Power Electronics in Renewable Systems

Power Electronics in Renewable Systems

Guest Editors

Shan He
Zhiqing Yang
Wenzhao Liu
Zian Qin

Basel • Beijing • Wuhan • Barcelona • Belgrade • Novi Sad • Cluj • Manchester

Guest Editors
Shan He
Department of Energy
Aalborg University
Aalborg
Denmark

Zhiqing Yang
School of Electric Engineering and Automation
Hefei University of Technology
Hefei
China

Wenzhao Liu
Vestas Wind Systems
Aarhus
Denmark

Zian Qin
Department of DC Systems Energy Conversion and Storage
Delft University of Technology
Delft
The Netherlands

Editorial Office
MDPI AG
Grosspeteranlage 5
4052 Basel, Switzerland

This is a reprint of the Special Issue, published open access by the journal *Electronics* (ISSN 2079-9292), freely accessible at: https://www.mdpi.com/journal/electronics/special_issues/0N641V3WB6.

For citation purposes, cite each article independently as indicated on the article page online and as indicated below:

Lastname, A.A.; Lastname, B.B. Article Title. *Journal Name* **Year**, *Volume Number*, Page Range.

ISBN 978-3-7258-5229-1 (Hbk)
ISBN 978-3-7258-5230-7 (PDF)
https://doi.org/10.3390/books978-3-7258-5230-7

Contents

electronics

MDPI

Article

High-Redundancy Design and Application of Excitation Systems for Large Hydro-Generator Units Based on ATS and DDS

Xiaodong Wang [1], Xiangtian Deng [1], Xuxin Yue [2,*], Haoran Wang [2], Xiaokun Li [2] and Xuemin He [2]

[1] School of Automation Wuhan, University of Technology Wuhan, Wuhan 430070, China; 363482@whut.edu.cn (X.W.); dengxt@whut.edu.cn (X.D.)
[2] Three Gorges Intelligent Control Technology Co., Ltd., Wuhan 430070, China; wang_haoran@ctg.com.cn (H.W.); li_xiaokun@ctg.com.cn (X.L.); he_xuemin@ctg.com.cn (X.H.)
* Correspondence: yue_xuxin@ctg.com.cn; Tel.: +86-13339990726

Abstract: The large-scale integration of stochastic renewable energy sources necessitates enhanced dynamic balancing capabilities in power systems, positioning hydropower as a critical balancing asset. Conventional excitation systems utilizing hot-standby dual-redundancy configurations remain susceptible to unit shutdown events caused by regulator failures. To mitigate this vulnerability, this study proposes a peer-to-peer distributed excitation architecture integrating asynchronous traffic shaping (ATS) and Data Distribution Service (DDS) technologies. This architecture utilizes control channels of equal priority and achieves high redundancy through cross-communication between discrete acquisition and computation modules. This research advances three key contributions: (1) design of a peer-to-peer distributed architectural framework; (2) development of a real-time data interaction methodology combining ATS and DDS, incorporating cross-layer parameter mapping, multi-priority queue scheduling, and congestion control mechanisms; (3) experimental validation of system reliability and redundancy through dynamic simulation. The results confirm the architecture's operational efficacy, delivering both theoretical foundations and practical frameworks for highly reliable excitation systems.

Keywords: self-shunt excitation system; peer-to-peer distributed system; distributed data exchange technology; asynchronous traffic shaping; data distribution service

1. Introduction

Hydropower, as a clean energy source, plays a vital role in conventional power systems and is the world's second-largest source of electricity generation. Amid the global shift toward low-carbon energy, the growing integration of variable renewable energy sources such as wind and solar power has highlighted hydropower's unique ability to stabilize the grid. As a result, it has become the key element in coordinating the balancing of dynamic supply and demand in modern power grids. Currently, large hydro-turbine units primarily utilize self-excited static excitation systems, where excitation power is generated by transformers connected at the generator terminals. This power is then supplied to the generator's field windings through controlled rectification. However, traditional excitation system designs lack distributed control modules within both the excitation power unit and the regulating unit. Failures in excitation regulation units, therefore, necessitate the immediate shutdown of the entire system, highlighting the reliability limitations of centralized control systems. Improving the stability and dependability of excitation systems remains the main goal driving technological progress in this area.

Compared to centralized control architectures, distributed architectures have been widely adopted to enhance system reliability. In the context of distributed system architecture design, Reference [1] introduces a distributed application architecture based on the GridAPPS-D platform, which coordinates the control of distributed energy resources by standardizing APIs and hierarchical architecture. The authors of [2] designed a distributed control architecture for parallel Buck-Boost converters, employing adaptive droop control parameters. The local controllers operate independently without real-time central communication, thereby reducing the communication burden. In harmonic compensation and power control, the master–slave distributed coordination strategy presented in Reference [3] empowers local controllers with autonomous decision-making capabilities, effectively mitigating the risks associated with single-point failures. The modular generation unit design in Reference [4] supports localized fault isolation, preventing system-wide collapse. Regarding reactive power compensation, Reference [5] systematically summarizes microgrid reactive compensation technologies, providing theoretical foundations for decentralized reactive control modules in distributed architectures. The authors of [6] propose a chain-connected STATCOM employing a redundant master sub-controller architecture, where sub-controllers can be dynamically dispatched to maintain DC bus voltage balance during main controller failures, significantly enhancing fault tolerance. In terms of microgrid integration and device interconnection, Reference [7] combines network reconfiguration (NR) and soft open points (SOPs) to overcome traditional distribution network capacity limitations and enhance fault recovery flexibility. The authors of [8] introduce a modular multi-port DC power electronic transformer (MDCPET) that employs an input-independent-output-series topology to eliminate single-point failures inherent in centralized designs, supporting "plug-and-play" expandability and improving system robustness. Regarding synchronization and coordinated control, Reference [9] proposes a distributed phase-locked loop (PLL) with a GPS-synchronized parallel mechanism to reduce reliance on communication links, ensuring frequency stability even during localized communication disruptions. The authors of [10] utilize virtual synchronous generator (VSG) control to emulate the inertial characteristics of traditional synchronous generators, enabling autonomous cooperative control of inverters. This approach minimizes dependence on central controllers, strengthening frequency stability and reliability in islanded microgrids. In terms of intelligent control and cybersecurity, Reference [11] employs a hierarchical communication architecture to distribute control functions to local controllers, reducing the risk of system-wide collapse caused by cyberattacks targeting central nodes. Regarding voltage regulation and novel devices, Reference [12] utilizes a consensus algorithm to coordinate multiple electric springs with BC (ESBC) to suppress voltage fluctuations. Its distributed cooperative strategy ensures bus voltage stability even under variations in load characteristics. The authors of [13] propose a novel decentralized stochastic recursive gradient (DSRG) method for solving the optimal power flow (OPF) problem in multi-area power systems. This method achieves global optimization through local decisions and minimizes information exchange, avoiding the data sharing requirements of traditional centralized approaches. The core steps of DSRG include initialization, local gradient calculation, stochastic update, recursive aggregation, and a consensus algorithm, with convergence checks ensuring the validity of the optimization results. The authors of [14] propose a hybrid approach combining an artificial neural network (ANN) with a backstepping controller to enhance the maximum power point tracking (MPPT) performance of photovoltaic generation systems (PVGSs). This method employs a hybrid strategy of particle swarm optimization (PSO) and genetic algorithm (GA) to optimize the weights and biases of the ANN for accurate prediction of the PV reference voltage. The backstepping controller ensures system stability and dynamic tracking performance by recursively constructing Lyapunov functions. Although distributed architectures and coordinated strategies have become a key research focus in the industry, studies specifically targeting distributed excitation systems remain scarce.

To further enhance the redundancy of excitation systems, an effective and practical technical approach involves eliminating hierarchical differences between control channels and ensuring that all control nodes operate in real-time equivalence, thereby establishing a functionally equivalent redundant control architecture within the system. Concurrently, this peer-to-peer redundant control framework requires seamless data flow across acquisition and computing units. Consequently, developing a robust distributed data distribution mechanism with strong real-time capabilities and high reliability becomes pivotal to realizing an equivalent distributed excitation control system.

Time-Sensitive Networking (TSN), with its microsecond-level transmission precision and predictable communication performance, demonstrates unique advantages in precision industrial control scenarios. Notably, in industrial control environments such as excitation systems, where stringent real-time responsiveness and substantial electromagnetic interference coexist, the collaborative architecture design of TSN and Data Distribution Service (DDS) has achieved significant technological breakthroughs: TSN establishes a hard real-time transmission channel at the physical layer, providing deterministic latency guarantees for data flow. DDS, as a middleware framework, offers semantic-rich data orchestration and intelligent scheduling capabilities at the logical layer. This vertically integrated "deterministic pipeline + semantic interconnection" architecture fundamentally addresses longstanding challenges in distributed excitation control systems, including the real-time coordination of multi-node operations and reliable data interaction under complex electromagnetic conditions [15]. Meanwhile, DDS implements intelligent data filtering and dynamic routing optimization at the application layer [16]. This cross-layer collaborative mechanism effectively resolves the latency bottleneck inherent in traditional industrial communication architectures, signaling that network switching technology research is evolving towards innovative integration of TSN and DDS. In studies of Time-Sensitive Networking (TSN) traffic scheduling mechanisms, asynchronous traffic shaping (ATS) technology has garnered significant attention due to its inherent independence from global clock synchronization. The authors of [17] broke through traditional synchronization constraints by innovatively developing an Urgency-Based Scheduling (UBS) algorithm. Through formal temporal modeling and large-scale simulation verification, its tight delay boundaries satisfy the hybrid transmission requirements of high-bandwidth sensor data and control flows in in-vehicle backbone networks, providing a low complexity deterministic transmission solution for automotive active safety scenarios. This pioneering work inspired subsequent theoretical advancements, notably the Pi-regulation theoretical framework proposed in [18]. By integrating network calculus with max-plus algebra, this framework has fundamentally revealed the mathematical equivalence between min-plus and max-plus algebras in traffic constraint formulations while establishing a minimal interleaving regulator model that provides theoretical grounding for asynchronous traffic shaping techniques, such as UBS. In collaborative research on heterogeneous shaping mechanisms, Reference [19] established, for the first time, a quantitative analysis model for hybrid TAS-ATS-CBS architectures. By constructing a worst-case latency upper-bound calculation framework, it revealed the gap in temporal analysis for ATS-CBS hybrid queue scenarios, providing both experimental evidence and a guiding framework for researchers and practitioners to select TSN sub-protocols tailored to application requirements.

As the number of communication nodes increases, the logical structure of distributed systems becomes more complex, rendering traditional communication middleware inadequate. Consequently, the data-centric DDS holds high application prospects. For example, in the power systems domain, Reference [20] proposed a DDS-based publish/subscribe protocol to implement dynamic monitoring in digital substations. By partitioning a multi-domain network structure to reduce discovery traffic and designing mapping rules between IEC 61850 [21] and DDS objects, the study experimentally validated the dynamic identification capability of IED devices (within 0.7 s), suitable for dynamic configuration scenarios such as microgrids. The authors of [22] addressed large-scale DDS data transmission scenarios by proposing a hierarchical priority scheduling and parallel processing mechanism. It categorizes monitoring tasks into three levels—basic storage, simple error detection, and complex data mining—and ensures real-time performance through multi-queue allocation. The system supports in-depth error analysis and monitoring while maintaining low latency. In intelligent transportation, Reference [23] developed a DDS and cloud computing-based validation framework for autonomous driving systems. By integrating FMI co-simulation, 3D vehicle simulators, and HILS hardware interfaces, DDS facilitates communication between distributed modules. This framework enables efficient resource allocation and multi-scenario validation, thereby lowering the cost and complexity associated with traditional hardware-in-the-loop simulations. To address emergency event response requirements, Reference [24] introduced an emergency QoS mechanism into the DDS publish/subscribe middleware, designing multi-level scheduling queues to prioritize high-urgency data. By optimizing scheduling algorithms, inefficiencies in traditional single-queue insertion are resolved, ensuring immediate responses to critical tasks.

The integration of DDS with ATS can enhance the efficiency and reliability of data distribution while improving system flexibility and security. However, research on the convergence of DDS and ATS remains limited, both domestically and internationally. It was not until 2023 that the Object Management Group (OMG) released a test version of the DDS extension specification 1.0, based on TSN [25]. Current studies on the integration of DDS and TSN are still in their initial stages.

To achieve cross-redundant links in a peer-to-peer distributed excitation system, the key lies in realizing high-precision data synchronization and efficient data distribution. This study investigates dynamic multi-priority queue scheduling algorithms and communication congestion control strategies that integrate ATS and DDS technologies, thereby completing a high-availability design to provide critical technical support for implementing peer-to-peer distributed excitation systems.

The primary contributions of this study include the following: (1) proposing a peer-to-peer distributed redundant architecture and quantitatively verifying its reliability advantages; (2) conducting research on core technologies for peer-to-peer distributed excitation systems using real-time data exchange methods based on ATS and DDS fusion; (3) validating the system's effectiveness through real-time simulation.

2. High-Redundancy Architecture Design for Peer-to-Peer Distributed Excitation Systems

2.1. Traditional Distributed Excitation System Redundant Architecture

Excitation systems employ various excitation methods. The self-shunt static excitation system typically comprises an excitation power unit and an excitation regulation unit, supplemented by auxiliary components such as a de-excitation device. A simplified system architecture is illustrated in Figure 1.

The traditional distributed excitation system features a regulating cabinet containing two regulating unit control channels and power cabinets, each housing one power unit control channel; a fiber-optic communication network diagram is shown in Figure 2. While the entire system incorporates five control channels capable of excitation control, the power unit control channels lack full excitation functionality, consequently necessitating operation under a two-tier priority architecture, where the two excitation regulator control channels constitute high-priority control and the three power unit control channels serve as low-priority control. System operation defaults to high-priority dual-channel control, with low-priority power unit control channels assuming authority only upon complete failure of both high-priority channels; within any single priority group, control channels dynamically determine control authority through real-time communication status detection, prioritizing selection of the channel maintaining normal communication with the maximum number of power cabinets and, when multiple channels exhibit identical communication integrity levels, executing switching according to the preset control channel priority sequence.

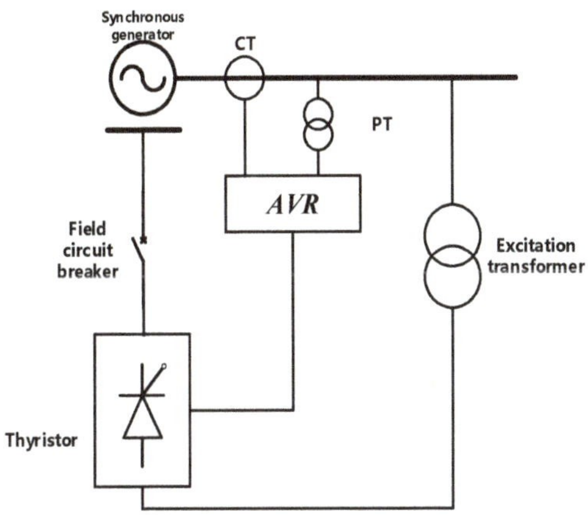

Figure 1. Schematic diagram of fundamental principles for self-shunt static excitation system.

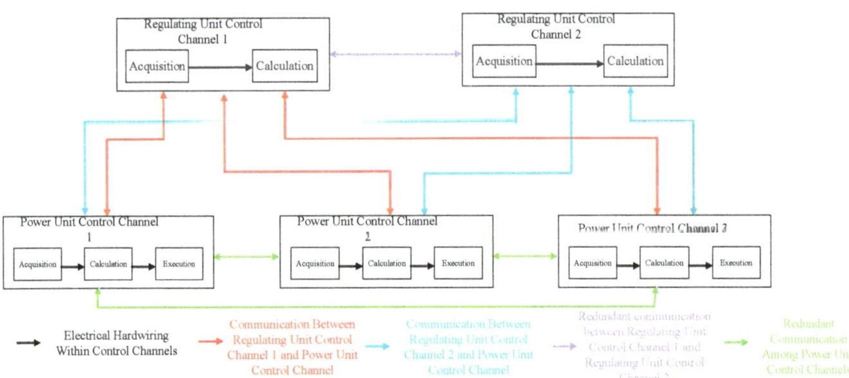

Figure 2. Traditional distributed excitation system fiber-optic network.

Focusing on single control channel failures, for the sake of analytical simplicity, it is assumed that the self-failure probabilities of the three internal modules within a control channel and the probability of communication abnormalities are all equal to p.

For the power unit control channel, the acquisition failure probability comprises two components: the self-failure of the acquisition module and the communication failure between the acquisition module and the calculation module. Let the acquisition failure probability of the power unit be P_1; then, the failure probability is expressed as:

$$P_1 = 1 - (1 - p)^2 \quad (1)$$

Calculation failure includes only self-failure, with a probability of p.

Execution failure also comprises two components: the self-failure of the execution module and the communication failure between the execution module and the calculation module. The failure probability is, thus, expressed as:

$$P_2 = 1 - (1 - p)^2 \tag{2}$$

Since the necessary and sufficient condition for a control channel to remain operational is that no failures occur, the failure probability P_3 of a single power unit control channel is expressed as:

$$P_3 = 1 - (1 - p)^5 \tag{3}$$

For the regulating unit control channel, the probabilities of acquisition failure and calculation failure are consistent with the failure probability analysis of the power unit control channel presented earlier. The necessary and sufficient condition for a regulating unit control channel to maintain operation is the absence of all failures; consequently, the failure probability P_4 of a single regulating unit control channel is expressed as:

$$P_4 = 1 - (1 - p)^3 \tag{4}$$

2.2. Peer-to-Peer Distributed Excitation System Redundant Architecture

The redundant architecture of the peer-to-peer distributed excitation system proposed in this study innovatively establishes module-level redundancy mechanisms within control channels. In this architecture, all control channels adopt identical hardware and software configurations. By creating cross-channel data interaction interfaces, the computing modules in each control channel can access monitoring data from the data acquisition modules of other control channels, while the actuation modules in each channel also receive control command output from the computing modules of different channels. A schematic diagram of this redundant architecture is shown in Figure 3.

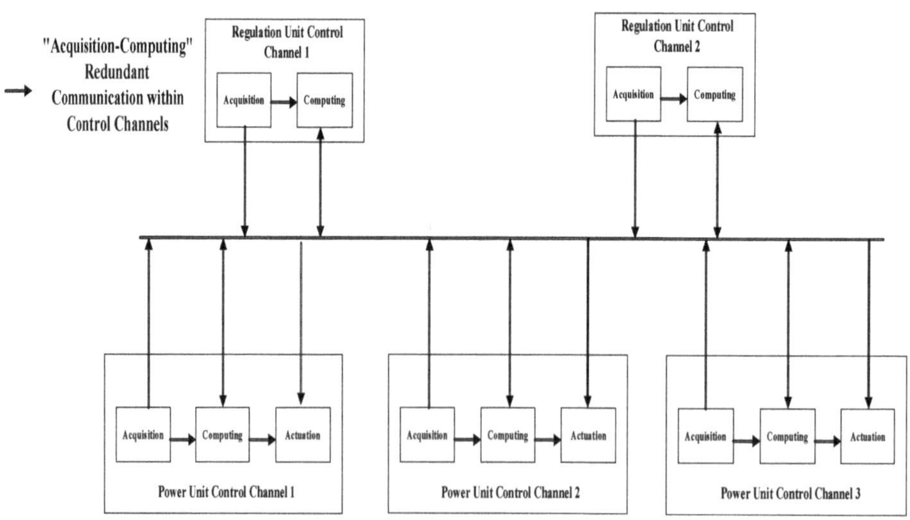

Figure 3. Cross-redundant architecture for peer-to-peer distributed excitation systems.

For the reliability analysis of a single power unit control channel, it is assumed that the probabilities of module self-failure and communication failure are both p. Accordingly, the failure probability P_5 of a single power unit control channel in the peer-to-peer distributed excitation system is:

$$\begin{aligned} P_5 = p + 2(p^{10} - 2p^9 - 2p^6 + 4p^5) - 2p(p^{10} - 2p^9 - 2p^6 + 4p^5) \\ -(p^{10} - 2p^9 - 2p^6 + 4p^5)^2 + p(p^{10} - 2p^9 - 2p^6 + 4p^5)^2 \end{aligned} \quad (5)$$

The exit paths for the deactivation of the excitation regulating unit control channel include failure of the local calculation module or the calculation module's inability to acquire data, and its failure probability is expressed as:

$$P_6 = p + 4p^5 - 6p^6 + 2p^7 - 2p^9 + 3p^{10} - p^{11} \quad (6)$$

Separately plot the comparative curves of the failure probability function of a single power unit control channel between the traditional distributed excitation system and the peer-to-peer distributed excitation system, as well as the comparative curves of the failure probability function of a single regulating unit control channel between the traditional distributed excitation system and the peer-to-peer distributed excitation system. As shown in Figures 4 and 5, within the probability interval [0, 1], the failure probability curve of the control channels of the same type in the peer-to-peer distributed excitation system always lies below that of the traditional system, and its vertical coordinate values are significantly reduced. The curve comparison results indicate that the peer-to-peer distributed excitation system designed in this study has obvious advantages in terms of reliability indicators. Through the quantitative comparative analysis of the probability function curves, the effectiveness of this architecture in optimizing the control system's failure rate is verified.

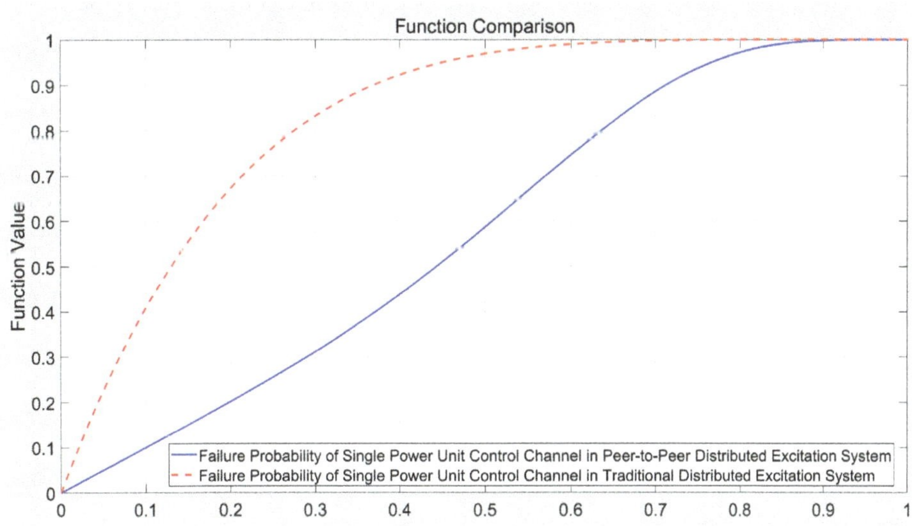

Figure 4. Comparison diagram of failure probability functions for single power unit control channel.

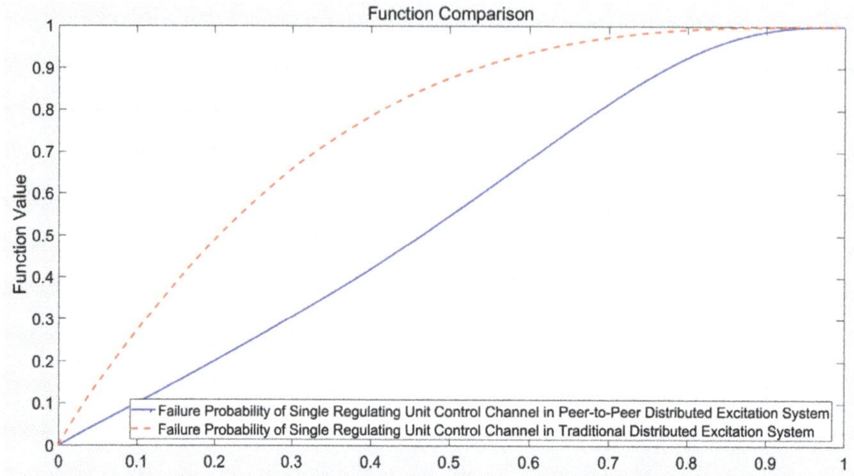

Figure 5. Comparison diagram of failure probability functions for single regulating unit control channel.

3. Research on Data Exchange Technologies and High-Availability Function Design for Peer-to-Peer Distributed Excitation Systems

3.1. Introduction to Asynchronous Traffic Shaping (ATS) Technology

The IEEE 802.1Qcr protocol [26], officially released in 2020, introduced an asynchronous traffic shaping (ATS) mechanism that breaks through traditional synchronous scheduling constraints. This mechanism utilizes an Urgency-Based Scheduler (UBS) to establish a hierarchical queue management architecture, primarily comprising traffic shaping queues and shared queues. As illustrated in Figure 6, during initial data processing, packets undergo traffic filtering and classification modules, where they are categorized into two types: guaranteed-transmission packets and best-effort packets. The UBS core components include multiple shaping queues and priority queues. The shaping queues utilize token bucket traffic policing technology to enforce bandwidth control. Simultaneously, by monitoring the dwell time of packets in rate adjustment queues, the system dynamically decides whether to reallocate guaranteed packets to corresponding hierarchical processing queues.

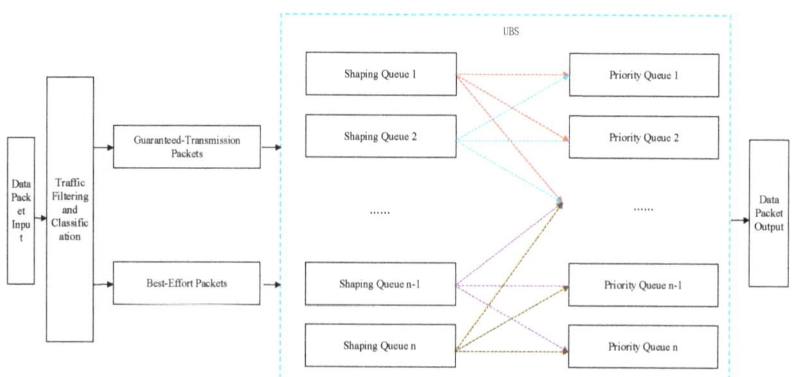

Figure 6. UBS queue structure diagram.

In the field of Time-Sensitive Networking (TSN), the asynchronous traffic shaping (ATS) mechanism eliminates dependency on network-wide clock synchronization systems by employing autonomous time reference maintenance and event-driven calibration technologies, establishing a decentralized temporal coordination framework. Through distributed timestamp management and dynamic priority allocation, it provides an innovative solution for deterministic transmission in heterogeneous network environments. The workflow involves three sequential phases: input ports first perform traffic shaping and classification via the Per-Stream Filtering and Policing (PSFP) mechanism to preliminarily filter and regulate data streams, directing qualified traffic into independent shaping queues configured with algorithms like Length-Based Rate Queuing (LRQ) or Token Bucket Emulation (TBE) to smooth traffic rates and suppress burst interference. Subsequently, the scheduler dynamically maps stream classifications to internal priorities by calculating an urgency metric derived from the ratio of a frame's remaining delay budget to a predefined threshold, ensuring highly time-sensitive traffic is prioritized for shared queues. Finally, shared queues execute conflict-free asynchronous transmission using a strict priority (SP) scheduling mechanism, optimized by real-time monitoring of output port link conditions to select optimal transmission windows. This technology excels in large-scale network deployments or environments where precise clock synchronization is impractical.

3.2. Introduction to Data Distribution Service

The DDS (Data Distribution Service) protocol is a data-centric, agentless middleware technology based on the publish/subscribe model, designed to provide efficient, reliable, and real-time data communication services for distributed systems. The DDS protocol defines data sharing as a domain where publishers send data to the domain space, and subscribers receive the required data from the domain space. The data in the domain space itself are stored in the local memory of all participating nodes. Therefore, during the data exchange process, it is as if the locally stored data are being operated on, resulting in very low latency for data publishing and receiving. This means that the protocol meets reasonable real-time requirements. At the same time, the nodes in the protocol's networks are distributed, with no master–slave relationships between them, functioning as peer nodes, and capable of supporting one-to-one, one-to-many, and many-to-many communication methods.

The DDS standard provides a set of QoS (Quality of Service) mechanisms. Some of the main strategies, such as the reliability mechanism, control the reliability of data transmission, including the following options: Best Effort: make every effort but no guarantee of data delivery; option Reliable: ensures reliable data delivery, where lost data will be retransmitted, suitable for scenarios requiring high reliability; the Durability mechanism controls whether data persist after publication, ideal for scenarios that require historical data access; the History mechanism controls the storage method of data samples, suitable for scenarios that require data caching; the latency budget mechanism defines the maximum allowable delay for data transmission, helping to optimize network and system performance to ensure data arrive within the specified time. The above mechanism allows developers to control the reliability, real-time performance, storage methods, and resource management of data transmission, thereby adapting to different application scenarios.

Within the DDS architecture, clock management and synchronization constitute critical components, particularly in distributed real-time systems. They are essential for ensuring data consistency, temporal accuracy of events, and fulfillment of QoS requirements. Given that the DDS standard itself does not mandate a specific clock synchronization protocol, it permits the integration of external clock synchronization solutions to support time management in distributed systems. The primary integrated synchronization approach utilizes the IEEE 1588 protocol [27] to achieve sub-microsecond synchronization accuracy. Hardware timestamping is utilized to minimize latency impacts from operating system scheduling and protocol stack processing. The master clock periodically broadcasts synchronization messages, and the slave clock adjusts its local time accordingly. The primary workflow for clock synchronization is as follows:

1. The master clock broadcasts a Sync message, marking the transmission time T_1;
2. The slave clock receives the Sync message, recording the reception time T_2;
3. The slave clock sends a Delay Request message, marking the transmission time T_3;
4. The master clock receives the Delay Request message, recording the reception time T_4;
5. Calculate the clock offset;

$$offset = \frac{(T_2 - T_1) + (T_4 - T_3)}{2} \tag{7}$$

6. Adjust the local clock to align it with the master clock.

3.3. ATS and DDS Integration Technology

In the joint scheduling system of ATS and DDS, DDS is responsible for QoS management at the application layer, such as data reliability and delay budgeting, while ATS is responsible for network traffic scheduling, such as traffic priority scheduling and traffic shaping. To ensure efficient data transmission between ATS and DDS during cross-layer collaboration, it is necessary to establish mapping rules between DDS QoS and ATS mechanisms, ensuring that ATS network scheduling can meet the real-time requirements of the DDS application layer.

The designed peer-to-peer distributed excitation system leverages the integration of ATS and DDS technologies, employing a five-layer software architecture, as depicted in Figure 7. The Driver Layer facilitates resource access via hardware interfaces, including AD drivers and memory drivers. Building upon the operating system kernel, the Transport Layer achieves efficient cross-node data distribution and real-time communication through the synergistic fusion of DDS middleware and ATS capabilities, specifically multi-priority traffic scheduling and traffic shaping. The Control Layer orchestrates multi-node resources via distributed data exchange management and communication management modules, thereby enabling peer-to-peer coordination. Finally, the application layer implements core algorithms such as excitation control and limit protection, establishing a decentralized control loop.

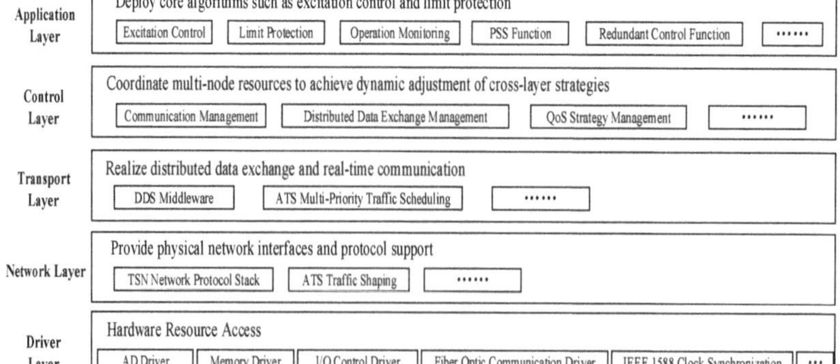

Figure 7. System software framework diagram.

To meet these requirements, traffic scheduling must adopt a multi-priority scheduling mechanism. This is achieved through proxy nodes to ensure prioritized transmission of HRT (hard real-time traffic) data for low-latency demands, while simultaneously balancing SRT (soft real-time traffic) performance and mitigating the impact of BE (best-effort) traffic on network performance. Thus, the overall design architecture is depicted in Figure 8.

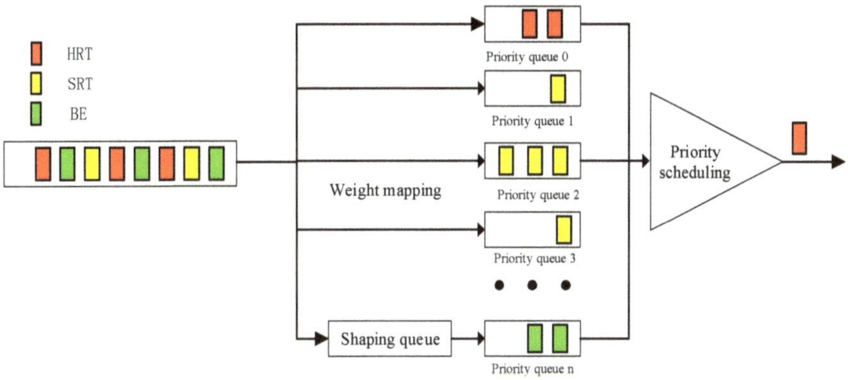

Figure 8. Traffic scheduling overall architecture.

Among them, HRT traffic has the highest priority, mapped to high-priority traffic in ATS, ensuring low-latency transmission, suitable for critical task flows. SRT traffic is mapped to medium-priority traffic according to the latency budget QoS policy, with the system calculating priority weights based on the latency budget. BE data have the lowest priority and are mapped as best-effort traffic in ATS. The transmission rate is controlled through traffic shaping to prevent network congestion. The mapping rules are shown in Table 1.

Table 1. Mapping of data types to scheduling strategies.

Business Type	DDS QoS Configuration	ATS Scheduling Strategy	Mapping Rules	Target Effect
HRT	1. LatencyBudget: <1 ms 2. Reliability: Reliable	ATS UBS Queue scheduling	ATS traffic scheduling ensures the highest priority transmission of traffic, free from interference by low-priority traffic.	Low latency and highest priority, ensuring data delivery within the specified latency budget.
SRT	1. LatencyBudget: 1~10 ms 2. Reliability: Reliable	ATS UBS Queue scheduling	In DDS, the latency budget is converted into allocation weights for queue priorities.	Avoid the impact of burst traffic on the network and ensure low-latency transmission of SRT traffic.
BE	1. LatencyBudget:default 2. Reliability: Best Effort	Traffic Shaping Low-Priority Queuing	BE traffic is transmitted when the network is idle. ATS allocates according to available bandwidth and may discard data if the network becomes congested.	BE traffic is transmitted when the network is idle. TSN allocates according to available bandwidth and may discard data if the network becomes congested.

The specific implementation logic of the mapping rules between DDS QoS and ATS mechanisms is visually elaborated in the Traffic Scheduling Priority Mapping Process, as shown in Figure 9. Initiated with DDS Topic Tasks, this process explicitly delineates the end-to-end decision-making workflow—from QoS policy extraction to ATS priority queue allocation—encompassing the dynamic logic of priority determination and queue mapping for HRT/SRT/BE traffic categories based on QoS parameters (e.g., latency budget). It serves as a critical enabler for translating the theoretical mapping rules into actionable network scheduling practices. The remaining flows along with the integrated architecture of ATS and DDS systems can be found in Appendix B.

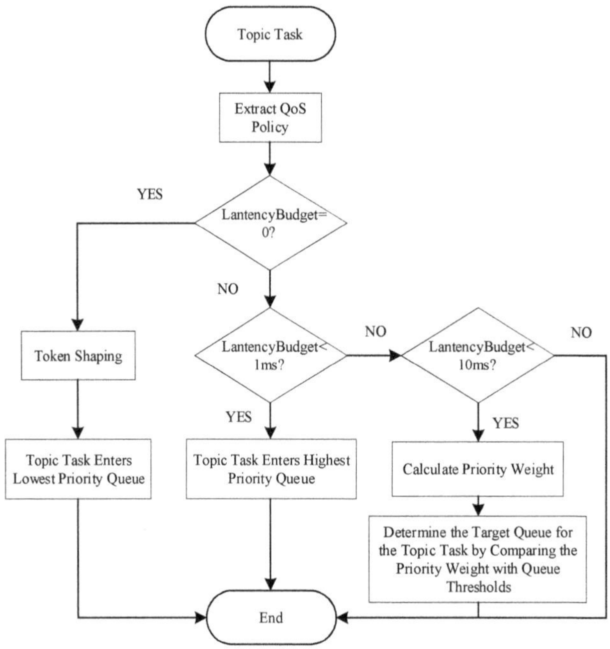

Figure 9. Implementation flow of mapping DDS QoS policies to ATS traffic priorities.

3.4. *Multi-Priority Queue Scheduling Algorithm Introduction*

To ensure HRT traffic is transmitted with the highest-priority sequence, SRT traffic is prioritized based on the latency budget to transmit data with tighter deadlines first; BE traffic is rate-controlled into the lowest-priority queue for transmission. The queue scheduling model must adopt a strict priority queue mechanism. Based on these scheduling requirements, this study employs the strict priority queue with the Delay Bound (SP^2DB) scheduling model [28]. Refer to Appendix A for definitions of critical variables.

Suppose there exists a priority scheduling queue pool composed of k ($k > 1$) multi-level sequential queues I with fixed capacities. The priorities of these multi-level sequential queues descend from I_1, I_2, ..., I_k, for any queue I_i ($1 < i < k$), and its queue length and capacity limit are L_i and C_i, respectively. When a data packet attempts to enter the strict priority queue, the system verifies whether the real-time detected queue length has reached its capacity limit. If exceeded, the tail data discarding policy is executed. When a data packet exits the strict priority queue, forwarding strictly adheres to the queue priority order.

Furthermore, assume there exists a sequence of N incoming data packets $\{p_1, p_2, \ldots, p_N\}$, each carrying its desired queuing delay upper bound d_i, where i ranges from 1 to N. Let the arrival time of packet p_i at the strict priority queue be t_a^i, and its departure time be t_b^i. If $t_b^i - t_a^i, \leq d_i$, p_i is deemed successfully scheduled; otherwise, it is marked as failed. Based on this, the scheduling target—packet acceptance rate S_d—for the queue model is defined by Equation (8):

$$S_d = \frac{\sum_{i=1}^{N} P_s(i)}{N} \quad (8)$$

Let $P_s(i)$ indicate whether the i-th data packet is successfully scheduled, where $P_{suc}(i) = 1$ for success and 0 otherwise. The scheduling objective of the queue model is to maximize the packet acceptance rate S_d.

Assume the current cached packets in queues I_1, I_2, and I_3 are n, z, and m, respectively. For a pending data packet $p(x)$ with a preset maximum cache depth threshold x (where $n + z \leq x < n + z + m$), the actual queuing length depends on its injection target: if injected into I_1, the actual queuing length is n; if injected into I_2, the queuing length becomes $n + z$ due to I_1's priority scheduling; if injected into I_3, the queuing length is $n + z + m$.

Analysis shows that $p(x)$ can be successfully scheduled when injected into either I_1 or I_2. Moreover, injecting $p(x)$ into I_2 enables I_1 to prioritize packets with stricter delay requirements, thereby maximizing S_d and achieving optimal resource allocation under system delay constraints.

However, when the newly added data packet $p(b)$ preemptively enters queue I_1, if $x < n+z + b$ at this time, the data packet $p(x)$ will fail to be scheduled. Therefore, the scheduling algorithm must account for the impact of high-priority preemption on queuing delays.

As illustrated by the analysis above, the key to improving the packet acceptance rate lies in minimizing the queuing length of data packets under the influence of new packet preemption, thereby maximizing the likelihood of successful scheduling.

Define the maximum port data packet forwarding rate as R, with a fixed data transmission time interval Δt. This is equivalent to assigning each data packet a processing time of Δt in the priority scheduling queue. Let the maximum queuing length of data packets be l, and map the queuing delay upper bound d in the strict priority queue to this maximum queuing length l:

$$l = \frac{d}{\Delta t} \quad (9)$$

Define the queue limit q_i of queue I_i as the queue length within ΔT after the data packet p enters queue I_i, which serves as the reference value for predicting the queue depth upon the packet's entry. Based on a dynamic comparison between this estimated value and the data packet's maximum allowable queue length l, the system executes the hierarchical queue dynamic selection mechanism. By sequentially comparing the relationship between the maximum queuing time length and the queue limit from low to high priority if $q_i \leq l_i$, it enters this queue; otherwise, it jumps to a queue with higher priority until reaching the highest-priority tier—an appropriate queue limit must be identified to improve the packet acceptance rate.

The SP^{2}DB−EQ algorithm [28] analyzes the traffic characteristics of the historical monitoring period. Using the inbound traffic intensity from the preceding t statistical periods, it calculates the traffic intensity V_i of queue I_i during statistical period T_{t+1}. Define the arrival time of packet p at the switch during T_{t+1} as T_s. The duration from T_s to the end of T_{t+1} is ΔT. In this dynamic evaluation model, the system treats continuously arriving packets within the ΔT window as the critical factor affecting delay accumulation for p.

As shown in Figure 10, this is the calculation framework for the queue limit q_i, assuming the transmission time interval does not exceed ΔT.

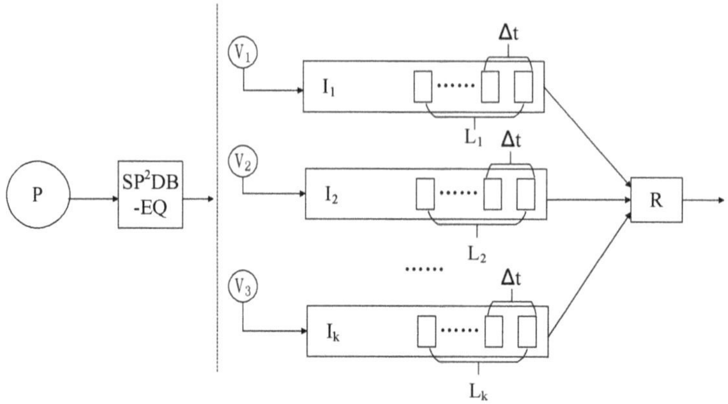

Figure 10. Queue boundary calculation.

When the data packet p is assigned to the transmission channel of the highest-priority queue I_1, the incoming data packet sequence in the subsequent statistical period T will not interfere with the queuing duration of p. In this case, the channel admission threshold is calculated based solely on the real-time depth parameter of the current buffer queue, i.e., $q_1 = L_1$.

If the data packet p is assigned to a non-highest-priority queue, it must wait until all buffered data packets in I_1 are processed. At this point, a dynamic transmission model for I_1 must be established: starting from the injection of the current data packet into other queues until either the queued data packets in I_1 are fully cleared or the monitoring statistical period T_{t+1} terminates. During this dynamic process, data packets are continuously injected and forwarded. Key evaluation parameters include the following: traffic intensity V_1 entering I_1 during this process, queue length L_1, duration of the dynamic process (JT_1), total data packet count in the dynamic process (JP_1)

When $V_1 < R$, the term $(R-V_1)$ represents the net egress forwarding rate:

$$JT_1 = \begin{cases} \frac{L_1}{R-V_1}, \frac{L_1}{R-V_1} < \Delta T \\ \Delta T, \frac{L_1}{R-V_1} \geqslant \Delta T \end{cases}, JP_1 = JT_1 \times V_1 + L_1 \tag{10}$$

When $V_1 > R$, L_1 gradually increases. Denote the time it takes for the packet queue length to grow from L_1 to the maximum queue capacity C_1 as T_1^C.

$$T_1^C = \frac{C_1 - L_1}{V_1 - R} \tag{11}$$

If $T_1^C < \Delta T$, then within the future ΔT period, the queue I_1 will reach the maximum capacity C_1. At this time, the duration of the dynamic process and the total number of data packets in this dynamic process are expressed as:

$$JT_1 = \Delta T, JP_1 = JT_1 \times R + C_1 \tag{12}$$

Conversely, if $T_1^C \geq \Delta T$, within the future ΔT period, queue I_1 will gradually increase but will not reach its maximum capacity C_1. In this case, the duration of the dynamic process and the total number of data packets within this dynamic process are defined as:

$$JT_1 = \Delta T, JP_1 = JT_1 \times V_1 + L_1 \tag{13}$$

If the traffic intensity V_1 is equal to the maximum rate value R, then within the future ΔT period, the actual length of the queue I_1 will remain unchanged at L_1. In this scenario, both the duration of the dynamic process and the total number of data packets in this process align with Equation (13).

Furthermore, if data packet p enters queue I_2, it must wait for the completion of the dynamic process in I_1. Only when $V_1 < R$ does I_2 gain the opportunity to forward p. During the time required to send out L_2 data packets, $L_2/(R - V_1)V_1$ packets enter I_1 and are forwarded. If $V_1 \geq R$, I_2 will persistently idle until the condition improves. The queue boundary q_2 is then defined as:

$$q_2 = \begin{cases} JP_1 + L_2 + \frac{L_2}{R - V_1} V_1, V_1 < R \\ JP_1 + L_2, V_1 \geqslant R \end{cases} \tag{14}$$

When the data packet p is assigned to a queue with a priority lower than I_2, the system triggers the queue-emptying and waiting mechanism. It must wait until all historical data packets in the queue channels I_1 and I_2 are fully forwarded. Define the dynamic sending process of I_2: starting from the injection of the current data packet until the channel becomes completely idle. During this period, data packets arrive at both I_1 and I_2; queued packets in I_1 and I_2 are transmitted. The traffic intensity of packets arriving at the queue during this process is $V_1 + V_2$. After a duration JT_1, there will be $JT_1 \times V_2$ packets arriving at queue I_2, and its actual queue length becomes L_2'. Let JT_2 denote the duration of the dynamic process for I_2 and JP_2 represent the total number of packets in the dynamic process of I_2. Assuming the duration does not exceed ΔT, then:

$$L_2' = \begin{cases} L_2 + V_2 \times JT_1, L_2 + V_2 JT_1 < C_2 \\ C_2, L_2 + V_2 JT_1 \geqslant C_2 \end{cases} \tag{15}$$

Based on the foregoing analysis, the expression for the queue boundary q_i ($1 \leqslant I \leqslant m$) can be derived as:

$$q_i = \begin{cases} \left(\sum\limits_{j=0}^{i-1} JP_j \right) + L_i + \frac{L_i}{R - \sum_{j=0}^{i-1} V_j} \sum\limits_{j=0}^{i-1} V_j, \sum\limits_{j=0}^{i-1} V_j < R \\ \left(\sum\limits_{j=0}^{i-1} JP_j \right) + L_i, \sum\limits_{j=0}^{i-1} V_j \geqslant R \end{cases} \tag{16}$$

In the dynamic transmission process of queue I_i, the traffic intensity arriving at the queue is the sum of all individual traffic intensities. Let the queue length be L_i'. Denote JP_i as the total number of data packets transmitted during the dynamic process of I_i, and JT_i as the duration of this process. Assuming JT_i does not exceed ΔT, the queue length is defined as:

$$L_i' = \begin{cases} L_i + V_i \sum\limits_{j=1}^{i-1} JT_j, L_i + V_i \sum\limits_{j=1}^{i-1} JT_j < C_i \\ C_i, L_i + V_i \sum\limits_{j=1}^{i-1} JT_j \geqslant C_i \end{cases} \tag{17}$$

If the traffic intensity is less than the switch egress forwarding rate, then:

$$JT_i = \begin{cases} \dfrac{L_i'}{R - \sum_{j=1}^{i} V_j}, \dfrac{L_i'}{R - \sum_{j=1}^{i} V_j} < \Delta T - \sum\limits_{j=1}^{i-1} JT_j \\ \Delta T - \sum\limits_{j=1}^{i-1} JT_j, \dfrac{L_i'}{R - \sum_{j=1}^{i} V_j} \geqslant \Delta T - \sum\limits_{j=1}^{i-1} JT_j \end{cases}, JP_i = JT_i \times \sum_{j=1}^{i} V_j + L_i' \tag{18}$$

If the traffic intensity exceeds the switch egress forwarding rate, when the queue length grows from L_i' to the maximum queue capacity C_i, the elapsed time is denoted as T_i^C:

$$T_i^C = \frac{C_i - L_i'}{\sum_{j=1}^{i} V_j - R} \tag{19}$$

$$JT_i = \Delta T - \sum_{j=1}^{i-1} JT_j, JP_i = \begin{cases} JT_i \times R + C_i, T_i^C < JT_i \\ JT_i \times \sum\limits_{j=1}^{i} V_j + L_i', T_i^C \geqslant JT_i \end{cases} \tag{20}$$

If the traffic intensity equals the switch egress forwarding rate, then:

$$JT_i = \Delta T - \sum_{j=1}^{i-1} JT_j, JP_i = JT_i \times \sum_{j=1}^{i} V_j + L_i' \tag{21}$$

3.5. Communication Blocking Regulation Strategies

In the peer-to-peer distributed excitation system, redundancy functionality is implemented through cross-communication links. Compared to traditional architectures, the number of internal communication nodes significantly increases, with simultaneous and substantial improvements in both communication latency sensitivity and data throughput requirements. To achieve highly reliable peer-to-peer distributed system configurations, the critical approach is to adopt robust distributed communication technologies. Through congestion control strategies, the synchronization and coordination of internal link states across control units are ensured.

Based on this analysis, priority indices must be computed for each incoming data stream and inserted into strict priority queues. During dequeuing, the highest-priority data are constantly retrieved from the queue.

Let the attributes of data stream $\mathrm{f_i}$ include data type $Type(f_i)$, QoS policy set $QoS(f_i) = \{Deadline, Latencybudget, Reliability, Durability, \ldots\}$. Define the strategy mapping function $\Phi(f_i)$ as the mapping of the service flow to its corresponding service category. Then,

$$\Phi(f_i) = \begin{cases} HRT & Deadline_i \leq T_{\mathrm{hrt}} or Type(f_i) \in \mathcal{T}_{\mathrm{critical}} \\ SRT & LatencyBudget_i \leq T_{\mathrm{srt}} \\ BE & else \end{cases} \tag{22}$$

Among them, T_{hrt} is the hard real-time deadline threshold, T_{srt} is the soft real-time delay budget threshold, and $\mathcal{T}_{\mathrm{critical}}$ is the set of critical data types. These parameters are mapped to their corresponding priority queues via priority mapping.

For dynamic priority calculation, let the current scheduling time of data packet i be t and its delay upper bound be d_i. The remaining maximum waiting time is defined as:

$$s_i(t) = d_i - t \tag{23}$$

Since the allowable maximum waiting time is inversely proportional to priority, the dynamic priority index is computed by incorporating the estimated service time ω_i and ε denotes the permissible error on the maximum waiting time:

$$I_i(t) = \frac{1}{d_i - t - w_i + \varepsilon} \tag{24}$$

After implementing real-time preemption for key traffic flows through the dynamic priority adjustment mechanism, the system further detects congestion states via a congestion detection mechanism. Subsequently, a token bucket-based dynamic flow control algorithm is employed to regulate the growth of low-priority traffic, thereby preventing congestion in high-priority data transmission.

The congestion detection employs a two-stage congestion discrimination method, performing judgments based on queue backlog status and delay proximity estimation. The queue performance analysis for DDS nodes must consider the following parameters: current queue backlog length $L_i(t)$; average packet arrival rate $\lambda_i(t)$; queue service rate $\mu_i(t)$; average queuing delay $D_i(t)$. Based on these parameters, the congestion judgment thresholds are defined as follows:

$$C_i(t) = \begin{cases} 1, L_i(t) > L_{high} \\ 0, L_i(t) < L_{low} \\ C_i(t-1), otherwise \end{cases} \tag{25}$$

In the congestion detection mechanism, L_{high} and L_{low} represent the upper and lower backlog thresholds, respectively. The queuing delay $D_i(t)$ is estimated using the current queue backlog length $L_i(t)$ and service rate $\mu_i(t)$ and then compared against these thresholds to determine congestion status.

For traffic flow control, the token bucket mechanism is implemented: a token bucket is configured for each flow fi, with a maximum bucket capacity Bi. Let $T_i(t)$ denote the current token count and $r_i(t)$ the allocated transmission rate. The token update rule is defined as:

$$T_i(t+\Delta t) = min\{\ B_i\ , T_i(t) +\ r_i(t)\cdot \Delta t\ \} \tag{26}$$

The sending rate is dynamically adjusted based on congestion status. The calculation formula for rate adjustment is defined as follows:

$$r_i(t+\Delta t) = \begin{cases} r_i(t)\cdot(1-\alpha), & C_i(t) = 1 \\ \min\{r_i(t)+\beta, r_{i,\max}\}, & C_i(t) = 0 \\ r_i(t), & else \end{cases} \tag{27}$$

Among them, α∈(0, 1) is the multiplicative decreasing factor, β > 0 is the additive increment, and $r_{i,max}$ is the maximum allocable rate.

3.6. High-Availability Functional Design

In the peer-to-peer distributed excitation system, each controller's acquisition module communicates with the computation modules in other controllers. In contrast, each computation module communicates with the execution modules in other controllers. Probe-live communication links exist between modules of the same type—all peer agent nodes of the same type share equal priority. When a node operates normally, other nodes remain capable of receiving domain-specific data. If a node fails, its responsibilities are reassigned to peer nodes of the same type, and data consistency and reliability are ensured through backups of DDS node-published production data. Additionally, online agent nodes proactively send heartbeat DDS messages to standby peer nodes at regular intervals to maintain a keep-alive mechanism. When an online agent node encounters communication link failures or internal malfunctions, standby agents initiate an election mechanism to designate a new primary agent, which subsequently assumes data forwarding duties. Figure 11 illustrates the architecture of the hot backup module system.

In this system, the online agent node broadcasts predefined topic information to the local area network (LAN) segment. This information serves as heartbeat liveness detection messages from the online agent node. By continuously transmitting these heartbeat signals, the online agent node declares its normal and active operational status to other components within the LAN, particularly standby peer agent nodes of the same type. The standby agent nodes persistently monitor the primary agent's heartbeat detection messages and use a timer to record the intervals between received heartbeats. If the timer exceeds a predefined threshold without detecting any heartbeat from the online agent node, the standby agent node concludes that the primary node has failed or become unavailable. At this point, the standby agent node replaces it as the new online agent node. Figure 12 illustrates the timing diagram for the hot backup functionality.

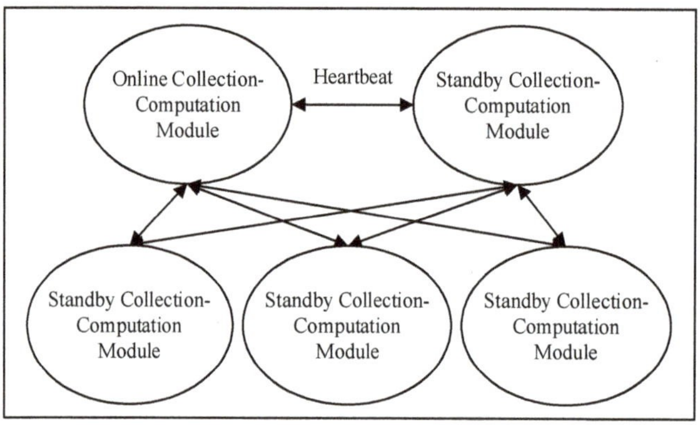

Figure 11. Hot backup module system architecture.

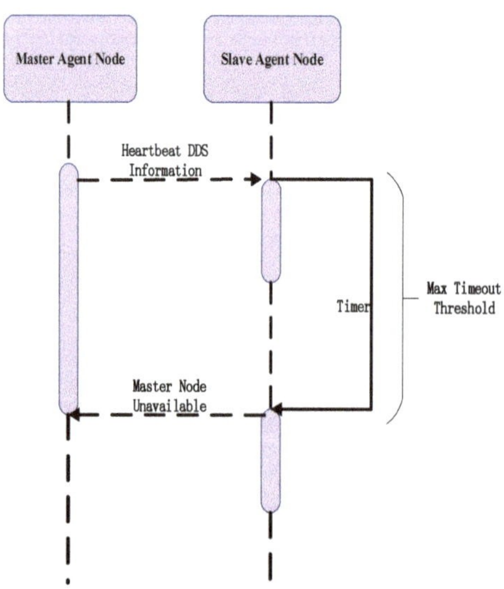

Figure 12. Sequence diagram of hot backup function.

In the peer-to-peer distributed excitation system designed in this study, all five controllers operate at an equal hierarchical level. Each controller integrates signal acquisition and execution modules; however, outputs from homogeneous modules across different controllers may exhibit discrepancies due to interference or communication delays. To ensure system reliability, an optimization strategy is embedded within the redundancy framework to preserve superior data while eliminating inferior inputs. This involves filtering out transient interference or faulty signals and selecting the most representative data from multi-source inputs that align with actual operational conditions. By doing so, both high fault tolerance and precise, rapid excitation regulation are maintained.

To fulfill these requirements, homogeneous modules across the five controllers execute time-synchronized parallel signal output and backup. The system first performs data pre-screening to discard outliers violating physical constraints. It then employs a 5-choose-3 majority voting mechanism: if three or more modules produce values within a $\pm\alpha\%$ error margin, the majority consensus is adopted. These modules are ranked by predefined priority, with the highest-priority module designated as the online module. If no clear majority emerges, dynamic clustering analysis is activated: adjacent signal differences are computed to identify densely clustered intervals, their median is derived, and the module closest to this median is selected as the online module. Figure 13 illustrates this workflow.

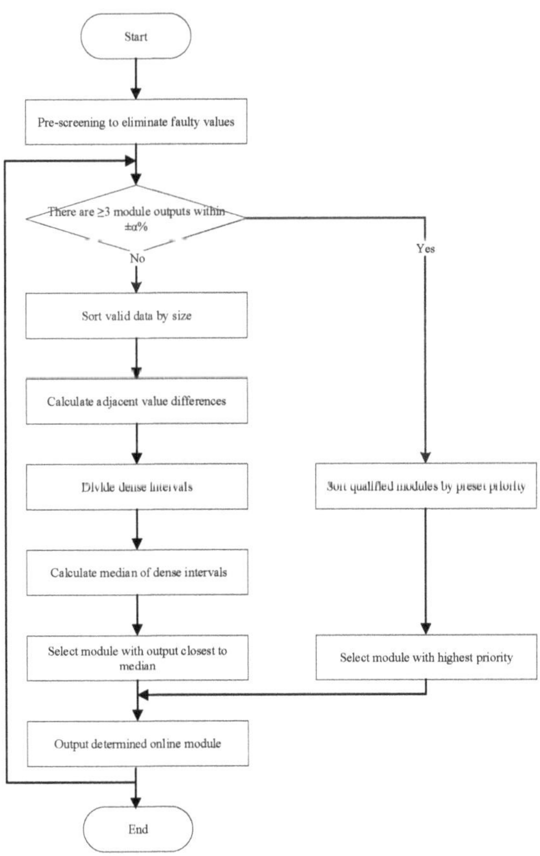

Figure 13. Flowchart of optimization design.

4. Experimental Verification of the Peer-to-Peer Distributed Excitation System

4.1. Performance Testing of Distributed Real-Time Data Distribution System

Configure one switch and one host. The development board is equipped with five on-board DDS nodes (A, B, C, D, E). The development board is connected to the switch, allowing data transmission through it. The test network topology is shown in Figure 14.

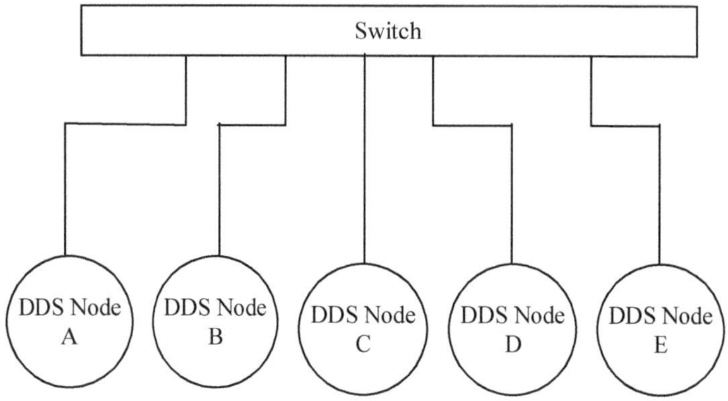

Figure 14. DDS test network topology.

To evaluate the computational overhead and resource requirements for implementing ATS and DDS, under data distribution traffic of 1000 packets per second (pps), the CPU utilization stabilizes within the range of 30–35%. Under this service load, the processor is capable of operating the DDS middleware, TSN traffic scheduling module, and operating system kernel efficiently and stably. The end-to-end latency of DDS is maintained at the sub-millisecond level. The total memory usage is approximately 200 megabytes (MB), accounting for about 2.6% of the current onboard memory.

To evaluate the performance of the multi-priority scheduling and flow control mechanisms in the DDS system proposed in this study, we aim to verify whether the scheduling mechanism can prioritize high-priority data streams (HRT/SRT) under high-load or resource-constrained network conditions, thereby ensuring compliance with both soft and hard real-time latency requirements. The test focuses on the mechanism's ability to mitigate congestion at DDS node transmitters under mixed traffic loads (HRT/SRT/BE), analyzing latency and packet loss metrics.

Without enabling dynamic priority scheduling or flow control policies, all nodes' service flow publishers and subscribers are activated, allowing low-priority background traffic (customizable rates) to exchange data between nodes. The scheduling mechanism on each node is configured to wait for publishers to transmit fixed-size high-priority data. Publishers iteratively attempt to send 64-byte test samples, await acknowledgments, and measure end-to-end latency. The latency results (Table 2) are collected under varying background traffic bandwidths (Mbits/sec).

From the analysis of latency results, as the bandwidth increases, the maximum deviation of the tested service flow's latency gradually rises, with the average latency increase remaining within approximately 10%. However, when the bandwidth reaches 450 Mbps, the system encounters severe congestion, resulting in significant deviations in maximum latency and a sharp increase in packet loss rates.

Table 2. Latency performance test data without priority scheduling and flow control policies.

Mbits/s	Latency (Mean/μs)	Latency (Min/μs)	Latency (Max/μs)	Packet Loss Rate (%)
50	1126.72	746.59	2452.98	0
100	1115.36	783.65	2189.75	0
150	1163.68	772.84	2509.56	0
200	1179.33	770.82	2643.75	0
250	1216.47	771.96	2782.27	0
300	1182.14	796.69	2846.58	0
350	1311.15	793.42	3359.79	0
400	1372.43	751.41	3916.54	1
450	2301.89	802.08	8923.24	12

Building upon the previous configuration, dynamic priority scheduling and flow control strategies were enabled to evaluate latency performance under identical network conditions. The experimental results are summarized in Table 3.

Table 3. Latency performance test data with priority scheduling and flow control policies enabled.

Mbits/s	Latency (Mean/μs)	Latency (Min/μs)	Latency (Max/μs)	Packet Loss Rate (%)
50	824.36	569.78	984.38	0
100	818.90	678.16	1012.25	0
150	833.19	698.45	1029.36	0
200	819.25	670.12	1049.35	0
250	846.71	679.24	989.23	0
300	882.29	690.30	1046.82	0
350	851.11	641.32	1102.93	0
400	875.68	651.82	1046.48	0
450	904.18	701.38	1151.76	0

The test results indicate that, after implementing priority scheduling and flow control strategies, the average latency of the tested service flow remains stable as the bandwidth increases, showing no significant upward trend. These experimental outcomes validate the effectiveness of the distributed communication technology, which combines the ATS and DDS frameworks, along with the associated QoS enforcement strategies proposed in this study.

4.2. *Calculation Module Redundant Switching Test*

In practical engineering, active fault injection for redundancy validation is infeasible due to safety risks and potential economic losses. Leveraging a high-fidelity dynamic simulation platform, this study replicates real-world grid conditions, including electromagnetic transient processes and complex operational scenarios, while enabling controlled fault injection. The photograph of the dynamic simulation platform can be found in Appendix C. Based on this platform, we systematically test the redundancy mechanisms of the designed peer-to-peer distributed excitation system. This approach validates the dynamic response of redundancy strategies under multi-failure scenarios, providing dual assurance (theoretical and practical) for system reliability and fault tolerance.

As shown in Figure 15, the experiment validates the redundancy functionality of computational modules between Control Channel 1 and Control Channel 2 in the regulator cabinet by actively switching the online module priority between the two channels. During the transition from Channel 1's computational module to Channel 2's, the generator terminal voltage exhibits no abrupt fluctuations. Similarly, when switching back to Channel 1's module, voltage stability is maintained. The results demonstrate that both control channels' computational modules achieve seamless control handover during real-time switching, with a voltage fluctuation deviation rate of no more than 0.36%. This confirms the effectiveness of the redundancy mechanism between the computational modules of Control Channel 1 and Control Channel 2.

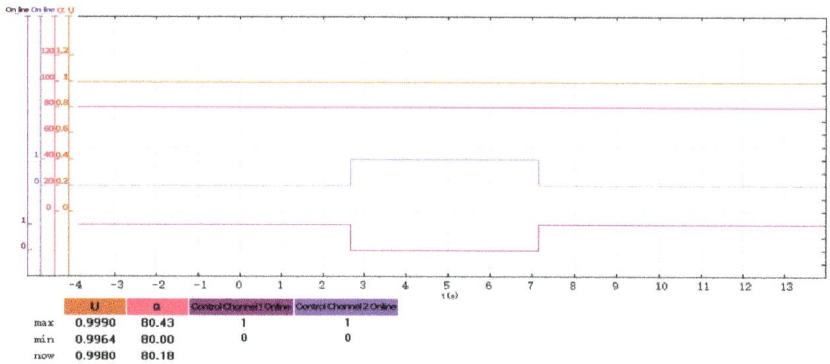

Figure 15. The online calculation module switches between Control Channel 1 and Control Channel 2 of the regulation cabinet.

As shown in Figure 16, the experiment validates the redundancy mechanism between computational modules of Control Channel 1 (Regulation Cabinet) and Power Cabinet 1 (denoted as SCR1 in the legend) by actively switching the online priority of computational modules across channels. When the active module transitions from Control Channel 1 (Regulation Cabinet) to Power Cabinet 1's Control Channel, the generator terminal voltage exhibits no abrupt fluctuations. Similarly, switching back to Control Channel 1's computational module maintains voltage stability. The results confirm that the computational modules of both control channels achieve seamless control handover during real-time switching, with a voltage fluctuation deviation rate not exceeding 0.34%, thereby verifying the redundancy functionality between the modules of these channels.

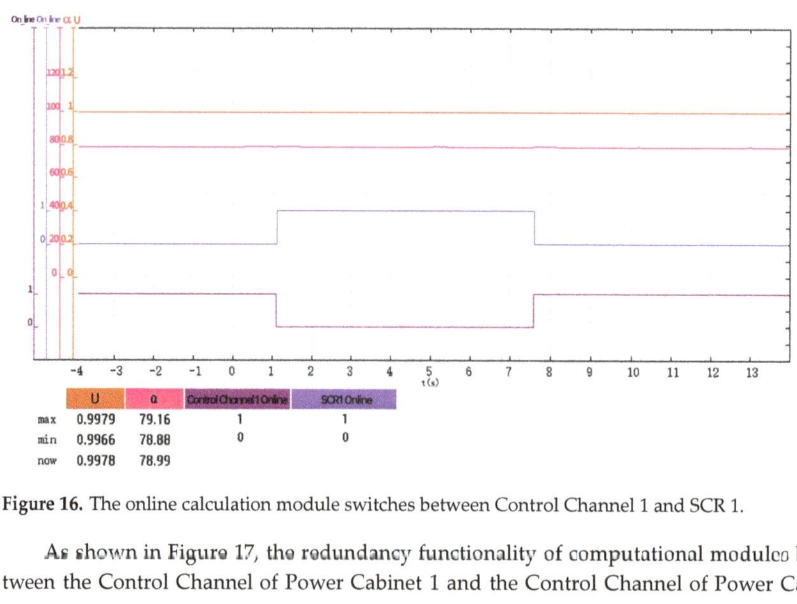

Figure 16. The online calculation module switches between Control Channel 1 and SCR 1.

As shown in Figure 17, the redundancy functionality of computational modules between the Control Channel of Power Cabinet 1 and the Control Channel of Power Cabinet 2 was experimentally validated. The results indicate that no abrupt fluctuations in generator terminal voltage occurred during the switching process, thereby confirming the redundancy capability between the computational modules of these two power cabinet control channels.

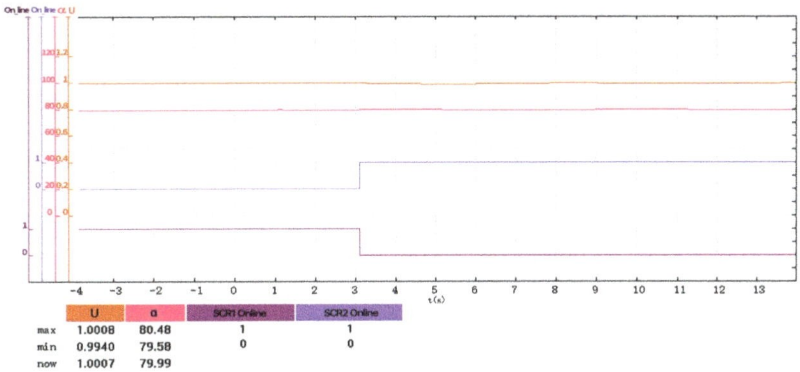

Figure 17. The online calculation module switches between SCR 1 and SCR 2.

The test items and summary are shown in Table 4.

Table 4. Summary of computing module redundancy switching test results.

No.	Test Item	Maximum Voltage Deviation Rate at Generator Terminals
1	Switching from Channel 1 to Channel 2 and back to Channel 1	0.36%
2	Switching from Channel 1 to Power Cabinet 1 and back to Channel 1	0.34%
3	Switching from Power Cabinet 1 to Power Cabinet 2	0.6%

The test results for the redundant switching of computational modules demonstrate that the maximum deviation rate of the generator terminal voltage is effectively controlled within 2%, strictly complying with the national standard's technical specification limit (a maximum allowable fluctuation of 5% during redundancy switching). These experimental data validate the effectiveness of the 2 + 3 redundant architecture implemented in the computational modules.

4.3. Redundant Switching Test of Acquisition Module

In the peer-to-peer distributed excitation system, the synchronization signal, as the core acquisition parameter with the highest requirements for real-time performance and stability, directly impacts the system's dynamic response characteristics through its redundant reliability. This experiment validates the effectiveness of the system's redundancy design through redundant switching of synchronization signals. As shown in Figure 18, when the Phase A synchronization source in the control channel of Power Cabinet 1 is disconnected, the system seamlessly switches to the Phase A synchronization source output by the acquisition unit of Power Cabinet 2 through its redundancy mechanism. The current-sharing coefficient measured during the test is 99.45%.

As shown in Figure 19, when the B-phase synchronization source of Power Cabinet 1's control channel is disconnected, the system automatically switches the synchronization signal to the B-phase synchronization source output by the acquisition unit of Power

Cabinet 2, utilizing a redundant design. The current-sharing coefficient during the test is 99.44%.

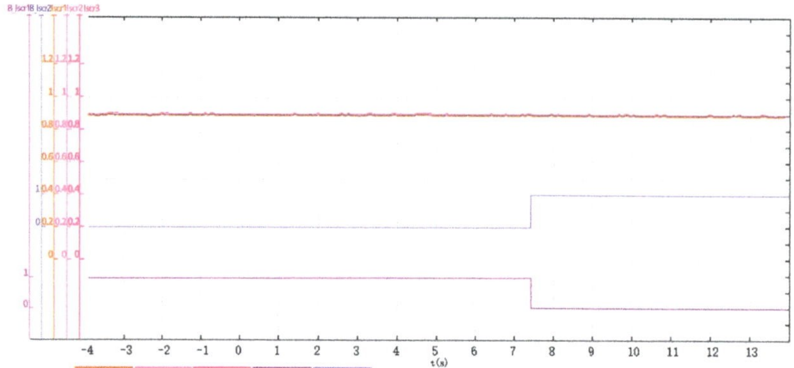

	Iscr1	Iscr2	Iscr3	A_scr1	A_scr2
max	0.8982	0.9080	0.9027	1	1
min	0.8879	0.8981	0.8925	0	0
now	0.8957	0.9058	0.9001		

Figure 18. Redundancy test for fault switching of Phase A synchronous sources between the Control Channel of Power Cabinet 1 and the Control Channel of Power Cabinet 2.

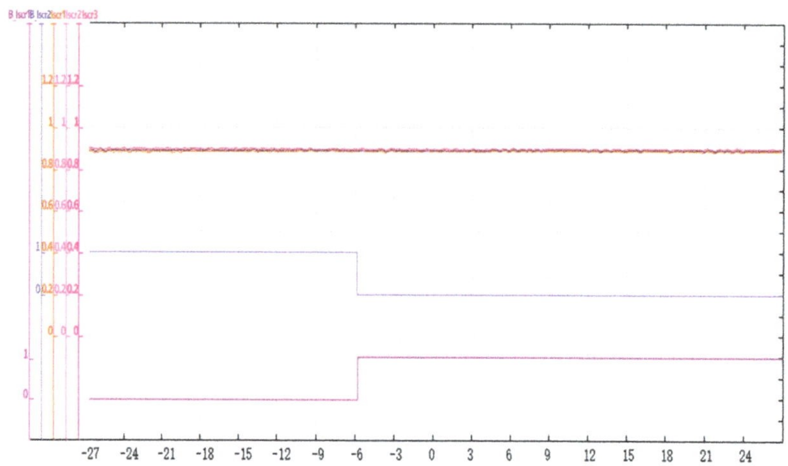

	Iscr1	Iscr2	Iscr3	B_scr1	B_scr2
max	0.8969	0.9070	0.9008	1	1
min	0.8807	0.8936	0.8870	0	0
now	0.8891	0.8987	0.8933		

Figure 19. Redundancy test for fault switching of Phase-B synchronous sources between the Control Channel of Power Cabinet 1 and the Control Channel of Power Cabinet 2.

Synchronized signal source acquisition redundancy switching test items and a summary are provided in Table 5.

The results of the synchronous signal acquisition redundancy switching test show that the system's current-sharing coefficient stays stable at ≥98% during the switching process. Throughout the test, all five computational modules, non-faulty acquisition modules, and power unit execution modules remain operational. This confirms the high availability of both the mutual redundancy among acquisition modules and the cross-redundancy between "acquisition-computation" modules.

Table 5. Synchronization source acquisition redundancy switching test results summary table.

No.	Test Item	Theoretical Synchronization Signal Switching Results	Actual Synchronization Signal Switching Test Results	Current Sharing Coefficient	Active Computing Module
1	Isolate Phase A synchronization source of Power Cabinet 1	Switch to Phase A of Power Cabinet 2	Switch to Phase A of Power Cabinet 2	99.45%	Computing Module 1 operating as primary
2	Isolate Phase B synchronization source of Power Cabinet 1	Switch to Phase B of Power Cabinet 2	Switch to Phase B of Power Cabinet 2	99.44%	Computing Module 1 operating as primary

Under no-load rated operating conditions of the dynamic simulation unit, the excitation system operates in Automatic Voltage Regulation (AVR) mode. By sequentially disconnecting the input signal sources of the generator terminal voltage acquisition modules in the regulation cabinet and power cabinets, the system's redundancy reliability is validated. As shown in Figure 20, disconnecting the generator terminal voltage input source (PT1) of Control Channel 1 in the regulation cabinet results in a maximum voltage deviation rate of 0.71%, while Control Channel 1 remains fully operational throughout the test.

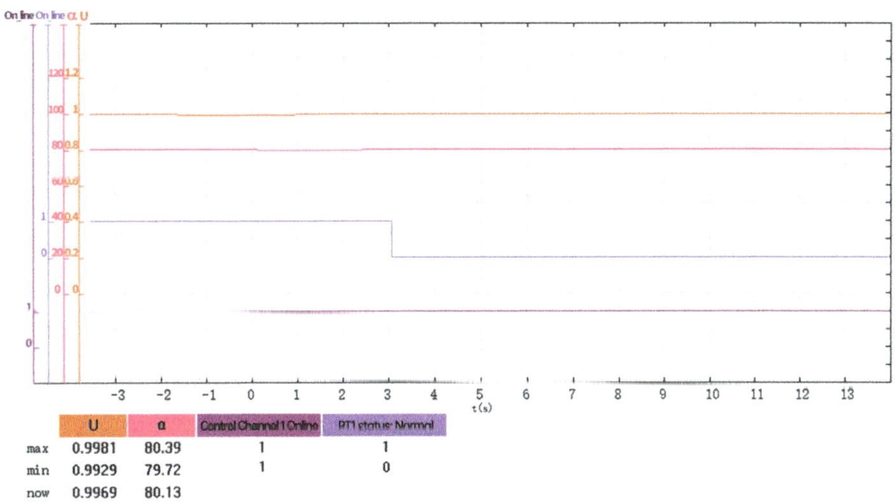

Figure 20. Redundancy test for acquisition fault of generator terminal voltage in Control Channel 1 of the regulation cabinet.

The test data for redundant switching of generator terminal voltage sampling signals indicate that the maximum voltage deviation is strictly controlled within 1%, meeting the national standard's technical requirement (fluctuation rate < 5% during redundancy switching). Throughout the test, all five computational modules, acquisition modules with non-disrupted signal sources, and power unit execution modules functioned normally, while preset Control Channel 1 remained fully operational. These results further validate the mutual redundancy between acquisition units and the cross-redundancy between "acquisition-computation" modules in the system.

4.4. Redundancy Switching Test During Load Rejection Transient

To validate the effectiveness of redundant switching functionality during transient processes, a load rejection test was conducted, focusing on the impact of online control channel switching on generator terminal voltage stability. As shown in Figure 21, a load rejection disturbance at 1.8 s triggered voltage fluctuations and an increase in unit frequency. Subsequently, during the voltage recovery phase, an online control channel switchover was performed. Throughout the switching process, the voltage recovery transitioned smoothly and stabilized at the rated value, with no secondary fluctuations observed due to the switching action.

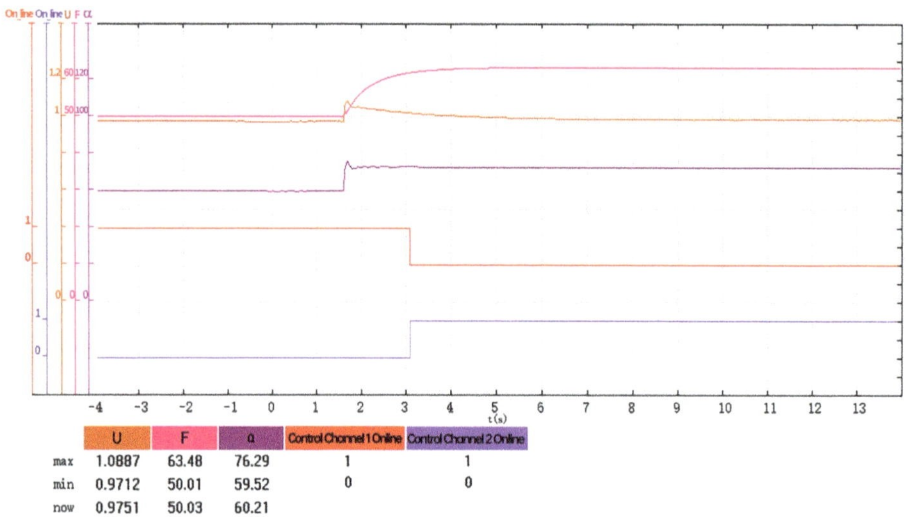

Figure 21. Redundancy test for online control channel switching during transient processes.

5. Discussion

With the ongoing progress of the energy transition, the widespread integration of renewable energy sources like wind and solar power presents significant challenges to the flexible regulation capabilities of power systems. Hydroelectric units, leveraging their quick response times and power regulation advantages, have become essential in modern power systems for stabilizing fluctuations and maintaining grid reliability. As the primary control equipment of synchronous generators, excitation systems directly influence power angle stability and voltage support capacity through dynamic adjustments of excitation currents. However, traditional distributed excitation systems face limitations in redundancy design. To overcome these challenges, this study combines ATS and DDS technologies to create a distributed real-time data exchange framework. All control nodes have identical configurations at both the hardware and software levels. When adding a new controller, it only needs to be connected to the ATS switches and DDS networks. Then, the domain discovery mechanism built into DDS automatically detects topology and synchronizes data, removing the need for changes to central controller settings. This allows for online, dynamic system expansion. Experimental results showed coordinated operation among five controllers, with the system theoretically scalable to N nodes. Without extra hardware costs, the reliability of the excitation system has been effectively improved through software and protocol modifications.

Validation through a dynamic simulation platform demonstrates the effectiveness of the peer-to-peer distributed excitation system's redundancy features, with key metrics including a voltage deviation of $\leq$1% and a current-sharing coefficient of $\geq$98%. These findings provide empirical evidence of operational reliability in high-renewable-penetration grids scenarios.

Author Contributions: Conceptualization, H.W.; Methodology, X.W., X.D., X.Y. and H.W.; Software, X.H.; Validation, X.H.; Formal analysis, X.D.; Investigation, X.W.; Resources, X.D.; Data curation, X.W.; Writing—original draft, X.W.; Writing—review & editing, X.W.; Supervision, X.D., X.Y. and X.L.; Project administration, X.L. All authors have read and agreed to the published version of the manuscript.

Funding: This work is supported by the Three Gorges Intelligent Control Technology Co., Ltd. Research Project (NBZZ202400220).

Data Availability Statement: The data that support the findings of this study are available on request from the corresponding author. The data are not publicly available due to privacy or ethical restrictions.

Acknowledgments: I sincerely thank the editors for their insightful guidance and extend gratitude for their invaluable academic support throughout this research.

Conflicts of Interest: Authors Xuxin Yue, Haoran Wang, Xiaokun Li and Xuemin He were employed by the company Three Gorges Intel-ligent Control Technology Co., Ltd. The remaining authors declare that the research was conducted in the absence of any commercial or financial relationships that could be construed as a potential conflict of interest.

Appendix A

Table A1. Variable description summary table.

Variable Name	Description
I_i	multi-level sequential queues
C_i	capacity limit
L_i	queue length
q_i	queue limit
R	the maximum port data packet forwarding rate
V	Traffic intensity
JT_i	Duration of the dynamic process
JP_i	Total data packet count in the dynamic process
T_{hrt}	the hard real-time deadline threshold
T_{srt}	the soft real-time delay budget threshold
$\mathcal{T}_{\text{critical}}$	the set of critical data types
d_i	delay upper bound
ω_i	the estimated service time
ε	Allowable error in the maximum waiting time
$r_i(t)$	the allocated transmission rate
B_i	maximum bucket capacity
$T_i(t)$	the current token count
α	the multiplicative decreasing factor
β	the additive increment

Appendix B

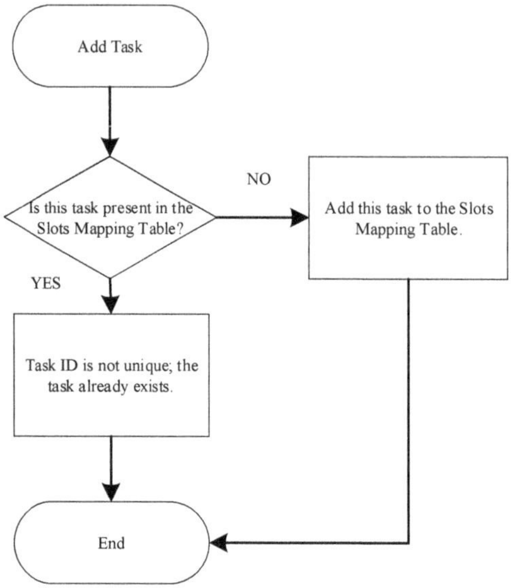

Figure A1. Flowchart for priority task registration process.

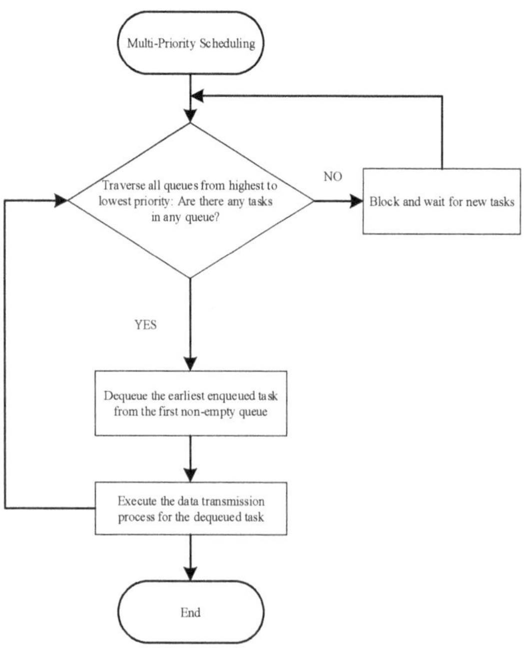

Figure A2. Flowchart of priority task execution process.

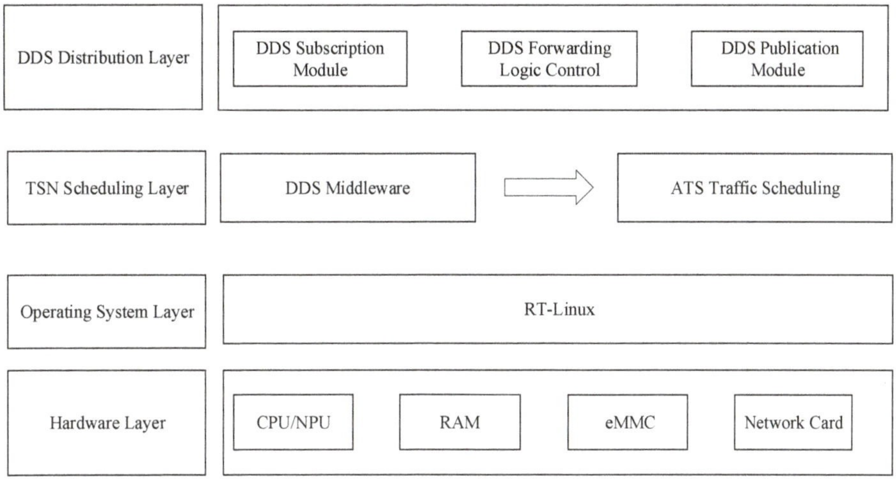

Figure A3. ATS and DDS system architecture diagram.

Appendix C

Figure A4. Physical diagram of the dynamic simulation platform.

The physical diagram of the dynamic simulation platform utilized in the dynamic simulation test comprises a small dynamic simulation test cabinet, a high-voltage low current test cabinet, a low-voltage high-current test cabinet, a dynamic simulation unit, and a reactor, among other components. The text "配电网" in the diagram denotes "Distribution Network", and "有电危险" denotes "Danger Electrical Hazard".

References

1. Sharma, P.; Reiman, A.P.; Anderson, A.A.; Poudel, S.; Allwardt, C.H.; Fisher, A.R.; Slay, T.E.; Mukherjee, M.; Dubey, A.; Ogle, J.P.; et al. GridAPPS-D Distributed App Architecture and API for Modular and Distributed Grid Operations. *IEEE Access* **2024**, *12*, 39862–39875.
2. Mesbah, M.A.; Sayed, K.; Ahmed, A.; Aref, M.; Gaafar, M.A.; Mossa, M.A.; Almalki, M.M.; Alghamdi, T.A.H. A distributed architecture of parallel buck-boost converters and cascaded control of DC microgrids-real time implementation. *IEEE Access* **2024**, *12*, 47483–47493.
3. dos Santos Alonso, A.M.; Brandao, D.I.; Caldognetto, T.; Marafão, F.P.; Mattavelli, P. A selective harmonic compensation and power control approach exploiting distributed electronic converters in microgrids. *Int. J. Electr. Power Energy Syst.* **2020**, *115*, 105452.

4. Naderi, Y.; Hosseini, S.H.; Zadeh, S.G.; Mohammadi-Ivatloo, B.; Vasquez, J.C.; Guerrero, J. An overview of power quality enhancement techniques applied to distributed generation in electrical distribution networks. *Renew. Sustain. Energy Rev.* **2018**, *93*, 201–214.
5. Gayatri, M.T.L.; Parimi, A.M.; Kumar, A.V.P. A review of reactive power compensation techniques in microgrids. *Renew. Sustain. Energy Rev.* **2018**, *81*, 1030–1036.
6. Li, S.; Geng, H.; Yang, G. A distributed control system of cascade static synchronous compensator. *Trans. China Electrotech. Soc.* **2017**, *32*, 153–159.
7. Qi, Q.; Wu, J. Increasing distributed generation penetration using network reconfiguration and soft open points. *Energy Procedia* **2017**, *105*, 2169–2174.
8. Zhu, X.; Hou, J.; Liu, L.; Zhang, B.; Wu, Y. A modular multiport DC power electronic transformer based on triple-active-bridge for multiple distributed DC units. *IEEE Trans. Power Electron.* **2024**, *39*, 15191–15205.
9. Bellini, A.; Bifaretti, S.; Giannini, F. A robust synchronization method for centralized microgrids. *IEEE Trans. Ind. Appl.* **2014**, *51*, 1602–1609.
10. Kandula, A.; Verma, V.; Solanki, S.K.; Solanki, J. Comparative analysis of self-synchronized virtual synchronous generator control and droop control for inverters in islanded microgrid. In Proceedings of the 2019 North American Power Symposium (NAPS), Wichita, KS, USA, 13–15 October 2019; IEEE: Piscataway, NJ, USA, 2019; pp. 1–5.
11. Palahalli, H.; Hemmati, M.; Gruosso, G. Analysis and design of a smart controller for managing penetration of renewable energy including cybersecurity issues. *Electronics* **2022**, *11*, 1861.
12. Zheng, Y.; Zhang, C.; Hill, D.J.; Meng, K. Consensus control of electric spring using back-to-back converter for voltage regulation with ultra-high renewable penetration. *J. Mod. Power Syst. Clean Energy* **2017**, *5*, 897–907.
13. Hussan, U.; Wang, H.; Ayub, M.A.; Rasheed, H.; Majeed, M.A.; Peng, J.; Jiang, H. Decentralized stochastic recursive gradient method for fully decentralized OPF in multi-area power systems. *Mathematics* **2024**, *12*, 3064.
14. Hussan, U.; Waheed, A.; Bilal, H.; Wang, H.; Hassan, M.; Ullah, I.; Peng, J.; Hosseinzadeh, M. Robust Maximum Power Point Tracking in PV Generation System: A Hybrid ANN-Backstepping Approach With PSO-GA Optimization. *IEEE Trans. Consum. Electron.* **2025**. [CrossRef]
15. Maletić, Ž.; Mlađen, M.; Ljubojević, M. A survey on the current state of time-sensitive networks standardization. In Proceedings of the 2023 10th International Conference on Electrical, Electronic and Computing Engineering (IcETRAN), East Sarajevo, Bosnia and Herzegovina, 5–8 June 2023; IEEE: Piscataway, NJ, USA, 2023; pp. 1–6.
16. Huang, T.; LV, J.J.; Zhu, H.L.; Zhang, H.Y.; Gu, Q.M. Automotive In-Vehicle Time-Sensitive Networking: The State of the Art and Prospect. *J. Beijing Univ. Posts Telecommun.* **2023**, *46*, 46.
17. Specht, J.; Samii, S. Urgency-based scheduler for time-sensitive switched ethernet networks. In Proceedings of the 2016 28th Euromicro Conference on Real-Time Systems (ECRTS), Toulouse, France, 5–8 July 2016; IEEE: Piscataway, NJ, USA, 2016; pp. 75–85.
18. Le Boudec, J.Y. A theory of traffic regulators for deterministic networks with application to interleaved regulators. *IEEE/ACM Trans. Netw.* **2018**, *26*, 2721–2733.
19. Zhao, L.; Pop, P.; Steinhorst, S. Quantitative performance comparison of various traffic shapers in time-sensitive networking. *IEEE Trans. Netw. Serv. Manag.* **2022**, *19*, 2899–2928.
20. Shin, I.; Choi, D.; Choi, H. Monitoring of digital substations using DDS. In Proceedings of the 2019 54th International Universities Power Engineering Conference (UPEC), Bucharest, Romania, 3–6 September 2019; IEEE: Piscataway, NJ, USA, 2019; pp. 1–5.
21. *IEC 61850*; Communication Networks and Systems for Power Utility Automation. Edition 2.0. International Electrotechnical Commission: Geneva, Switzerland, 2011–2013.
22. Son, M.Y.; Kim, D.S.; Cha, J.H. Efficient DDS monitoring system for large amount of data. In Proceedings of the 2018 14th IEEE International Workshop on Factory Communication Systems (WFCS), Imperia, Italy, 13–15 June 2018; IEEE: Piscataway, NJ, USA, 2018; pp. 1–4.
23. Cho, D.S.; Yun, S.; Kim, H.; Kwon, J.; Kim, W.-T. Autonomous driving system verification framework with FMI co-simulation based on OMG DDS. In Proceedings of the 2020 IEEE International Conference on Consumer Electronics (ICCE), Las Vegas, NV, USA, 4–6 January 2020; IEEE: Piscataway, NJ, USA, 2020; pp. 1–6.
24. Meng, Y.; Xingmin, W.; Wei, S.Z. Design and Implementation of Emergency QoS of Publish-Subscribe Middleware Based on DDS. In Proceedings of the 2019 3rd International Conference on Electronic Information Technology and Computer Engineering (EITCE), Xiamen, China, 18–20 October 2019; IEEE: Piscataway, NJ, USA, 2019; pp. 2024–2027.
25. Object Management Group. *DDS Extensions for Time Sensitive Networking*; 1.0 beta; Object Management Group: Milford, MA, USA, 2023.
26. *IEEE Std 802.1Qcr-2020 (Amendment to IEEE Std 802.1Q-2018 as amended by IEEE Std 802.1Qcp-2018, IEEE Std 802.1Qcc-2018, IEEE Std 802.1Qcy-2019, and IEEE Std 802.1Qcx-2020)*; IEEE Standard for Local and Metropolitan Area Networks--Bridges and Bridged Networks-Amendment 34: Asynchronous Traffic Shaping. IEEE: Piscataway, NJ, USA, 6 November 2020; pp. 1–151.

27. *IEEE Std 1588-2019 (Revision ofIEEE Std 1588-2008)*; IEEE Standard for a Precision Clock Synchronization Protocol for Networked Measurement and Control Systems. IEEE: Piscataway, NJ, USA, 16 June 2020; pp. 1–499.
28. Shi, X. Research on Delay-Constrained Queue Scheduling Method Based on Strict Priority. Master's Thesis, Huazhong University of Science and Technology, Wuhan, China, 2023.

electronics

MDPI

Article

Research on Nonlinear Pitch Control Strategy for Large Wind Turbine Units Based on Effective Wind Speed Estimation

Longjun Li [1], Xiangtian Deng [1], Yandong Liu [1,*], Xuxin Yue [2], Haoran Wang [2], Ruibo Liu [2], Zhaobing Cai [2] and Ruiqi Cai [2]

1 School of Automation, Wuhan University of Technology, Wuhan 430072, China; li_lng@whut.edu.cn (L.L.); dengxt@whut.edu.cn (X.D.)

2 Three Gorges Intelligent Control Technology Co., Ltd., Wuhan 430070, China; yue_xuxin@ctg.com.cn (X.Y.); wang_haoran@ctg.com.cn (H.W.); liu_ruibo@ctg.com.cn (R.L.); cai_zhaobing@ctg.com.cn (Z.C.); cai_ruiqi@ctg.com.cn (R.C.)

* Correspondence: 348818@whut.edu.cn; Tel.: +86-15908686016

Abstract: With the increasing capacity of wind turbines, key components including the rotor diameter, tower height, and tower radius expand correspondingly. This heightened inertia extends the response time of pitch actuators during rapid wind speed variations occurring above the rated wind speed. Consequently, wind turbines encounter significant output power oscillations and complex structural loading challenges. To address these issues, this paper proposes a novel pitch control strategy combining an effective wind speed estimation with the inverse system method. The developed control system aims to stabilize the power output and rotational speed despite wind speed fluctuations. Central to this approach is the estimation of the aerodynamic rotor torque using an extended Kalman filter (EKF) applied to the drive train model. The estimated torque is then utilized to compute the effective wind speed at the rotor plane via a differential method. Leveraging this wind speed estimate, the inverse system technique transforms the nonlinear wind turbine dynamics into a linearized, decoupled pseudo-linear system. This linearization facilitates the design of a more agile pitch controller. Simulation outcomes demonstrate that the proposed strategy markedly enhances the pitch response speed, diminishes output power oscillations, and alleviates structural loads, notably at the tower base. These improvements bolster operational safety and stability under the above-rated wind speed conditions.

Keywords: large wind turbines; pitch control; effective wind speed estimation; inverse system approach

1. Introduction

In the context of the accelerating global energy transition, wind energy has emerged as the renewable energy source with the greatest potential for large-scale development, playing a pivotal role in achieving a sustainable energy transformation. According to the Global Wind Energy Report 2024 published in July 2017 [1], by the end of 2023, the global wind power installed capacity had surpassed 1000 GW, with onshore wind installations reaching a historic milestone by exceeding 100 GW in annual capacity additions for the first time—representing a remarkable 54% year-on-year growth.

As countries accelerate carbon neutrality initiatives, wind turbine technology is rapidly evolving towards gigawatt-scale systems and intelligent operation. However, the increasing mass of supersized wind turbines and excessive blade lengths introduce critical load related challenges [2]. Furthermore, wind energy exhibits inherent intermittency, stochastic variability, and limited controllability as a non-dispatchable power generation source within electrical grids [3,4], compounded by the rotational inertia of massive turbine components and the nonlinear dynamics of pitch control systems (including rate limitations, hysteresis effects, and control delays) [5], which collectively hinder timely responses to wind speed variations. These technical constraints not only cause rotational speed overshoots and significant power fluctuations under extreme operating conditions but also induce structural vibrations that exacerbate tower load issues. In high-wind regions where power regulation and mechanical loading problems are particularly acute [6], manufacturers typically enhance the blade mechanical robustness through material overdesign, though this approach substantially escalates production costs. The effective implementation of nonlinear pitch control strategies offers a dual-benefit solution: achieving the simultaneous optimization of power regulation and load mitigation while reducing manufacturing costs and operational failures.

The wind speed data obtained from wind measurement radars is often integrated into the pitch control system of wind turbines to optimize the automated control strategy [7]. Refence [8] studied and analyzed the optimal distance of wind measurements by LiDAR and used radar-based wind speed feedforward control to reduce wind turbine loads. Reference [9] proposed an adaptive feedforward control algorithm for LiDAR-based on PI control, which improves the tracking speed and smoothes the output power fluctuation. Reference [10] used LiDAR-captured wind speed variations to implement an RBF neural network for the simultaneous optimization of the pitch angle and torque, addressing the objectives of power maximization and structural load mitigation. References [11–13] proposed a model predictive pitch control algorithm based on radar-measured wind using a linear or nonlinear model of wind turbines to improve the wind energy utilization factor of large wind turbines. Reference [14] compared radar-based linear and nonlinear predictive controls for pitch regulation, with the latter exhibiting an enhanced stabilization of the rotational speed and power output. Reference [15] examined the operational mechanisms of nacelle acceleration feedback versus lidar feedforward control. The results demonstrate that the LIDAR feedforward diminishes the rotor speed and thrust variability through pulsating load reductions. In contrast, nacelle feedback amplifies these effects. A combined implementation achieves a tower first-mode suppression alongside a decreased rotor speed and power fluctuations.

Highly accurate and wide-range LiDAR is used to measure wind speed, but the application of this technology significantly increases the installation and maintenance costs of large wind turbines [16], and wind speed estimation is gradually becoming a research focus as a low-cost means. Reference [17] employed support vector regression (SVR) to process and predict real-time wind speed data acquired by SCADA systems. The particle swarm algorithm was employed to optimize the SVR parameters in this process. Then, a feedforward controller for the pitch angle was designed based on the optimized SVR model to smooth the wind turbine output power and optimize wind turbine loads. Reference [18] firstly used an extended Kalman filter to estimate the wind speed in the past period of time and then used polynomial curve fitting to extrapolate the short-term predicted wind speed and optimized it using a neural network, and the results showed that the shorter the prediction time, the higher the accuracy. Reference [19] predicts the wind speed signal at the hub of a wind turbine by building a neural network physical model and wavelet analyzes it to filter out the high-frequency components to obtain a smooth low-frequency wind that plays a decisive role in wind speed predictions. Reference [20], addressing the vast number of wind turbines in wind farms and the intricate spatial and temporal correlation of wind speed, presented a correlation analysis method that integrates the whale optimization algorithm with the hybrid Copula function. On the foundation of conducting a correlation analysis on the wind speed of wind turbines, the study went on to establish a wind speed prediction model. Reference [21] proposed that Taylor's turbulence freezing assumption neglects the evolution of the wind field structure, which will affect the accuracy of LiDAR wind measurements, so the autoregressive moving average with an exogenous input (Armax) model is used to model the wind evolution process and a particle swarm optimization algorithm is used to estimate the model parameters and finally predicted the effective rotor plane wind speed in advance based on the developed model. Reference [22] established the correlation between the tip–speed ratio, pitch angle, and power coefficient (Cp) via a Gaussian process regression using empirical wind turbine data, subsequently designing a wind speed estimator integrating this regression model with an extended Kalman filter. Reference [23] developed a turbulence intensity identification method using wind turbine operational data, established a simplified turbine model derived from the measured rotor speed and electromagnetic torque, and implemented an immersion-invariant observer for real-time wind speed estimations. Reference [24] introduced an adaptive neuro-fuzzy inference system with hybrid intelligent learning for online effective wind speed estimations, utilizing the real-time feedback of the tip–speed ratio, rotor speed, and mechanical power. References [25,26] employed an unknown input disturbance observer to accurately estimate the aerodynamic torque and utilized intelligent algorithms to approximate wind turbine aerodynamic characteristics. The effective wind speed at the rotor plane was subsequently determined from the measured rotor speed and pitch angle. This derived wind speed was integrated into the control loop to compute optimal control commands. Reference [27] included the rotor plane effective wind speed among the system dynamics to formulate the state space equation for the wind turbine. An extended Kalman filter was implemented to estimate this wind speed. Building upon the estimated wind speed, an adaptive model predictive controller (MPC) was developed. This design enhanced the system robustness and minimized power output fluctuations under high-wind-speed conditions. Reference [28] quantified the turbine structural deflection via Kalman filtering, thereby inferring the rotor plane effective wind speed from blade and tower bending characteristics.

Currently, an effective wind speed estimation is the primary means of reducing costs for large wind turbines, but it is primarily used in meteorology and has not been utilized to optimize control strategies. Therefore, this paper addresses power output fluctuations and overload issues in regions with high wind speeds. An effective wind speed estimation is performed under step and turbulent wind conditions, and an inverse system method is employed based on this estimation to design a nonlinear pitch control strategy. This significantly enhances the pitch system's dynamic response, effectively mitigating speed and power fluctuations during high-wind operations while substantially reducing tower base load amplitudes. Simultaneously, the integration of the turbine output power with critical operational state parameters enables high-accuracy estimations of the rotor plane effective wind speed. This reduces the reliance on lidar and cuts system manufacturing costs directly. Additionally, the inverse system method's low computational complexity results in lower hardware performance requirements for controllers during the engineering implementation. This significantly improves the pitch control quality without incurring hardware upgrade costs. This strategy enhances operational safety under extreme wind conditions and reduces maintenance costs for operators by alleviating structural fatigue.

2. Wind Turbine System Modeling

2.1. Wind Turbine Aerodynamic Modeling

The transformation of wind kinetic energy into electrical power within wind turbines is fundamentally governed by aerodynamic principles, but due to the complex computational problems involving the precise description of impeller geometric features and turbulence field resolution, a simplified model is usually used to characterize its energy conversion mechanism. The mathematical characterization of the model is as follows:

$$\begin{cases} P = \frac{1}{2} C_p(\lambda, \beta) \cdot \rho \cdot S \cdot v^3 \\ \lambda = \frac{\omega \cdot R}{v} \end{cases} \tag{1}$$

where ω denotes the rotational mechanical angular velocity (rad/s) of the wind turbine, R represents the rotor radius, and $C_p(\lambda,\beta)$ signifies the wind energy conversion coefficient. The determination of C_p involves multifaceted dependencies on the wind speed, rotor geometry, generator rotational speed, and blade pitch angle. During operational implementations, this coefficient is typically characterized by the following expression:

$$\begin{cases} C_p(\lambda, \beta) = 0.5176\left(\frac{116}{\lambda_i} - 0.4\beta - 5\right)e^{\frac{-21}{\lambda_i}} + 0.0068\lambda \\ \frac{1}{\lambda_i} = \frac{1}{\lambda + 0.08\beta} - \frac{0.035}{\beta^3 + 1} \end{cases} \tag{2}$$

The above equation represents, in detail, the relationship between the wind energy utilization factor C_p of the wind turbine with the blade tip speed ratio λ and pitch angle β, as shown in Figure 1.

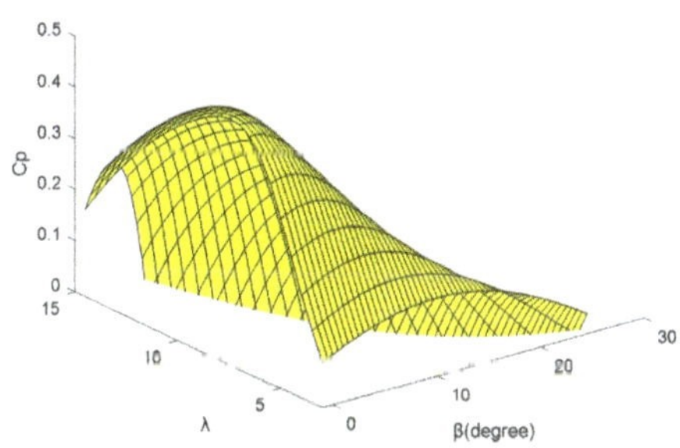

Figure 1. Wind energy utilization factor.

An analysis reveals that the operational characteristics vary significantly at specific blade pitch angles β, and there exists a uniquely determined optimal tip speed ratio λ_{opt} corresponding to the corresponding maximum wind energy utilization coefficient C_{pmax}, which allows the wind turbine to capture the most energy, and therefore, the tip speed ratio can be maintained as a constant by controlling the wind turbine speed, which leads to a stable and efficient operation of the wind turbine.

2.2. Wind Turbine Drive Train and Generator Modeling

In wind turbine drive train modeling, the system is typically conceptualized as multiple flexibly coupled rigid bodies. These representations are categorized as single-mass or multi-mass block models, depending on the number of rigid bodies in the system. In order to simplify the theoretical analysis and engineering calculations, this model adopts the equivalent inertia aggregation method: the equivalent inertia J_r on the low-speed side and the equivalent inertia J_g on the high-speed side are integrated and merged into a single equivalent inertia J_t that characterizes the overall characteristics of the drive system, and at the same time, the influence of the stiffness coefficient C_d and damping characteristics of the drive shaft system K is ignored, which results in the establishment of a single-block centralized parameter model as shown in Figure 2. While this omits high-frequency torsional oscillations, it retains a fidelity for pitch-dominated dynamics above the rated wind speed. Its operating equations are expressed as follows:

$$J_t\dot{\omega}_r = T_a - nT_e \tag{3}$$

$$\begin{cases} D_r\omega_r + J_r\dot{\omega}_r = T_a - T_{ls} \\ D_g\omega_g + J_g\dot{\omega}_g = \frac{T_{ls}}{n} - T_e \\ T_{ls} = K\gamma + C_d\dot{\gamma} \\ \gamma = \omega_r - \frac{\omega_g}{n} \end{cases} \tag{4}$$

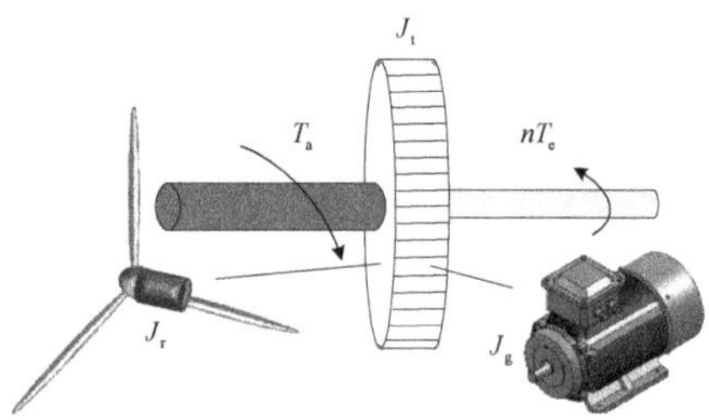

Figure 2. The equivalent single-mass block model of the driveline.

2.3. Modeling of Wind Turbine Pitch Systems

In large wind turbines, pitch actuators play an important role as an important component in all stages of operation. Due to the large inertia of the blades of a large wind turbine, there is a delay from the time the pitch controller issues a command to the time the actuator starts to perform its action. The time constant of this process is usually between 100 and 300 ms. Therefore, in the modeling process, the pitch actuator is usually equated to a first-order inertial link, whose transfer function can be expressed as follows:

$$\beta = \frac{k_1}{T_{act}s + 1}\beta_c \tag{5}$$

where β_c denotes the commanded pitch angle generated by the pitch controller, T_{act} represents the pitch actuator's inertial time constant, and k_1 signifies the gain of the inertial element, typically assigned a unit value.

Secondly, the pitch actuator also ensures the smooth operation of the wind turbine by limiting the amplitude and rate of the pitch change. The wind turbine pitch angle operates within a 0° to 90° range. The pitch rate varies by manufacturer specifications, with typical safety limits constrained to 6–8°/s. When the command value of the pitch angle exceeds the upper limit, the control command that exceeds the limit will not be responded to, so as not to cause adverse effects; the block diagram of the pitch actuator is shown in Figure 3.

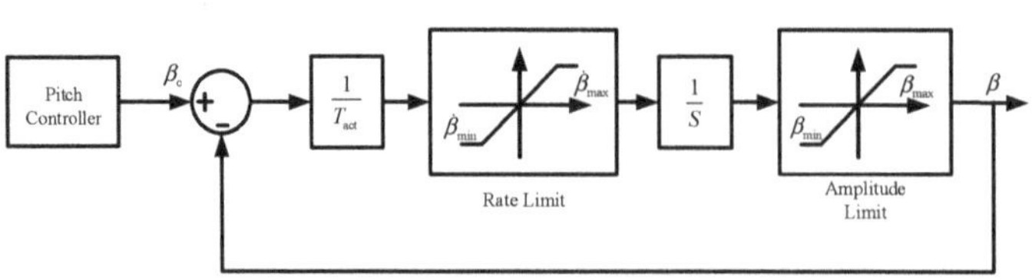

Figure 3. Block diagram of pitch actuator.

3. Wind Speed Estimation Based on Extended Kalman Filtering

This paper utilizes the extended Kalman filter (EKF) algorithm to estimate aerodynamic torque. The effective wind speed at the rotor plane is subsequently computed via a differential approach.

3.1. Extended Kalman Filter Theory

The extended Kalman filter is homologous to the classical Kalman filter in terms of its algorithmic architecture, which is divided into two parts: updating the timing and updating the measurements. Unlike the classical Kalman filter, the extended Kalman filter is a nonlinear approximation filter; i.e., it solves the nonlinear problem by local linearization on the basis of the classical algorithm. The extended Kalman filter can achieve better computational results when dealing with highly nonlinear systems, such as wind power systems. The prediction model of the extended Kalman filter can be expressed as follows:

$$x_{k+1} = \varphi(x_k, k) + \Gamma(x_k, k)w_k \tag{6}$$

$$z_{k+1} = h(x_{k+1}, k+1) + v_{k+1} \tag{7}$$

where k denotes the discrete moment; $\varphi(xk, k)$ denotes the nonlinear state function; $h(x_{k+1}, k+1)$ denotes the nonlinear measurement function; v_{k+1} denotes the measurement noise; w_k denotes the process evolution noise; $\Gamma(x_k, k)$ denotes the output matrix of the process evolution noise; x denotes the nonlinear state vector; and z denotes the nonlinear measurement vector.

Applying a first-order Taylor expansion to the nonlinear state function derives the prediction equation:

$$\begin{aligned} x_{k+1} &\approx \varphi(\hat{x}_k, k) + \frac{\partial \varphi}{\partial x_k}\bigg|_{x_k=\hat{x}_k}(x_k - \hat{x}_k) + \Gamma(\hat{x}_k, k)w_k \\ &= \frac{\partial \varphi}{\partial x_k}\bigg|_{x_k=\hat{x}_k} x_k + \varphi(\hat{x}_k, k) - \frac{\partial \varphi}{\partial x_k}\bigg|_{x_k=\hat{x}_k}\hat{x}_k + \Gamma(\hat{x}_k, k)w_k \end{aligned} \tag{8}$$

$$\hat{x}_{k|k-1} = \Phi_{k|k-1}\hat{x}_{k-1} \tag{9}$$

$$P_{k|k-1} = \Phi_{k|k-1}P_{k-1}\Phi_{k|k-1}^T + \Gamma_{k|k-1}Q_{k-1}\Gamma_{k|k-1}^T \tag{10}$$

$$\Phi_{k+1|k} = \frac{\partial \varphi}{\partial x_k}\bigg|_{x_k=\hat{x}_k} \tag{11}$$

Expanding the nonlinear measurement function, $h(x_{k+1}, k+1)$, via the first-order Taylor series updates the equation:

$$\begin{aligned} z_{k+1} &\approx h(\hat{x}_{k+1}, k+1) + \frac{\partial h}{\partial x_{k+1}}\bigg|_{x_{k+1}=\hat{x}_{k+1}}(x_{k+1} - \hat{x}_{k+1}) + v_k \\ &= \frac{\partial h}{\partial x_{k+1}}\bigg|_{x_{k+1}=\hat{x}_{k+1}} x_{k+1} + h(\hat{x}_{k+1}, k+1) - \frac{\partial h}{\partial x_{k+1}}\bigg|_{x_{k+1}=\hat{x}_{k+1}}\hat{x}_{k+1} + v_k \end{aligned} \tag{12}$$

$$K_k = P_{k|k-1}H_k^T\left(H_kP_{k|k-1}H_k^T + R_k\right)^{-1} \tag{13}$$

$$\hat{x}_k = \hat{x}_{k|k-1} + K_k\left(z_k - H_k\hat{x}_{k|k-1}\right) \tag{14}$$

$$P_k = (I - K_kH_k)P_{k|k-1}(I - K_kH_k)^T + K_kR_kK_k^T \tag{15}$$

$$H_{k+1} = \frac{\partial h}{\partial x_{k+1}}\bigg|_{x_{k+1}=\hat{x}_{k+1}} \tag{16}$$

3.2. Pneumatic Torque Estimation

Wind speed, as an external perturbation of the wind turbine, is difficult to integrate into the state space equations of the system to solve, so the aerodynamic torque is first estimated, and then the corresponding wind speed is calculated using the wind power curve. The aerodynamic torque, being solely governed by wind speed, is thus modeled via a first-order Markov process.

A Markov process is a class of stochastic processes, specifically for time series. If the probability of the present state depends only on the previous state, it is called a first-order Markov process, which is expressed in mathematical language as follows:

$$P(S_n = i_n | S_{n-1} = i_{n-1}, \ldots, S_0 = i_0) = P(S_n = i_n | S_{n-1} = i_{n-1}) \tag{17}$$

where S_n denotes the random variable S after the nth time step, and in represents the value of S taken.

Based on the above properties, we can get the value of the random variable taken at each moment:

$$\begin{gathered} P(S_n = i_n, S_{n-1} = i_{n-1}, \ldots, S_0 = i_0) = P(S_n = i_n | S_{n-1} = i_{n-1}) \\ \ldots P(S_1 = i_1 | S_0 = i_0) P(S_0 = i_0) \end{gathered} \tag{18}$$

Thus, the first-order Markov process for pneumatic torque can be expressed as follows:

$$\dot{T}_{\mathrm{a}} = \frac{-1}{T_{\mathrm{t}}} T_{\mathrm{a}} + \varepsilon \tag{19}$$

where T_t is the time constant of the pneumatic torque, and ε is the Gaussian white noise.

Based on the equivalent two-mass block model of the drive chain, the desired equation of the motion of the drive train is obtained as follows:

$$\begin{cases} \dot{\omega}_r = \frac{T_a}{J_r} - \frac{C_d \dot{\gamma} + K\gamma}{J_r} \\ \dot{\omega}_g = \frac{C_d \dot{\gamma} + K\gamma}{N J_g} - \frac{T_e}{J_g} \\ \dot{T}_a = -\frac{T_a}{T_t} \\ \dot{\gamma} = \omega_r - \frac{\omega_g}{N} \end{cases} \tag{20}$$

The corresponding state space equations are

$$\begin{cases} \dot{X} = AX + BU \\ Y = CX + \eta \end{cases} \tag{21}$$

Among them

$$X = \begin{bmatrix} \omega_r & \omega_g & T_a & \gamma \end{bmatrix}^{\mathrm{T}} \tag{22}$$

$$U = T_e \tag{23}$$

$$A = \begin{bmatrix} -\frac{C_d}{J_r} & \frac{C_d}{N J_r} & \frac{1}{J_r} & -\frac{K}{J_r} \\ \frac{C_d}{N J_g} & -\frac{C_d}{N^2 J_g} & 0 & \frac{K}{N J_g} \\ 0 & 0 & -\frac{1}{T_t} & 0 \\ 1 & -\frac{1}{N} & 0 & 0 \end{bmatrix} \tag{24}$$

$$B = \begin{bmatrix} 0 & -\frac{1}{J_g} & 0 & 0 \end{bmatrix}^{\mathrm{T}} \tag{25}$$

$$C = \begin{bmatrix} 1 & & & \\ & 0 & & \\ & & 0 & \\ & & & 0 \end{bmatrix} \tag{26}$$

where η is the measurement noise of the system. Taking the sampling period T_s = 0.001 s, using the bilinear transformation method to discretize the above state-space model, and considering the process noise of the system and the measurement noise of the sensor, the above expression is formalized as follows:

$$\begin{cases} X(k+1) = A_z X(k) + B_z U(k) \\ Y(k) = C_z X(k) + D_z U(k) \end{cases} \tag{27}$$

Among them

$$A_z = e^{AT} = \sum_{k=0}^{\infty} \frac{A^k T^k}{k!} \tag{28}$$

$$B_z = \int_0^T e^{A\tau} d\tau B = \sum_{k=0}^{\infty} \frac{A^k T^{k+1}}{(k+1)!} B \tag{29}$$

3.3. Wind Speed Estimation

Based on the extended Kalman filtering approach, the aerodynamic torque of the wind turbine, a strongly nonlinear system, has been accurately estimated in this paper. For the estimation of the effective wind speed, the Newton–Raphson algorithm has been used in the literature [29,30]. However, the wind turbine power curve is very complex and difficult to accurately represent by equations, making the derivation process in the Newton–Raphson algorithm challenging and prone to obtaining inaccurate results. This section implements a finite difference method approach to iteratively determine wind speed, eliminating derivational uncertainties in the solution process. The differential step size is set to 0.01 s. This value is determined through multiple trials and can strike a balance between the calculation accuracy and convergence speed, ensuring the stability of the iterative process and the accuracy of the results.

The expression of the wind turbine aerodynamic torque is given by

$$T_a = P/\omega_r \tag{30}$$

The effective wind speed was computed iteratively using the following formulation:

$$\hat{V}(t) = \text{argmin} S(t, v) \tag{31}$$

$$S(t, v) = (\hat{T}(t) - J(v))^2 \tag{32}$$

where $\hat{V}(t)$ denotes the optimal effective wind speed, $\hat{T}(t)$ represents the Kalman-filtered aerodynamic torque estimate, $J(v)$ defines the optimization criterion, and $S(t,v)$ quantifies the variance between the estimated aerodynamic torque and the objective function.

The optimization criterion corresponds to the measured aerodynamic torque.

$$J(v) = T_a \tag{33}$$

Ultimately, effective wind speed estimates are iteratively computed:

$$\hat{v}_{n+1} = \hat{v}_n - J_n^{-1} K_n \tag{34}$$

where J_n denotes the discrepancy between the aerodynamic torque computed at the current wind speed and its Kalman-filtered estimate, combined with K_n, this yields the differential wind speed component:

$$J_n = \hat{T}_a - \frac{1}{2}\rho\pi R^3 C_p(\lambda, \beta) V^2/\lambda \tag{35}$$

$$K_n = -\rho\pi R^3 \hat{v}_n C_p(\hat{\lambda}_n)/\lambda + \frac{1}{2}\rho\pi\hat{\omega}_r R^4 \frac{\partial C_q(\hat{\lambda}_n)}{\partial\lambda} \tag{36}$$

We repeat the above iterative process until the convergence condition is met and calculate the effective wind speed of the impeller surface. The flowchart of the algorithm is shown in Figure 4.

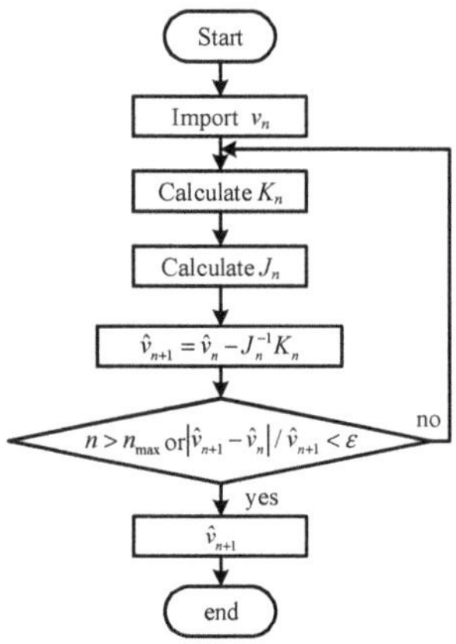

Figure 4. Effective wind speed iterative calculation process.

4. Nonlinear Pitch Control Strategy

The realization of the wind speed integration to the pitch control system can be mainly divided into two types of technical paths: one is to maintain the original closed-loop pitch control architecture on the basis of an external wind speed disturbance compensation mechanism to achieve feedforward control; the other is based on the dynamic model of the wind turbine reconfiguration of the entire closed-loop control strategy of the systematic design approach. Compared with the systematic design method that needs to reconstruct the control framework, the feedforward compensation scheme is widely used in engineering practice due to its unique robustness advantage—adjusting the system's closed-loop pole distribution only through a perturbation compensation without changing the zero-point configuration, which fundamentally ensures the stability of the control system. This paper applies the inverse system control theory to develop a disturbance-compensating pitch strategy, leveraging a precise wind speed estimation via the inverse system method.

4.1. Fundamentals of Inverse Systems Theory Analysis

The inverse system method is a kind of method to convert a nonlinear system into a linear, decoupled standard form (i.e., pseudo-linear system) by constructing an "inverse system" model of the control object and then utilizing the design theory of the linear system to carry out the comprehensive design of the system. In the research field of nonlinear system control, the inverse system approach opens up completely new control strategies for researchers.

In general, for reversible systems, the steps to realize their inverse systems are as follows:

1. Find the inverse system $\hat{\Sigma}$ of the original system Σ and its initial value.
2. Find the inverse of the α-order integral system $\hat{\Sigma}_\alpha$ and its initial value.
3. Cascade the system Σ with its α-order integral inverse system to form a pseudo-linear composite system $\Sigma\hat{\Sigma}_\alpha$, which successfully decouples and linearizes the controlled object.
4. Considering each subsystem contained in the above pseudo-linear composite system $\Sigma\hat{\Sigma}_\alpha$ as a controlled object, the design of the target control system is carried out by utilizing the design method of a single-variable linear system, such as the frequency response method or the root trajectory correction method.

Figure 5 shows the control loop designed based on the inverse system method.

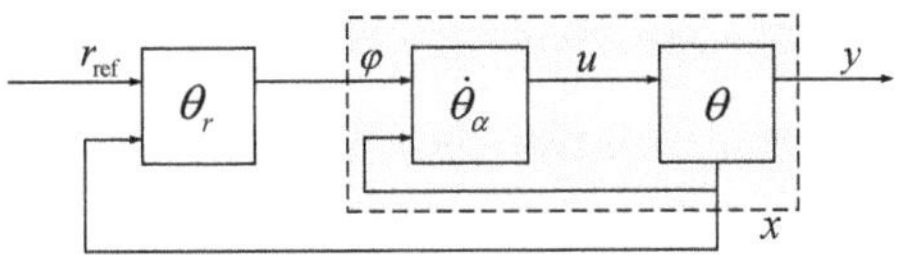

Figure 5. The schematic diagram of the inverse system approach.

Where r_{ref} is the reference value of the input signal, θ_r is the linear controller of the pseudo-linear system after cascading within the dotted box, $\dot{\theta}_\alpha$ is its α-order derivative, u is the system input, y is the system output, and φ is the error.

4.2. Pitch Controller Design

The large inertia of a large wind turbine causes the response speed of the electromagnetic torque and pitch control system to be much greater than that of the large wind turbine itself. The main manifestation is that the inertia lag time constant generated by the equivalent moment of the inertia J_t of the low-speed and high-speed axes of the drive chain far exceeds the time constant T_{act} of the pitch actuator and the electromagnetic time constant of the generator. In the singular regeneration theory, the system dynamics are divided into fast and slow two modes, ignoring the changes in the "fast mode" system, and it only studies the "slow mode" system; the fast system is considered to quickly enter the quasi-steady state [31], and the wind generator system model is downgraded and simplified to

$$\dot{\omega}_r = \frac{1}{J_t}\left(\overline{T}_a - \overline{T}'_e\right) \tag{37}$$

$$\overline{T}_a = \frac{1}{2}C_q(\lambda, \beta_c)\pi R^3 \rho v^2 \tag{38}$$

where $\overline{T}'_e$ represents the quasi-steady state value of the generator's electromagnetic torque, then $\overline{T}_a$ represents the quasi-steady state value of the aerodynamic torque; the value of which is determined by β_c, which is the given value of the pitch angle before the pitch actuator.

Under the high-wind-speed section above the rated operating condition studied in this paper, the generator's electromagnetic torque has reached its maximum value and cannot continue to increase, so the wind turbine speed is controlled only by changing the applied aerodynamic torque through the pitch, and the electrical dynamics of the generator's electromagnetic torque can be neglected under this operating condition, i.e.,

$$T'_e = \overline{T}'_e = K_g T_{eN} \tag{39}$$

where T_{eN} is the rated electromagnetic torque of the generator.

Using the wind turbine rotational speed and its derivatives, the rotor-effective wind speed, and the previously estimated pitch angle input β_c substituted into Equation (37) produces

$$\beta = h^{-1}(\omega_r, v, \dot{\omega}_r) \tag{40}$$

This equation constitutes the inverse system for the reduced-order model (3–56).

Figure 6 shows the torque coefficient curve C_q for a wind turbine. If the blade tip speed ratio λ is certain, the relationship between C_q and the pitch angle β can be described by a parabolic equation in the two-dimensional plane. Thus, for a given C_q, there exists one to two values of β corresponding to it.

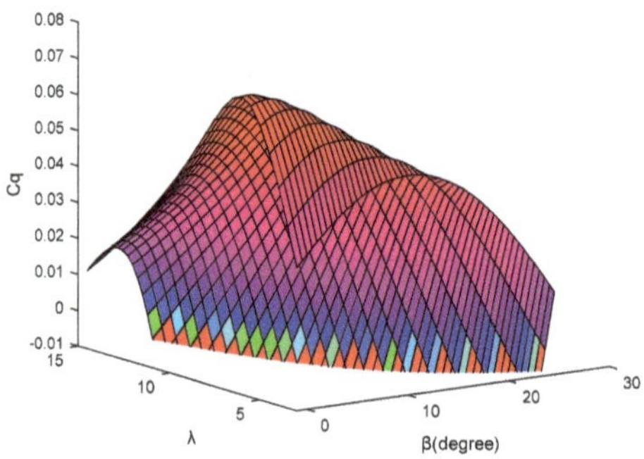

Figure 6. Wind turbine torque coefficient curve.

Continuing with the proof of the uniqueness of the above pitch angle solution, it is necessary to use the torque–speed curves of a typical pitch wind turbine system shown in Figure 7. In the figure, different pitch angles and wind speeds correspond to different curves, with a wind speed of $v_1 > v_2$ and pitch angle of $\beta_1 < \beta_2$. When the wind turbine operates at the intersection of curve ODB and curve OEC at A, it satisfies $T_a = T_A$ and $\omega_r = \omega_A$. At this point, at a certain wind speed and rotational speed—i.e., the above-mentioned wind turbine torque coefficient C_q is certain—there are two solutions for β, which are β_1 and β_2, respectively.

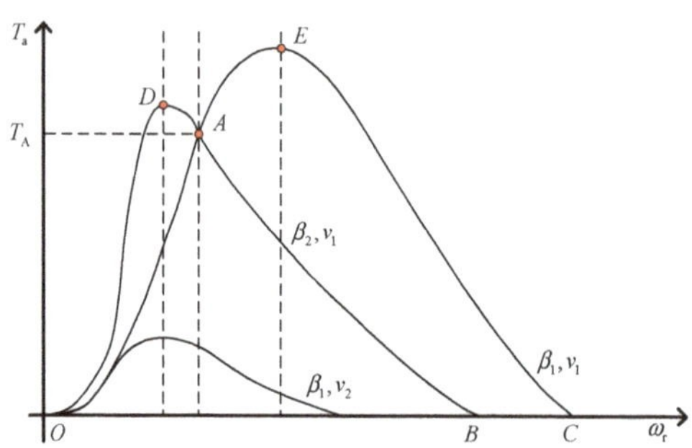

Figure 7. Wind turbine speed–torque curve.

The analysis of the stability of the working point A shows that when the rotational speed ω_r appears as a positive (negative) perturbation, the aerodynamic torque Ta on the curve segment OAE will increase (decrease) accordingly, and since the electromagnetic torque that reaches the limit value has been kept unchanged, it is known from the rotor motion equation that the rotational speed ω_r continues to increase (decrease), which results in the system getting further and further away from the working point A. On the contrary, in the curve segment DAB, no matter whether the rotational speed ω_r is increasing—or on the contrary, in the curve section DAB, no matter if the rotational speed ω_r increases or decreases—the aerodynamic torque will always be opposite to the trend of its change, i.e., $dT_a/d\omega_r < 0$, thus suppressing the change in the rotational speed. In summary, there exists a unique stable inverse system of system (40), and its corresponding pitch angle can be derived from the current speed–torque curve of the wind turbine.

If system (39) is cascaded with its inverse system (40) to form a "pseudo-linear" system, as shown in the dashed box in Figure 8, it is possible to use a linear system analysis to dynamically configure the cascaded system.

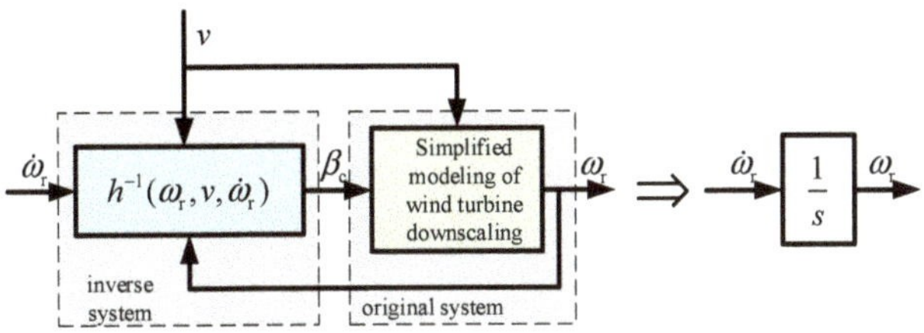

Figure 8. A schematic diagram of the inverse system approach to wind turbine modeling.

Where $\dot{\omega}_r$ is the rate of change in the rotor speed, ω_r is the rotor speed, v is the wind speed, $h^{-1}(\omega_r, v, \dot{\omega}_r)$ is the inverse system, β_c is the pitch angle setpoint before the variable pitch actuator, and s is the Laplace operator.

$$\ddot{\omega}_r + a_1\dot{\omega}_r + a_0\omega_r = r(t) \tag{41}$$

where a_1 and a_0, determined via pole placement, yield the reference input $r_{ref} = a_0\omega_{ref}$ for the inverse system pitch controller, and where ω_{ref} denotes the rated turbine speed. Figure 9 presents the proposed feedback-compensated control framework based on effective wind speed estimations within the inverse system methodology.

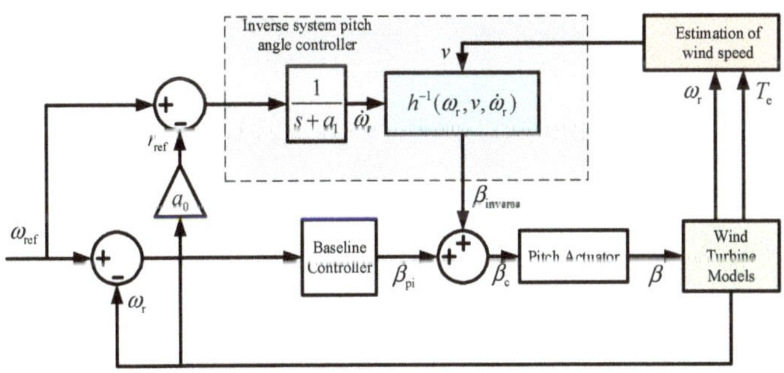

Figure 9. Block diagram of pitch control based on inverse system method for effective wind speed estimation.

Where ω_{ref} is the rotor speed reference value, r_{ref} is the given inverse system pitch angle, v is the wind speed, ω_r is the rotor speed, T_e is the electromagnetic torque, $\dot{\omega}_r$ is the rotor speed change rate, $h^{-1}(\omega_r, v, \dot{\omega}_r)$ is the inverse system, a_1 and a_0 are the inverse system controller parameters, β_{pi} is the PI controller output pitch angle value, $\beta_{inverse}$ is the inverse system output pitch angle value, β_c is the pitch angle setpoint before the variable pitch actuator, and s is the Laplace operator.

5. Simulation Verification

To validate the feasibility of the control strategy, joint simulations are conducted using the FAST nonlinear wind turbine platform and Matlab/Simulink. FAST provides a variety of standard wind turbine models, and this paper adopts an onshore 5 MW large-scale wind turbine for the study. Its detailed parameters are shown in Table 1.

Table 1. System design parameters.

Name	Parameters
Rated power	5 MW
Number of blades	3
Impeller diameter	126 m
Hub diameter	3 m
Cut-in wind speed	3 m/s
Rated wind speed	11.4 m/s
Cutting out wind speed	25 m/s
Rated electromagnetic torque	43,093.55 N·m
Rated speed	12.1 rpm
Rotor mass	110,000 kg
Blade initial pitch angle	0 deg
Drive train damping	6.22 × 106 N·m/(rad/s)
Drive train stiffness	8.67 × 108 N·m/rad

5.1. Simulation Results and Analysis of Wind Speed Estimation

In this section, firstly, the real-time estimation of the effective wind speed at the rotor plane of the wind turbine is carried out, and two wind conditions, the step wind changing from 13 m/s to 16 m/s and the turbulent wind with an average wind speed of 15 m/s and a turbulence degree of 10%, are selected for the experiment, where the actual wind speed is taken to be the wind speed captured at the hub in the FAST platform, and the maximum number of iterations in the iterative solution of the calculation of the effective wind speed nmax is taken to be 500 times, and the convergence condition ε is taken as 1×10^{-6}; the errors of the estimated and actual wind speeds for the two wind conditions are shown in Figure 10a,b, where (a) is the comparison of the estimated wind speeds, and (b) is the estimation error.

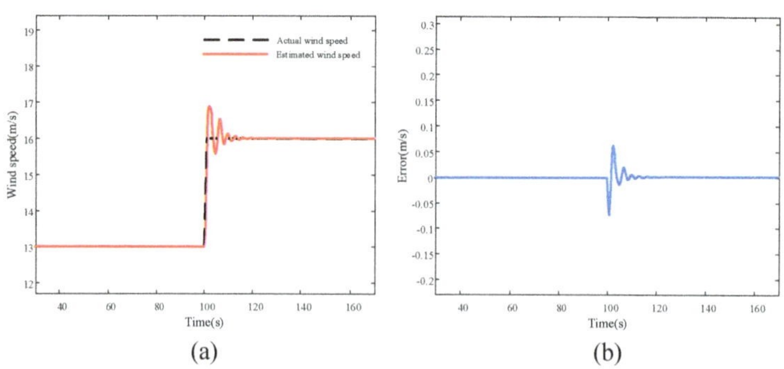

Figure 10. Estimation of step wind speed. (**a**) Estimated wind speed. (**b**) Estimation error.

When the wind speed changes, the pitch angle is limited by the rate of change, and it takes time from action to stabilization, which causes the rotational speed to fluctuate when the pitch angle changes. The effective wind speed iteration calculation needs to call the wind turbine torque coefficient curve to check the current rotational speed ω and pitch angle β corresponding to the value of the wind speed, while the rotational speed and pitch angle changes are closely related to the performance of the pitch controller. Figure 10a shows that the effective wind speed estimate lags behind the actual wind speed and has an error when the wind speed changes suddenly. Figure 10b shows that the error is up to 8% when the wind speed changes, and the error tends to be 0 when the wind speed is stable.

Figure 11b shows that there is no stable working point of the wind turbine in the turbulent wind field, and it is difficult to respond to the high-frequency component of the wind speed, so the error between the estimated effective wind speed and the actual wind speed is very large in high-frequency changes, close to 15%; in the low-frequency wind speed changes that the wind turbine can respond to, the error stabilizes at about 5%. It can be seen that the effective wind speed estimation algorithm proposed in this paper can better estimate the actual wind speed at the hub of the wind turbine.

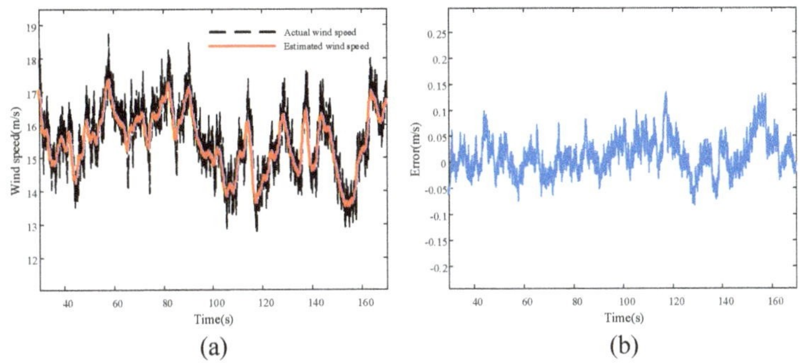

Figure 11. Estimation of turbulent wind speed. (**a**) Estimated wind speed. (**b**) Estimation error.

5.2. Pitch Control Simulation Results and Analysis

To assess the performance of the effective wind speed estimation-based inverse system pitch control, simulations were performed under both wind scenarios. Figure 12 presents the corresponding pitch angle variations, with (a) step-change and (b) turbulent wind conditions.

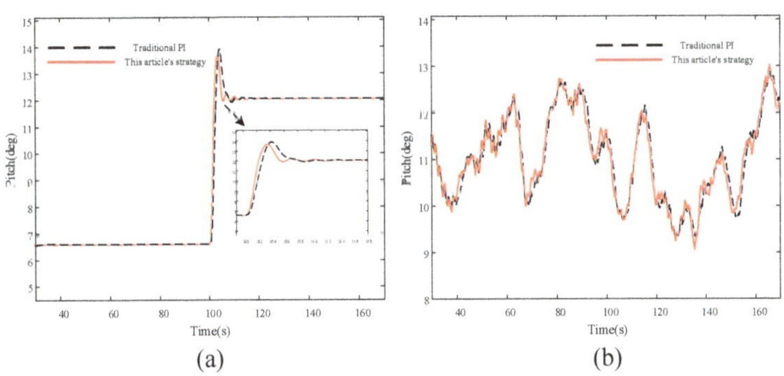

Figure 12. Comparison of pitch angle simulation results. (**a**) Step wind condition. (**b**) Turbulent wind conditions.

As shown in Figure 12a, the pitch angle adjusts 30% faster than conventional control under step-changing winds. Figure 12b reveals an enhanced wind speed tracking in turbulent conditions. Therefore, the feedback compensation pitch control loop that incorporates the inverse system method based on effective wind speed estimations can improve the response speed of the pitch control system to a certain extent, compared with traditional pitch controllers based on speed feedback.

Figures 13–15 present the simulated wind turbine speed, generator output power, and tower bottom pitching moment under both wind conditions, with (a) step-change and (b) turbulent scenarios.

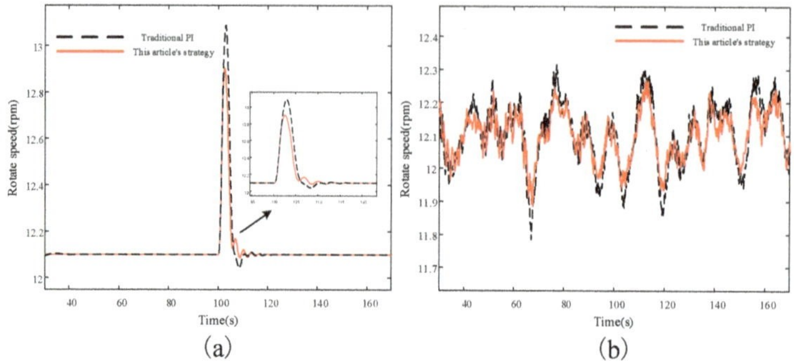

Figure 13. Comparison of speed simulation results. (**a**) Step wind condition. (**b**) Turbulent wind conditions.

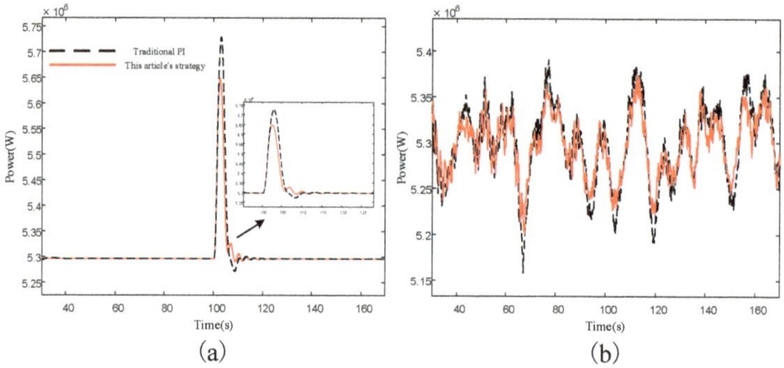

Figure 14. Comparison of output power simulation results. (**a**) Step wind condition. (**b**) Turbulent wind conditions.

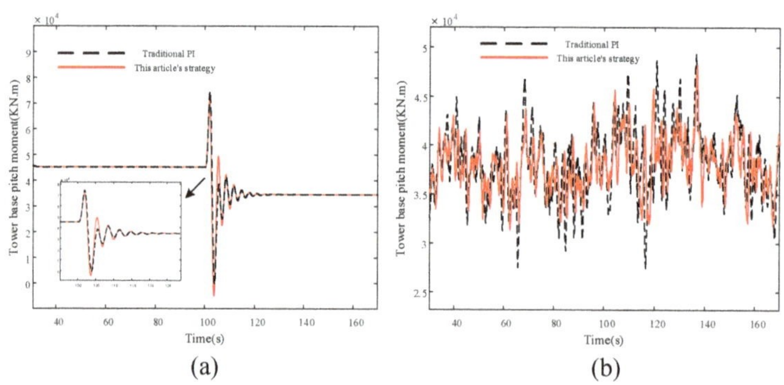

Figure 15. Comparison of tower bottom pitching moment simulation results. (**a**) Step wind condition. (**b**) Turbulent wind conditions.

From the above simulation results, it can be seen that the evaluation criteria for the control effect are different between the step wind condition and turbulent wind condition. Therefore, the rated power and the tower bottom pitching moment before the step wind are used as the benchmarks to calculate the overshooting amount when the step wind arrives and as the evaluation index to judge the control effect; in the turbulent wind condition, the mean squared deviation (MSD) of the output power and the tower bottom pitching moment during the simulation time are used as the evaluation index for the control effect; in the turbulent wind condition, the mean squared deviation of the output power and tower bottom pitching moment during the simulation time are used as the evaluation index for the control effect. In turbulent wind conditions, the mean square difference between the output power and the tower bottom pitching moment during the simulation time is used as the evaluation index of the control effect.

Tables 2 and 3 present comparative data on the output power and tower bottom pitching moment between the proposed method and traditional PI control, corresponding to step and turbulent wind conditions, respectively.

Table 2. Step wind condition output power and tower bottom pitching moment data.

Control Strategy	Output Power (W)		Tower Bottom Pitching Moment (kN·m)	
	Maximum Values	Overshoot	Maximum Values	Overshoot
PI	5.74×10^6	8.32%	7.5×10^4	63.10%
Proposed method	5.65×10^6	6.61%	7.0×10^4	52.13%

Table 3. Turbulent wind condition output power and tower bottom pitching moment data.

Control Strategy	Output Power (W)	Tower Bottom Pitching Moment (kN·m)
	Mean Square Error	Mean Square Error
PI	5.610×10^4	3.367×10^3
Proposed method	3.387×10^4	3.137×10^3

Figure 13 demonstrates that the inverse system-based feedback-compensated pitch control (utilizing the effective wind speed estimation) reduces the wind turbine speed overshoot during sudden wind changes more effectively than the conventional control, due to faster pitch responses in both step and turbulent conditions. The simulation results in Figure 14 reveal a reduced generator power overshoot under step winds and a closer alignment with rated values in turbulent winds, enhancing the power smoothness and grid delivery quality. As shown in Figure 15, the proposed strategy decreases the wind thrust on turbines during wind transients by improving the pitch response speed, achieving a tower base load reduction.

From Tables 2 and 3, we can observe the specific data, under the step wind condition, and in the method proposed in this paper the output power of the overshoot decreased by 22%, compared to the traditional pitch control in the bottom of the tower pitching moment, and the overshoot decreased by 17%, with an obvious output power smoothing and load shedding effect; the turbulent wind conditions, related to the mean square deviation as an evaluation index, also significantly demonstrate the advantages of the method proposed in this paper. Table 4 qualitatively compares the proposed pitch control strategy with the conventional PI control across critical performance metrics.

Table 4. Comparative performance analysis.

Evaluation Metric	Traditional PI Control	Proposed Method
Rotational Speed Stability	Medium	High
Power Fluctuation	Medium	Low
Tower Bottom Pitching Moment Fluctuation	Medium	Low
Model Complexity	Low	Medium

6. Conclusions

This study presents a nonlinear pitch control strategy based on an effective wind speed estimation designed to enhance the operational stability of wind turbines operating under complex above-rated wind speed conditions. The core objective is the mitigation of power fluctuations and the prevention of structural overloads. The strategy achieves this by integrating the real-time effective wind speed estimation, derived via a differential computation following aerodynamic torque estimations using an extended Kalman filter applied to the drivetrain, with an inverse system methodology for the pitch angle synthesis. This integrated approach enables the determination of the optimal real-time pitch angle based on the rotor plane wind speed estimation.

Simulation results substantiate the efficacy of the proposed control system. The strategy demonstrates an enhanced pitch system response, effectively suppressing fluctuations in both the rotational speed and power output within high-wind-speed regimes. Furthermore, a significant reduction in the tower base loading is achieved. Collectively, these improvements enhance operational safety during extreme wind conditions and reduce long-term maintenance costs by mitigating structural fatigue damage. The strategy also holds the potential to lower manufacturing expenses by diminishing the necessity for overdesigned turbine components.

The core advantages of the control strategy proposed in this paper, including the stable power regulation and structural load alleviation, have been successfully validated through simulations encompassing both step and turbulent wind inputs. Future work will focus on bridging the gap between the simulation and practical application by implementing and testing the proposed control framework on the wind turbine under real-world environmental conditions.

Author Contributions: Methodology, L.L.; Software, Y.L.; Validation, L.L.; Formal analysis, L.L.; Investigation, Y.L. and R.C.; Resources, Y.L. and R.L.; Data curation, Z.C.; Writing—original draft, L.L.; Writing—review & editing, L.L. and X.D.; Supervision, X.D.; Project administration, H.W.; Funding acquisition, X.Y. All authors have read and agreed to the published version of the manuscript.

Funding: This work is supported by the Hubei Provincial Key Research and Development Program Project (2023DJC201) and the Three Gorges Intelligent Control Technology Co., Ltd. Research Project (TGIC-KY-2024-04-00).

Data Availability Statement: The data that support the findings of this study are available on request from the corresponding author. The data are not publicly available due to privacy or ethical restrictions.

Acknowledgments: We thank the editor and the anonymous reviewers for their constructive comments that helped to improve our work.

Conflicts of Interest: Author Xuxin Yue, Haoran Wang, Ruibo Liu, Zhaobing Cai and Ruiqi Cai were employed by the company Three Gorges Intelligent Control Technology Co., Ltd. The remaining authors declare that the research was conducted in the absence of any commercial or financial relationships that could be construed as a potential conflict of interest.

References

1. GWEC. *Global Wind Report 2024;* Global Wind Energy Council: Abu Dhabi, United Arab Emirates, 2024.
2. Alhrshy, L.; Lippke, A.; Jauch, C. Variable Blade Inertia in State-of-the-Art Wind Turbine Structural-Dynamics Models. *Energies* **2023**, *16*, 6061. [CrossRef]
3. Lu, P.; Shi, Y.P. China's wind power industry enters a stage of high-quality market-oriented development. *Wind Energy* **2023**, *5*, 72–76.
4. Wan, Y. Cooperative Control for DFIG-Based Wind Turbine Generation System Covering All Operating Regions. *IEEE Trans. Sustain. Energy* **2025**, *16*, 945–954. [CrossRef]

5. Lasheen, A.; Elshafei, L.A. Wind-turbine collective-pitch control via a fuzzy predictive algorithm. *Renew. Energy* **2016**, *87*, 298–306. [CrossRef]
6. Abdin, E.S.; Xu, W. Control design and dynamic performance analysis of a wind turbine-induction generator unit. *IEEE Trans. Energy Convers.* **2000**, *15*, 91–96. [CrossRef]
7. Scholbrock, A.; Fleming, P.; Schlipf, D.; Wright, A.; Johnson, K.; Wang, N. Lidar-enhanced wind turbine control: Past, present, and future. In Proceedings of the 2016 American Control Conference (ACC), Boston, MA, USA, 6–8 July 2016; pp. 1399–1406.
8. Haizmann, F.; Schlipf, D.; Raach, S.; Scholbrock, A.; Wright, A.; Slinger, C.; Medley, J.; Harris, M.; Bossanyi, E.; Cheng, P.W. Optimization of a feed-forward controller using a CW-lidar system on the CART3. In Proceedings of the 2015 American Control Conference, Chicago, IL, USA, 1–3 July 2015; pp. 3715–3720.
9. Wang, N.; Johnson, K.E.; Wright, A.D. FX-RLS-Based Feedforward Control for LIDAR-Enabled Wind Turbine Load Mitigation. *IEEE Trans. Control. Syst. Technol.* **2012**, *20*, 1212–1222. [CrossRef]
10. Han, B.; Zhou, L.; Zhang, Z. LIDAR-assisted radial basis function neural network optimization for wind turbines. *IEEJ Trans. Electr. Electron. Eng.* **2018**, *13*, 195–200. [CrossRef]
11. Mirzaei, M.; Soltani, M.; Poulsen, N.K.; Niemann, H.H. An MPC approach to individual pitch control of wind turbines using uncertain LIDAR measurements. In Proceedings of the 2013 European Control Conference (ECC), Zurich, Switzerland, 17–19 July 2013; pp. 490–495.
12. Mirzaei, M.; Hansen, M.H. A LIDAR-assisted model predictive controller added on a traditional wind turbine controller. In Proceedings of the 2016 American Control Conference (ACC), Boston, MA, USA, 6–8 July 2016.
13. Schlipf, D.; Pao, L.Y.; Cheng, P.W. Comparison of feedforward and model predictive control of wind turbines using LIDAR. In Proceedings of the 2012 IEEE 51st IEEE Conference on Decision and Control (CDC), Maui, HI, USA, 10–13 December 2012; pp. 3050–3055.
14. Schlipf, D.; Grau, P.; Raach, S.; Duraiski, R.; Trierweiler, J.; Cheng, P.W. Comparison of linear and nonlinear model predictive control of wind turbines using LIDAR. In Proceedings of the 2014 American Control Conference, Portland, OR, USA, 4–6 June 2014; pp. 3742–3747.
15. Atsushi, Y.; Iman, Y.; Takeshi, I. Reduction in the Fluctuating Load on Wind Turbines by Using a Combined Nacelle Acceleration Feedback and Lidar-Based Feedforward Control. *Energies* **2020**, *13*, 4558. [CrossRef]
16. Li, D.Y.; Cai, W.C.; Li, P.; Jia, Z.J.; Chen, H.J.; Song, Y.D. Neuroadaptive Variable Speed Control of Wind Turbine with Wind Speed Estimation. *IEEE Trans. Ind. Electron.* **2016**, *63*, 7754–7764. [CrossRef]
17. Jiao, X.; Yang, Q.; Xu, B. Hybrid Intelligent Feedforward-Feedback Pitch Control for VSWT with Predicted Wind Speed. *IEEE Trans. Energy Convers.* **2021**, *36*, 2770–2781. [CrossRef]
18. Hur, S.H. Short-term wind speed prediction using Extended Kalman filter and machine learning. *Energy Rep.* **2020**, *7*, 1046–1054. [CrossRef]
19. Gao, Y.; Zhong, H.Y.; Chen, X.Y.; Geng, A.C.; Zhang, L.; Lei, C.J. Ultra-short-term wind speed prediction based on neural network and wavelet analysis. *Renew. Energy* **2016**, *34*, 705–711.
20. Wang, D.F.; Zhang, Z.Y.; Gu, Z.Y.; Huang, Y. Wind farm wind speed prediction based on hybrid Copula function and whale optimization algorithm. *Electr. Power Sci. Eng.* **2022**, *38*, 33–40.
21. Cao, S.Q.; Hao, W.J.; Wang, H.; Sun, Z.H.; Zhou, J.Y. Effective wind speed prediction at wind turbine hub based on LiDAR-Armax. *Laser Optoelectron. Prog.* **2020**, *57*, 228–234.
22. Hou, W.L.; Ang, L.; Fan, Z.M. Real-time rotor effective wind speed estimation using Gaussian processregression and Kalman filtering. *Renew. Energy* **2021**, *169*, 670–686.
23. Wang, Y.Z.; Cai, X.; Wu, A.; Xu, B.F.; Lin, S.F. Turbulence intensity identification and load reduction of wind turbine under extreme turbulence. *Ocean. Eng.* **2022**, *257*, 111710. [CrossRef]
24. Asghar, A.B.; Liu, X. Adaptive neuro-fuzzy algorithm to estimate effective wind speed and optimal rotor speed for variable-speed wind turbine. *Neurocomputing* **2018**, *272*, 495–504. [CrossRef]
25. Deng, X.; Yang, J.; Sun, Y.; Song, D.; Xiang, X.; Ge, X.; Joo, Y.H. Sensorless effective wind speed estimation method based on unknown input disturbance observer and extreme learning machine. *Energy* **2019**, *186*, 115790. [CrossRef]
26. Deng, X.; Yang, J.; Sun, Y.; Song, D.; Yang, Y.; Joo, Y.H. An effective wind speed estimation based extended optimal torque control for maximum wind energy capture. *IEEE Access* **2020**, *8*, 65959–65969. [CrossRef]
27. Tian, D.; Zhou, C.K.; Tang, S.Z.; Zhou, Q.; Huang, M.Y.; Deng, Y. MPPT method for large wind turbines based on adaptive model predictive control. *Acta Energiae Solaris Sin.* **2023**, *44*, 501–508.
28. Simley, E.; Pao, Y.L. Evaluation of a wind speed estimator for effective hub-height and shear components. *Wind Energy* **2016**, *19*, 167–184. [CrossRef]
29. He, Y.L.; Huang, S.; Du, J.; Su, D.X.; Li, J. Feedforward-based pitch control for wind turbines. *Power Syst. Prot. Control* **2012**, *40*, 15–20.

30. Wang, X.L.; Tang, H.M.; Bao, G.Q.; Zhang, X.Y.; Liang, C. Disturbance feedforward and predictive feedback composite control for load suppression in wind turbines. *Trans. China Electrotech. Soc.* **2016**, *31*, 230–235.
31. Dai, S.Z. Singular perturbation theory. *Syst. Eng. Electron.* **1988**, *2*, 1–12.

electronics

MDPI

Article

A Two-Stage Fault Reconfiguration Strategy for Distribution Networks with High Penetration of Distributed Generators

Yuwei He [1], Yanjun Li [1], Jian Liu [1], Xiang Xiang [1], Fang Sheng [1], Xinyu Zhu [1], Yunpeng Fang [1] and Zhenchong Wu [2,*]

[1] Hangzhou Power Supply Company, State Grid Zhejiang Electric Power Co., Ltd., Hangzhou 310016, China; sjtu_ywhe@163.com (Y.H.); eeliyanjun@163.com (Y.L.); lj13588101881@gmail.com (J.L.); xiangxiang_gf@163.com (X.X.); shengfangcyj@gmail.com (F.S.); zxinyu0316@163.com (X.Z.); fangyunpeng@126.com (Y.F.)

[2] College of Electrical Engineering, Zhejiang University, Hangzhou 310027, China

* Correspondence: wuzhenchong@zju.edu.cn

Abstract: In distribution networks with high penetration of distributed generators (DGs), traditional fault reconfiguration strategies often fail to achieve maximum load recovery and encounter operational stability challenges. This paper proposes a novel two-stage fault reconfiguration strategy that addresses both the fault ride through capability and output uncertainty of DGs. The first stage introduces a rapid power restoration reconfiguration model that integrates network reconfiguration with fault ride-through, enabling DGs to provide power support to the distribution network during faults, thereby significantly improving the recovery rate of lost loads. An AdaBoost-enhanced decision tree algorithm is utilized to accelerate the computational process. The second stage proposes a post-recovery optimal reconfiguration model that uses fuzzy mathematics theory and the transformation of chance constraints to quantify the uncertainty of both generation and load, thereby improving the system's static voltage stability index. Case studies using the IEEE 69-bus system and a real-world distribution network validate the effectiveness of the proposed strategy. This two-stage strategy facilitates short-term rapid load power restoration and enhances long-term operational stability, improving both the resilience and reliability of distribution networks with high DG penetration. The findings of this research contribute to enhancing the fault tolerance and operational efficiency of modern power systems, which is essential for integrating higher levels of renewable energy.

Keywords: distributed generator; network reconfiguration; fault ride-through; uncertainty; decision tree

1. Introduction

The primary task of a distribution network is to deliver electricity to end-users. As society progresses, users' demands for the quality and reliability of power supply have been steadily increasing. Distributed generators (DGs) provide a broader range of energy sources for distribution networks. However, as the penetration level of DGs increases, maintaining smooth and uninterrupted operation of distribution networks presents significant challenges. Renewable energy devices typically lack the rotational inertia inherent to conventional thermal power units, resulting in a weakened frequency regulation capability of the grid. To address frequency stability issues, grid-forming inverter technologies, such as virtual synchronous generators, virtual impedance control, virtual inertia control, and virtual oscillator control, have been developed [1–3].

The power output of distributed renewable energy sources is inherently volatile, which may lead to local voltage rises or sags, resulting in voltage violations. To mitigate these effects, DGs can be effectively coordinated with other flexibility resources across multiple time scales through mechanisms such as virtual power plants (VPPs) and microgrids [4,5]. This coordination enables aggregated resources to participate as a unified entity in electricity market operations, facilitating optimized dispatch. Furthermore, by deploying energy storage systems, fluctuations in distributed generation output can be smoothed, thereby achieving better power matching between generation and load [6]. In distribution networks with high DG penetration, bidirectional power flows become common, rendering traditional grid protection schemes insufficient. Adaptive protection schemes or novel intelligent algorithms need to be developed to address these new protection challenges [7]. Islanded microgrid technologies offer a promising solution by enabling parts of the distribution system to operate independently from the main grid following a fault [8].

In addition, fault reconfiguration serves as an essential means to ensure supply reliability and maintain smooth, uninterrupted operation of the system. Fault reconfiguration technology refers to an optimization technique whereby, upon a fault occurring in the distribution network, the system's topology is modified to transfer the lost load in non-faulted areas to other feeders or generators that still have power supply capability, enabling rapid restoration of power supply [9].

The integration of a substantial number of DGs into distribution networks has introduced significant challenges to the fault reconfiguration process [10,11]. On one hand, DGs have transformed the distribution network from a traditional single-source structure to a multi-source structure, providing more flexible and diverse recovery options in case of faults. DGs with frequency regulation and voltage stabilization capabilities can support islanded operation of the distribution network during faults, while DGs with fault ride-through capabilities can forcibly remain connected to the grid for a short period to support system power balance [12,13]. On the other hand, the multi-source structure complicates fault reconfiguration, requiring consideration of factors such as power output characteristics and topological constraints. Moreover, the uncertainty of DG output necessitates dynamic adjustments to fault reconfiguration strategies [14].

The issue of fault reconfiguration for distribution networks with DGs has been extensively addressed in the existing literature, with the primary focus on fault ride-through (FRT) characteristics during fault events and the impacts of post-fault uncertainty. In [15], the authors investigate the relationship between fault ride-through characteristics of DGs and their power recovery support capabilities, proposing a multi-stage self-healing method based on the margin capacity of interconnection lines. In [16], the authors analyze the electrical output characteristics of inverter-based DGs during both normal operation and fault ride-through conditions. In [17], the authors propose a fault reconfiguration scheme for distribution networks that coordinates reclosing operations with DGs' LVRT capabilities. However, these studies have primarily focused on analyzing the impact of DGs' fault ride-through characteristics on fault reconfiguration, while offering limited incorporation of this condition within the reconfiguration algorithms themselves.

In [18], the authors emphasize that to ensure the long-term stability of distribution networks, fault reconfiguration strategies must account for dynamic variations in generation and load over multiple time periods. In [19], the authors use C-Vine Copula and conditional probability to characterize the dynamic uncertainty of renewable energy sources. In [20], the authors analyze the impact of the stochastic nature of renewable energy output on system stability, and propose a fault reconfiguration algorithm based on probabilistic power flow. In [21], a fault recovery model based on robust stochastic optimization is proposed to address the uncertainty of DG output. However, the methods still exhibit the following issues: they rarely account for the uncertainties associated with the output of DGs and load power, leading to lower reliability; and they primarily focus on load shedding and switch operation counts, with less consideration given to the impact on system stability. Additionally, the stability indicators for distribution networks are often nonlinear, making them difficult to handle [22].

In the absence of sufficient operational data and accurate forecasting methods, fuzzy mathematical theory provides an effective approach for analyzing uncertainty, and offers greater applicability in practical scenarios [23,24]. In [25], the authors investigate energy management in multi-energy microgrids using an optimization algorithm integrated with fuzzy decision-making techniques, demonstrating the effectiveness of fuzzy methods in handling uncertain conditions in complex energy systems. In [26], the authors propose a two-stage robust adaptive model, based on a predictive control approach, for the optimal scheduling of integrated energy systems, incorporating fuzzy-inspired uncertainty modeling and adaptive control to balance economy and robustness.

The fault ride-through capability and output uncertainty of DGs exhibit differences in time scales. During the fault recovery process, fault ride-through plays a role in the early stages of the fault, while output uncertainty affects the system's operation after the fault. Few studies simultaneously consider the impacts of both factors and achieve the dual objectives of rapid power restoration and long-term operational stability. Furthermore, due to the inclusion of fault ride-through control and output uncertainty of DGs, fault reconfiguration models tend to be nonlinear and complex. Traditional solution methods may suffer from slow computational speed [27], which makes them less suitable for fast power restoration requirements. To address this, heuristic algorithms, such as decision trees [28] and genetic algorithms [29], are often used to enhance solution speed. Nevertheless, these methods still face challenges, such as the risk of converging to local optima in complex scenarios and the requirements of high quality for training data.

To address the issues outlined above, this paper proposes a two-stage fault reconfiguration strategy that comprehensively considers both the fault ride-through capability and the output uncertainty of DGs, aiming to improve both the short-term power restoration performance and the long-term operational stability of the distribution network system. Specifically, the contributions of this paper are as follows:

1. Establishing a rapid power restoration reconfiguration model, in which different handling strategies are applied to various types of DGs during fault recovery by adding specific constraint conditions, handled based on the Big-M method, that account for the fault ride-through, improving the load recovery percentage for areas that have lost power supply;
2. Proposing a fast solution algorithm for the reconfiguration model based on an AdaBoost-enhanced decision tree, which accelerates the solving process while maintaining the same accuracy;
3. Establishing a post-recovery optimal reconfiguration model by assessing system stability using a linearized static voltage stability index, modeling the uncertainty of DG output and load power based on fuzzy mathematics theory, and applying fuzzy chance constraint transformation to facilitate the model's solution.

The main contents of the subsequent sections are as follows: Section 2 analyzes the role of DGs in distribution network fault recovery, including a classification-based treatment strategy for DGs, which considers fault ride-through capabilities and analysis of uncertainty and stochastic processes, and presents a two-stage fault reconfiguration strategy. Section 3 proposes a rapid power restoration strategy and introduces a fast solution algorithm based on AdaBoost-enhanced decision trees. Section 4 establishes a post-recovery optimal reconfiguration model, incorporating fuzzy representations of uncertainties and the transformation of fuzzy chance constraints. Section 5 conducts case studies using the IEEE standard system and a real-world distribution network example, comparing multiple algorithms to validate the effectiveness of the proposed approach. Finally, Section 6 concludes the paper.

2. The Impacts of DGs on Distribution Network Fault Recovery

2.1. Analysis of the Support Capability of DGs

2.1.1. Fault Ride-Through Characteristics of DGs

Considering the safety and stability of active distribution networks, DGs should have low-voltage ride-through (LVRT) capability [30]. IEEE 1547-2018 [31] establishes relevant requirements for the fault ride-through capability of DGs, classifying them into three response categories. DGs should have LVRT capability, meaning that when grid disturbances or faults cause voltage drops at the point of common coupling (PCC), DGs should continue to operate without disconnecting from the grid for a specified duration. This helps to prevent cascading failures that could lead to widespread power outages. The requirements for the LVRT capability of DGs are shown in Figure 1.

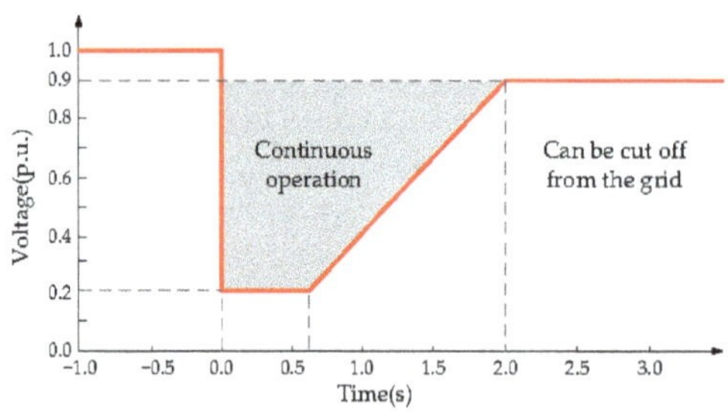

Figure 1. Standard requirements for LVRT capability of DG.

In the normal operation region ($U_{DG} \geq 0.9$ p.u.), the DG remains in a grid-connected state. When the voltage drop is relatively minor (0.2 p.u. $\leq U_{DG} \leq 0.9$ p.u.), the DG continues to operate without disconnecting from the grid, with an output power not less than 80% of that before the disturbance or fault occurred. When the voltage drop point is severe ($U_{DG} \leq 0.2$ p.u), the DG is disconnected from the system.

During the fault recovery process, the status of the DG may vary with voltage and frequency fluctuations at the PCC. Considering the limitations in the data processing capabilities of equipment and the real-time requirements for generating power restoration schemes, it is necessary to simplify the calculation of the DG's support capacity for engineering purposes.

When a fault causes a voltage drop at U_{DG}, the total output current may exceed the maximum allowable short-circuit current due to the control strategy. To ensure device safety, the DG output power is subject to the following constraint:

$$\begin{cases} P_{DG}^2 + Q_{DG}^2 \leq I_{max}^2 U_{DG}^2, & U_{DG} > 0.2 \text{ p.u.} \\ P_{DG} = 0\ ,\ Q_{DG} = 0, & U_{DG} \leq 0.2 \text{ p.u.} \end{cases} \tag{1}$$

where P_{DG} is the active power of the DG; Q_{DG} is the reactive power of DG. I_{max} is typically 1.2 to 2 times the rated current [32].

2.1.2. Handling Strategies for Different Types of DGs During Fault Recovery

Currently, inverter-based DGs, such as wind and solar DGs, can be categorized into grid-following and grid-forming types based on their grid connection control methods. Both grid-forming DGs and grid-following DGs with fault ride-through capabilities can provide active support to the system during fault recovery. Therefore, different handling strategies should be considered for different types of DGs.

(1) Directly decommissioned type

After fault isolation, if there is no grid-forming DG within the downstream area of the fault point to provide frequency or voltage support, DGs without fault ride-through capability should be decommissioned.

(2) Fault ride-through type

After a fault occurs, grid-forming DGs and grid-following DGs with fault ride-through capabilities begin fault ride-through. If the voltage meets the requirements, the DGs remain connected to the grid; otherwise, they are disconnected.

2.2. Analysis of the Output Uncertainty of DGs

For a distribution system that has just recovered from a fault and is in a relatively fragile state, the uncertainty in both power supply (from DGs) and load demand can exacerbate system instability, posing significant risks. The calculation based on the state of the system at the time of the fault will lead to neglect of the potential power change of the generation and load, and this prediction error will make the reconfiguration result no longer effective. After fault isolation and partial restoration, it is essential to readjust the distribution network to adapt to subsequent operational requirements and enhance its long-term operational capability.

The power output of DGs during a fault recovery period can be predicted using various forecasting methods. However, due to measurement errors during data collection, as well as the inherent limitations of predictive models in addressing complex scenarios, significant deviations may arise between the actual power output of the DG and its predicted value. This deviation can be expressed as follows:

$$\widetilde{P}^t_{\mathrm{DG}} = \overline{P}^t_{\mathrm{DG}} + \Delta^t_{\mathrm{DG}} \tag{2}$$

where $\widetilde{P}^t_{\mathrm{DG}}$ is the actual power value of the DG at time t; $\overline{P}^t_{\mathrm{DG}}$ is the predicted power value of the DG at time t; and Δ^t_{DG} denotes the deviation between the actual value and the predicted value of the DG.

The deviation between the predicted and actual output values of DGs is commonly described using probability distribution functions. The probability distribution functions for the output deviation of various types of DGs are as follows:

(1) Photovoltaic generation

The probability distribution function for the deviation between the predicted and actual output of a photovoltaic generator can be described by the TLS (t location-scale) distribution, as expressed mathematically below:

$$\phi(\widetilde{P}^t_{\mathrm{pv}}) = \frac{\Gamma(\frac{v+1}{2})}{\sigma_{\mathrm{pv}}\sqrt{v\pi}\Gamma(\frac{v}{2})}\left(\frac{v + \left(\frac{\widetilde{P}^t_{\mathrm{pv}} - \overline{P}^t_{\mathrm{pv}}}{\sigma_{\mathrm{pv}}}\right)^2}{v}\right)^{-\frac{v+1}{2}} \tag{3}$$

where $\widetilde{P}^t_{\text{pv}}$ and $\overline{P}^t_{\text{pv}}$ represent the actual and predicted power values of the photovoltaic system at time t, respectively; σ_{pv} and ν are the parameters for the probability distribution of the photovoltaic output deviation; and Γ is the Gamma function.

(2) Wind turbine generation

The probability distribution function for the deviation between the predicted and actual output of a wind turbine generator can be described by the Weibull distribution, with the following mathematical expression:

$$\overline{P}^t_{\text{wind}} = \begin{cases} 0, & v^t_{\text{wind}} \leq v_{\text{wind,ci}} \\ a + bv^t_{\text{wind}}, & v_{\text{wind,ci}} \leq v^t_{\text{wind}} \leq v_{\text{wind,e}} \\ P_{\text{wind,e}}, & v_{\text{wind,e}} \leq v^t_{\text{wind}} \leq v_{\text{wind,co}} \end{cases} \tag{4}$$

$$a = \frac{P_{\text{wind,e}} v_{\text{wind,ci}}}{v_{\text{wind,ci}} - v_{\text{wind,e}}}, \; b = \frac{P_{\text{wind,e}}}{v_{\text{wind,e}} - v_{\text{wind,ci}}} \tag{5}$$

$$\phi(\widetilde{P}^t_{\text{wind}}) = \frac{K}{C}\left(\frac{\overline{P}^t_{\text{wind}} - a}{bC}\right)^{K-1} \exp\left[-\left(\frac{\overline{P}^t_{\text{wind}} - a}{bC}\right)^{K}\right] \tag{6}$$

where $\widetilde{P}^t_{\text{wind}}$ and $\overline{P}^t_{\text{wind}}$ are the actual and predicted power values of the wind turbine at time t, respectively; v^t_{wind} is the wind speed at time t; $\nu_{\text{wind,ci}}$, $\nu_{\text{wind,co}}$, and $\nu_{\text{wind,e}}$ are the cut-in, cut-out, and rated wind speeds of the wind turbine, respectively; $P_{\text{wind,e}}$ is the rated output power of the wind turbine; K is the shape parameter of the Weibull distribution; and C is the scale parameter.

(3) Conventional DGs

Conventional DGs, such as gas turbines and diesel generators, generally produce stable power output when the fuel supply is sufficient. The power output deviations of such systems are often modeled using a discrete 0–1 probability distribution. In this model, the generator is assumed to either be in normal operation, producing the scheduled power output, or shut down, producing no power. The mathematical expression for this is as follows:

$$P(\widetilde{P}^t_{\text{DG,c}} = \overline{P}^t_{\text{DG,c}}) = p_{\text{normal}} \tag{7}$$

where $\widetilde{P}^t_{\text{DG,c}}$ and $\overline{P}^t_{\text{DG,c}}$ are the actual and scheduled power values of the conventional DG at time t, respectively, and p_{normal} is the probability that the generator is operating normally.

2.3. *Two-Stage Fault Reconfiguration Strategy for Distribution Networks with DGs*

The traditional fault reconfiguration strategy does not consider the changes brought about by DG integration. Firstly, the DG's fault ride-through capability provides additional power support. Secondly, the output uncertainty of DGs will make the restored system potentially unstable. Ignoring these factors will make the reconfiguration result unreliable. In the early stage of the fault, the primary task is to ensure the continuous power supply of important loads. After the fault is completely isolated and the system is restored to stability, the grid-connected DG can be re-connected to the system. At this point in time, the power grid can be reconfigured again to eliminate short-term overload and ensure the long-term stable operation of the system until the faulted line is repaired. It is difficult for existing methods to meet the requirements of load recovery speed and long-term stability at the same time.

The LVRT characteristics exhibit dynamic responses in a matter of seconds, while the output volatility involves longer time scales. Based on the above analysis, to maximize the support capability of DGs in the early stages of a fault and simultaneously consider the potential long-term stability impacts of their output uncertainty on the system, a staged optimization strategy must be adopted.

In response to the above problems, this paper proposes a two-stage fault reconfiguration strategy, as shown in Figure 2.

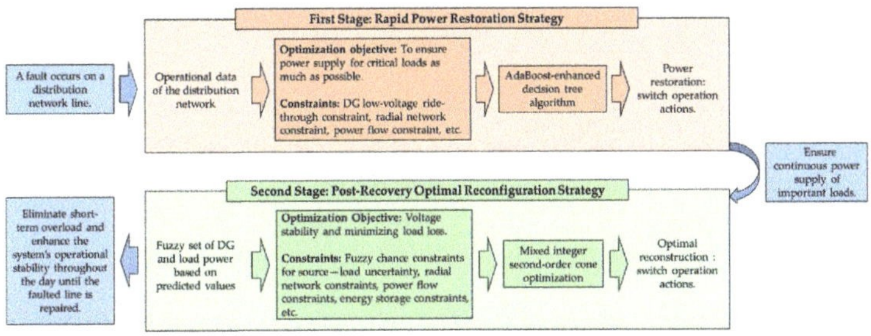

Figure 2. Two-stage fault reconfiguration strategy for distribution network with DGs.

The first stage is a rapid power restoration stage, considering the LVRT capability of DGs and aiming to minimize load shedding, and involves fast optimal calculations. It makes full use of the support capacity of the DG, and provides a very fast solution speed, which can provide a rapid power supply restoration scheme.

The second stage is a post-recovery optimal reconfiguration stage, aiming to optimize system stability and minimize load shedding. It focuses on dealing with the uncertainty of DGs and loads, improving the stability of the distribution network system for a long time.

3. Rapid Power Restoration Strategy

3.1. Rapid Power Restoration Reconfiguration Model Considering FRT Capability of DGs

3.1.1. Objective Function of Rapid Power Restoration Reconfiguration Model

After a fault occurs, we assume that the fault location has already been identified, and that the faulted lines have been isolated from the network. In the rapid power restoration stage following a fault, the system's primary objective is to restore power to the lost loads as quickly as possible by regulating tie switches and fully utilizing DGs and other devices capable of supporting the system's power supply. Thus, the optimization strategy for this stage is focused on ensuring continuous power supply to critical loads, with the objective function formulated as follows:

$$F_1 = \min \sum_{i \in \Omega_{\text{load}}} \omega_i (P_{\text{load},i,0} - P_{\text{load},i}) \tag{8}$$

In Equation (8), ω_i is the priority of load i, where a larger value indicates higher importance of the load; $P_{\text{load},i,0}$ is the pre-fault power of load i; $P_{\text{load},i}$ is post-fault power of load i; and Ω_{load} is the set of loads connected to the system.

3.1.2. Constraints of Rapid Power Restoration Reconfiguration Model

The rapid power restoration reconfiguration model must satisfy the following constraints:

(1) DG output power constraint

The constraint on the DG's output described in Equation (1) is a piecewise function, which is handled using the Big-M method in this paper:

$$\begin{cases} P_{\text{DG}}^2 + Q_{\text{DG}}^2 \leq I_{\max}^2 U_{\text{DG}}^2 \\ -M\alpha_{\text{on}} \leq P_{\text{DG}}^2 + Q_{\text{DG}}^2 \leq M\alpha_{\text{on}} \\ \alpha_{\text{on}} \leq U_{\text{DG}}/0.2 \end{cases} \tag{9}$$

α_{on} is a binary variable, where a value of 1 indicates that the DG is connected to the grid, and 0 indicates that the DG is disconnected; and M is a sufficiently large positive constant. In this paper, M was calibrated as the upper limit of the DG output power.

For DGs using a virtual impedance control strategy, they are handled by introducing the auxiliary binary variables α_{z1} and α_{z2}:

$$\begin{cases} Z_{DG} = \alpha_{z1} Z_0 + \alpha_{z2} Z_{virtual} \\ \alpha_{z1} + \alpha_{z2} = 1 \\ 0.9\alpha_{z1} + 0.2\alpha_{z2} \leq U_{DG} \leq \alpha_{z1} + 0.9\alpha_{z2} \end{cases} \tag{10}$$

where Z_{DG} is the additional impedance in series with the DG, Z_0 is the impedance of the DG during normal operation, and $Z_{virtual}$ is the virtual additional impedance of the DG.

(2) Radial network constraint

The distribution network should maintain radial operation, ensuring that the reconfigured network does not form unintended islands. In this paper, a description method based on the non-connectivity condition of the power supply loop is adopted [33]. Compared to the conventional spanning tree model, this method offers a reduced scale and complexity.

$$\begin{cases} \sum\limits_{l=1}^{L} y_l = N_b - 1 \\ \sum\limits_{m=1}^{M_h} y_{hm} \leq M_h - 1, h = 1, 2, ..., H \end{cases} \tag{11}$$

where l is the branch index of the distribution network; L is the total number of branches; N_b is the number of nodes; y_l is a binary variable representing the status of the l-th branch, where 1 indicates that the branch is closed and 0 indicates that the branch is open; M_h is the number of branches in the h-th power supply loop; H is the total number of power supply loops; and y_{hm} is a binary variable representing the status of the m-th branch in the h-th power supply loop.

(3) Other constraints

Other constraints are provided in Appendix A, including power flow constraints using second-order cone relaxation, line operational safety constraints to prevent line overloads, voltage constraints to prevent overvoltage at nodes, and energy storage output constraints.

3.2. Fast Solution Algorithm for Rapid Power Restoration Reconfiguration Model

3.2.1. Fast Solution Algorithm Based on Decision Trees

For the power restoration model established above, conventional mathematical optimization methods may struggle to meet the real-time solution requirements. To overcome this challenge, this paper proposes a fast solution algorithm based on decision trees. This method based on decision trees has the characteristics of rapid computation speed and strong interpretability, while also exhibiting excellent adaptability to a wide range of operating conditions [34]. The workflow of the proposed method is shown in Figure 3.

The fast algorithm based on decision trees consists of two main components: the "knowledge base" and the "inference machine". The "knowledge base" is a collection of instances, each containing the system's operational state, fault details, and corresponding restoration plans. The "inference machine" is trained on the "knowledge base" samples by a decision tree generation algorithm. It can quickly infer the power restoration operation solutions based on fault characteristics.

The instances in the "knowledge base" are derived from the calculation results of the proposed power restoration optimization model or manually set and verified reasonable solutions. The inference machine model is constructed using a decision tree algorithm improved by AdaBoost.

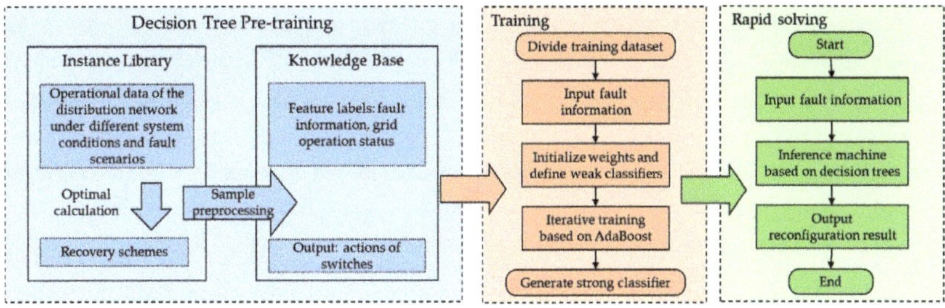

Figure 3. Flow chart of fast solution algorithm for power restoration strategy based on decision trees.

3.2.2. Steps of Solution Algorithm

The steps of the solution algorithm based on decision trees are as follows:

(1) Establishment of the instance library

Distribution network operation data are collected under different system conditions and fault scenarios, and reconfiguration problems are solved using optimization models to build the instance library for training. The feature labels for the instance library are selected as generation and load power, line status (on/off), energy storage status, and fault information. The solution, which is the output of the decision tree, corresponds to the switching actions of each connection switch.

(2) Construction of the knowledge base sample set

The feature labels serve as the classification criteria for the decision tree. To enhance the decision tree's accuracy, normalization and other preprocessing steps are applied to the feature labels in the instance library. For Boolean data (e.g., switch status), the value of 1000 represents a closed line, to mitigate the difference in magnitude between this and continuous data (e.g., power). For continuous data, normalization is performed using the rated values as the benchmark. Finally, the corresponding power restoration solution for each fault is added, completing the knowledge base sample set.

(3) Training of the inference machine based on the decision tree

The sample set of the knowledge base is divided into 70% for training and 30% for testing. The specific training algorithm is described in Section 3.2.3. After training, the decision tree inference machine for power restoration is generated. When a system fault is detected, the inference machine first processes the fault information and grid operation data according to the knowledge base standards. Then, the inference machine retrieves and infers the power restoration solution based on the trained decision tree, ultimately providing the recovery strategy.

3.2.3. AdaBoost-Enhanced Decision Tree

Ordinary decision tree training often employs algorithms like CART [34], but decision trees trained using CART typically serve as weak classifiers, and may lack accuracy when handling new fault scenarios. To address this issue, we propose an improved decision tree algorithm based on AdaBoost.

AdaBoost is capable of combining multiple weak classifiers into a strong classifier by following a specific rule [35]. It works by iteratively adjusting the weights of different samples, focusing more on those misclassified by previous iterations, thereby improving the overall decision accuracy. The specific process is outlined below:

(1) Assume there are n training samples, where the i-th sample has a feature label X_i and a decision result Y_i. Set the initial weight of each sample as $D_{0,i} = 1/n$.

(2) Iteratively train decision trees based on the CART algorithm to obtain m weak classifiers. In the CART algorithm, the Gini index is used as the criterion for node splitting [36].

At each iteration, the sample weights are updated, and the weight of the weak classifier in the current iteration is computed. Let E_j be the error rate of the weak classifier in the j-th iteration. The weight W_j of the classifier and the sample weights $D_{j,i}$ are updated as follows:

$$W_j = 0.5\ln(\frac{1-E_j}{E_j}) \tag{12}$$

$$D_{j,i} = \begin{cases} D_{j-1,i}e^{W_j}/Z_j \text{ , If the classification is correct} \\ D_{j-1,i}e^{-W_j}/Z_j\text{, If the classification is incorrect} \end{cases} \tag{13}$$

where Z_j is the normalization factor.

(3) After m iterations, m weak classifiers are generated. The final strong classifier is obtained by summing them with weights W_j. In this paper, the number of AdaBoost iteration rounds is 30 [37,38].

4. Post-Recovery Optimal Reconfiguration Strategy

During the rapid power restoration stage, DGs are allowed to operate under low voltage for a short period. However, prolonged operation in this fault ride-through state is not sustainable, as it poses the risk of overload and can undermine the long-term stability of the system. During the period when the faulted line has been isolated but not fully repaired (which typically ranges from several hours to a few days), the uncertainty of the output of DGs and load power fluctuations may also affect the stability of the system. Therefore, it is necessary to perform further optimal reconfiguration of the post-recovery system to ensure that it can operate safely and reliably over the long term.

4.1. DG and Load Uncertainty Models

4.1.1. Fuzzy Representation of Uncertainty Parameters

In Section 2.2, the probability distribution function for the output of DGs has been modeled. Similarly, the probability distribution function for load demand power can be described using the same approach.

$$\begin{cases} \widetilde{P}^t_{\text{load}} = \overline{P}^t_{\text{load}} + \Delta^t_{\text{load}} \\ \phi(\widetilde{P}^t_{\text{load}}) = \frac{1}{\sqrt{2\pi}\sigma_{\text{load}}}\exp[-\frac{(\widetilde{P}^t_{\text{load}}-\overline{P}^t_{\text{load}})^2}{2\sigma^2_{\text{load}}}] \end{cases} \tag{14}$$

where $\widetilde{P}^t_{\text{load}}$ and $\overline{P}^t_{\text{load}}$ represent the actual and predicted power values of the load at time t, respectively, and Δ^t_{load} denotes the deviation between the actual and predicted load power. σ_{load} is the parameter for the normal distribution function describing the load power deviation.

In practical distribution network engineering, parameters for the probability distribution function of DGs' output and load power are often difficult to obtain accurately, due to factors such as sustained faults, which can affect subsequent processing. To address this issue, fuzzy mathematics theory provides an operational method with good accuracy [21]. The triangular membership function, known for its computational efficiency and applicability in real-time systems, is used to describe the output of DGs and load power, as follows:

$$\mu(\widetilde{P}_i^t) = \begin{cases} \frac{\widetilde{P}_i^t - k_{i,1}^t \overline{P}_i^t}{\overline{P}_i^t - k_{i,1}^t \overline{P}_i^t}, & k_{i,1}^t \overline{P}_i^t \leq \widetilde{P}_i^t \leq \overline{P}_i^t \\ \frac{k_{i,2}^t \overline{P}_i^t - \widetilde{P}_i^t}{k_{i,2}^t \overline{P}_i^t - \overline{P}_i^t}, & \overline{P}_i^t \leq \widetilde{P}_i^t \leq k_{i,2}^t \overline{P}_i^t \\ 0, & \text{other} \end{cases} \tag{15}$$

where $\widetilde{P}_i^t$ is the actual output of the i-th DG at time t, and $\overline{P}_i^t$ is the predicted output of the i-th DG at time t. $k_{i,1}^t$ denote the positive scaling factors, and $k_{i,2}^t$ denote the negative scaling factors; they satisfy $0 < k_{i,1}^t < 1$, $k_{i,2}^t > 1$.

The triangular membership function parameters for each device can be derived through statistical fitting of the predicted data and actual measurement data. Taking photovoltaic generation as an example, the procedure is as follows:

First, a photovoltaic output forecasting model is used to obtain a time series of power fluctuation data at a specified time granularity (e.g., every 15 min). Next, the actual PV output power measurements at corresponding time points are collected, and the relative difference k_i^t between the actual measurement and predicted values is calculated. Then, the frequency distribution of k_i^t is obtained through statistical analysis. Finally, based on the frequency distribution of k_i^t, a triangular membership function is fitted to represent the uncertainty in the photovoltaic output.

Additionally, if the probability distribution function of the DG or the load power can be accurately described, the triangular membership function can also be directly fitted to the probability distribution function.

4.1.2. Transformation of Fuzzy Chance Constraints

Due to the incorporation of uncertainty, the deterministic model established in Section 3 for the rapid power restoration reconfiguration model requires revision. Specifically, the constraint conditions involving random variables, such as the output of DGs and load power (e.g., Equation (A13)), should be reformulated as fuzzy chance constraints:

$$\begin{cases} \mathrm{Cr}\{\widetilde{P}_{\mathrm{load},j,t} + \sum\limits_{k\in\Omega_{j,z}} P_{jk,t} - (\widetilde{P}_{\mathrm{DG},j,t} + P_{\mathrm{m},j,t} + P_{\mathrm{ess},j,t}) - \sum\limits_{i\in\Omega_{j,t}} (P_{ij,t} - r_{ij} I_{ij,t}^{\mathrm{sqr}}) \leq 0\} \geq \alpha \\ \mathrm{Cr}\{\widetilde{Q}_{\mathrm{load},j,t} + \sum\limits_{k\in\Omega_{j,z}} Q_{jk,t} - (\widetilde{Q}_{\mathrm{DG},j,t} + Q_{\mathrm{m},j,t}) - \sum\limits_{i\in\Omega_{j,t}} (Q_{ij,t} - x_{ij} I_{ij,t}^{\mathrm{sqr}}) \leq 0\} \geq \alpha \end{cases} \tag{16}$$

where $\widetilde{P}_{\mathrm{DG},j,t}$ and $\widetilde{Q}_{\mathrm{DG},j,t}$ are the actual active and reactive power of the DG at node j at time t; $\widetilde{P}_{\mathrm{load},j,t}$ and $\widetilde{Q}_{\mathrm{load},j,t}$ are the actual active and reactive power of the load at node j at time t; $I_{ij,t}^{\mathrm{sqr}}$ is the square of the current flowing through the line ij at time t; $P_{\mathrm{m},j,t}$ and $Q_{\mathrm{m},j,t}$ are the active and reactive power outputs of the main grid at node j at time t; $P_{\mathrm{ess},j,t}$ is the charging and discharging power of the energy storage at node j at time t (discharge is positive); $\mathrm{Cr}\{\cdot\}$ denotes the confidence level of the event that the load power is less than the power output of the source; and α is the confidence level of this event, and since the system optimal reconfiguration strategy aims to ensure reliable load supply, a higher value is chosen (e.g., α = 90%).

Based on the principle of fuzzy chance constraint equivalence transformation [25], when $\alpha > 0.5$, Equation (16) can be transformed as follows:

$$\begin{cases} \sum\limits_{k\in\Omega_{j,z}} P_{jk,t} - \sum\limits_{i\in\Omega_{j,t}} (P_{ij,t} - r_{ij} I_{ij,t}^{\mathrm{sqr}}) - P_{\mathrm{m},j,t} - P_{\mathrm{ess},j,t} + (2-2\alpha)\overline{P}_{\mathrm{load},j,t} + (2\alpha-1)k_{\mathrm{load},2}^t \overline{P}_{\mathrm{load},j,t} - (2-2\alpha)\overline{P}_{\mathrm{DG},j,t} - (2\alpha-1)k_{\mathrm{DG},1}^t \overline{P}_{\mathrm{DG},j,t} \leq 0 \\ \sum\limits_{k\in\Omega_{j,z}} Q_{jk,t} - \sum\limits_{i\in\Omega_{j,t}} (Q_{ij,t} - x_{ij} I_{ij,t}^{\mathrm{sqr}}) - Q_{\mathrm{m},j,t} + (2-2\alpha)\overline{Q}_{\mathrm{load},j,t} + (2\alpha-1)k_{\mathrm{load},2}^t \overline{Q}_{\mathrm{load},j,t} - (2-2\alpha)\overline{Q}_{\mathrm{DG},j,t} - (2\alpha-1)k_{\mathrm{DG},1}^t \overline{Q}_{\mathrm{DG},j,t} \leq 0 \end{cases} \tag{17}$$

where $\overline{P}_{DG,j,t}$ and $\overline{Q}_{DG,j,t}$ are the predicted active and reactive power of the DG at node j at time t; and $\overline{P}_{load,j,t}$ and $\overline{Q}_{load,j,t}$ are the predicted active and reactive power of the load at node j at time t.

By following the above steps, the uncertainty optimization is transformed into a deterministic optimization problem under a specific confidence level, thereby reducing the solving complexity.

4.2. Optimization Objective of Post-Recovery Optimal Reconfiguration Model

The uncertainty of generation and load leads to increasing power imbalances and voltage violations at nodes, which pose significant safety risks to the distribution networks [39]. Through optimal reconfiguration, the voltage distribution can be improved, thereby enhancing system stability. To address this, the optimization objective based on the static voltage stability index L_{ij} is given by the following:

$$L_{ij} = \frac{4[(x_{ij}P_{ij} - r_{ij}Q_{ij})^2 + (r_{ij}P_{ij} + x_{ij}Q_{ij})U_i^2]}{U_i^4} \tag{18}$$

where r_{ij} and x_{ij} are the resistance and reactance of branch ij; P_{ij} and Q_{ij} are the active and reactive power flowing into node j; U_i is the voltage at node i; and the voltage stability is indicated by L_{ij}, where a value less than 1 signifies stability, and the closer it is to 0, the better the system's voltage stability. For distribution networks with multiple branches, the maximum value of L_{ij} across all lines is selected as the overall voltage stability indicator for the system.

Equation (18) is nonlinear and requires simplification. In practical applications of distribution networks, the voltage magnitudes and phase angles at both ends of the line are approximately equal. Therefore, $x_{ij}P_{ij}$ and $r_{ij}Q_{ij}$ can be considered approximately equal [40], allowing Equation (18) to be simplified as follows:

$$\widetilde{L}_{ij} = 4(r_{ij}P_{ij} + x_{ij}Q_{ij}) - U_i^2 \tag{19}$$

where $\widetilde{L}_{ij}$ less than 0 indicates voltage stability, and the smaller the value, the better the voltage stability.

The quadratic voltage terms in Equation (19) can be handled using second-order cone relaxation. Thus, the optimization objective F_{2_1} is given by the following:

$$F_{2\_1} = \min\sum_{t=1}^{T} \widetilde{L}_{ij,t} \tag{20}$$

where $\widetilde{L}_{ij,t}$ represents the static voltage stability index of line ij at time t, and T is the number of scheduling periods.

To ensure the supply of critical loads, the optimization objective minimizing load shedding F_{2_2} is given by the following:

$$F_{2\_2} = \min\sum_{t=1}^{T} \sum_{i\in\Omega_{load}} \varpi_i(P_{load,i,t,0} - P_{load,i,t})\Delta T \tag{21}$$

ΔT denotes the length of each scheduling period, which depends on the granularity of the power forecast for the source and load. $P_{load,i,t,0}$ and $P_{load,i,t}$ are the predicted and actual power of load i at time t, respectively.

Since the multiple objective functions in the proposed optimal reconfiguration model have different dimensions and magnitudes, normalization is required for each objective function:

$$F'_i = \frac{F_{i,\max} - F_i}{F_{i,\max} - F_{i,\min}} \tag{22}$$

where F'_i is the normalized value of the objective; and $F_{i,\max}$ and $F_{i,\min}$ are the maximum and minimum values of the objective i, respectively.

After normalization, the linear weighted sum is applied to obtain the fault reconfiguration objective function F_2, which comprehensively considers the system's long-term voltage stability and load shedding:

$$F_2 = \min(\omega_1 F'_{2\_1} + \omega_2 F'_{2\_2}) \tag{23}$$

where ω_1 and ω_2 represent the weights of the two optimization objectives. In this paper, ω_1 is set to 0.4 and ω_2 is set to 0.6. Furthermore, the impact of different weight values on the optimization results is shown in Section 5.3.

4.3. Constraints of Post-Recovery Optimal Reconfiguration Model

The power flow constraints, radial network constraints, and other conditions of the post-recovery optimal reconfiguration model are fundamentally similar to those in the rapid power restoration reconfiguration model. The key distinction lies in the need to account for the power variations of DGs, loads, and energy storage devices over different time periods, where these variables are associated with distinct time instances. Additionally, some constraints must be transformed into fuzzy chance constraints, as detailed in Section 4.1.2.

The resulting model, which considers the uncertainty of the DG output, is a mixed-integer second-order cone optimization problem, and can be solved using solvers such as Gurobi.

5. Case Study

5.1. Case Study Model Description

In this study, two systems are used to analyze the performance of algorithms: an IEEE 69-bus distribution network with high penetration of DGs (as shown in Figure 4), and a certain 11-bus distribution network project (as shown in Figure 5). The DG penetration rates in these two systems are approximately 65% and 60%, respectively, and all connected DG units are equipped with fault ride-through capability. The specific parameters of the two systems, as well as the variations in power generation and load, are detailed in Appendix B.

In these cases, the faulty line exits operation. A variety of different fault cases were set up, as shown in Tables 1 and 2.

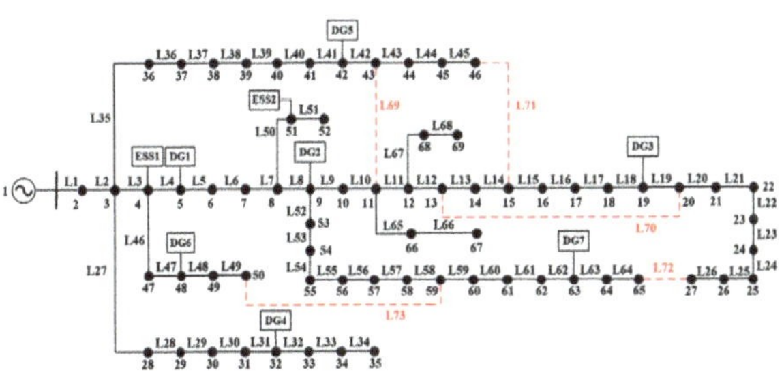

Figure 4. An IEEE 69-bus distribution network with DGs.

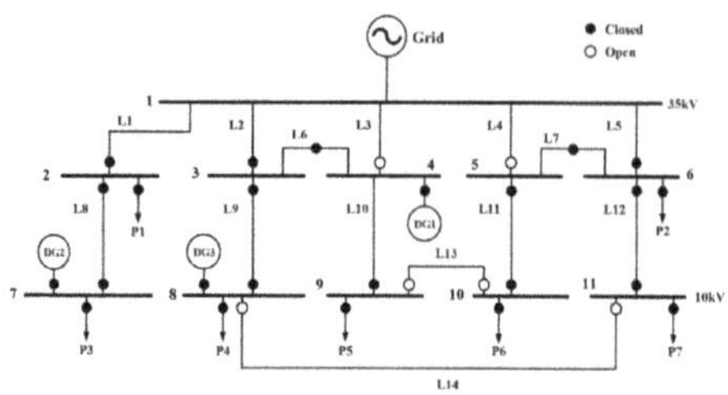

Figure 5. A real-world 11-bus distribution network.

Table 1. The fault cases of the 69-bus system.

Case Number	Faulty Line
1	L3
2	L10
3	L5 and L18
4	L38 and L64
5	L48; DG6 fault ride-through operation
6	L62; DG7 fault ride-through operation

Table 2. The fault cases of the 11-bus system.

Case Number	Faulty Line
1	L2
2	L9
3	L3 and L10
4	L4; DG1 fault ride-through operation

In Table 1, Case 1 simulates a fault occurring at the beginning of a branch line, resulting in the largest power outage area. Case 2 simulates a fault occurring in the middle section of a branch line, near a tie switch, which causes multiple power outage areas. Cases 3 and 4 simulate situations where multiple lines fail simultaneously. Cases 5 and 6 simulate line faults with DGs operating in fault ride-through mode. In Table 2, Cases 1–3 simulate single and multiple fault cases, while Case 4 simulates a line fault with a DG operating in fault ride-through mode.

5.2. *Analysis of Calculation Results for the Rapid Power Recovery Stage*

Based on the model and fault instance library established above, calculations for the first stage of rapid power recovery were performed. The results are shown in Tables 3 and 4.

Table 3. Load recovery results of 69-bus system.

Case Number	Reconfiguration Operation	Load Recovery Percentage		
		Primary Load	Secondary Load	Tertiary Load
1	L69 is closed	100%	100%	90.57%
2	L71 is closed	100%	100%	100%
3	L69 and L72 are closed	100%	100%	81.33%
4	L69 and L72 are closed	100%	100%	100%
5	L73 is closed	100%	100%	93.04%
6	L72 is closed	100%	100%	95.48%

Table 4. Load recovery results of 11-bus system.

Case Number	Reconfiguration Operation	Load Recovery Percentage		
		Primary Load	Secondary Load	Tertiary Load
1	L11 is closed	100%	100%	100%
2	L14 is closed	100%	100%	100%
3	L12 and L13 are closed	100%	100%	100%
4	L11 is closed	100%	100%	100%

From the results in Tables 3 and 4, it can be observed that the proposed algorithm successfully provides reasonable reconfiguration operations for different fault scenarios. The reconfiguration schemes differ across the various fault cases, indicating that the proposed algorithm is capable of analyzing different fault scenarios and providing tailored recovery solutions.

In all cases, the algorithm restores the power supply to all primary and secondary loads, ensuring that the most critical loads are prioritized for recovery. When the fault is relatively mild (e.g., Case 2 and Case 4 in Table 3), the algorithm restores power to all loads, as the system's damage is limited and the available resources are sufficient. However, in more severe fault scenarios (e.g., Case 1 and Case 3 in Table 3), the restoration of tertiary loads cannot be fully achieved. As the fault severity increases, the recovery rate for tertiary loads gradually decreases.

This demonstrates that the algorithm adapts to the severity of the fault and dynamically seeks to maximize load restoration, while dealing with the limitations imposed by the faulted lines. It is important to clarify that under more extreme fault conditions, the recovery rate for tertiary loads may be significantly lower, and even the restoration of primary and secondary loads might not reach 100%.

5.3. Analysis of Calculation Results for the Post-Recovery Optimal Reconfiguration Stage

The post-recovery optimal reconfiguration model in the second stage is solved using the Gurobi solver, with the calculation results presented in Tables 5 and 6. Loss of load refers to the total accumulated load shedding across all time periods. The worst $\widetilde{L}_{ij}$ refers to the maximum $\widetilde{L}_{ij}$ of all nodes in all periods (24 h) after reconfiguration, which can characterize the stability of the system.

Table 5. The results of the 69-bus system reconfiguration.

Case Number	Reconfiguration Operation	The Worst $\widetilde{L}_{ij}$	Loss of Load (MWh)
1	L69 is closed	−0.8898	1.4664
2	L69 is closed	−0.9264	0.4602
3	L72 and L73 are closed	−0.9479	1.2070
4	L69 and L72 are closed	−0.9251	0.6349

Table 6. The results of the 11-bus system reconfiguration.

Case Number	Reconfiguration Operation	The Worst $\widetilde{L}_{ij}$	Loss of Load (MWh)
1	L11 is closed	−0.9984	0
2	L14 is closed	−0.9919	0
3	L12 and L13 are closed	−0.9919	0

For the 11-bus system, the power supply is sufficient, and no load loss can be realized after the optimal reconfiguration. The value of the worst $\widetilde{L}_{ij}$ is small, meaning that the voltage stability of the system is relatively high. For the 69-bus system, there is still some load loss after the optimal reconfiguration, but the stability of the system is also guaranteed.

Taking Case 1 as an example, the static voltage stability index curves for each node at different times are shown in Figure 6.

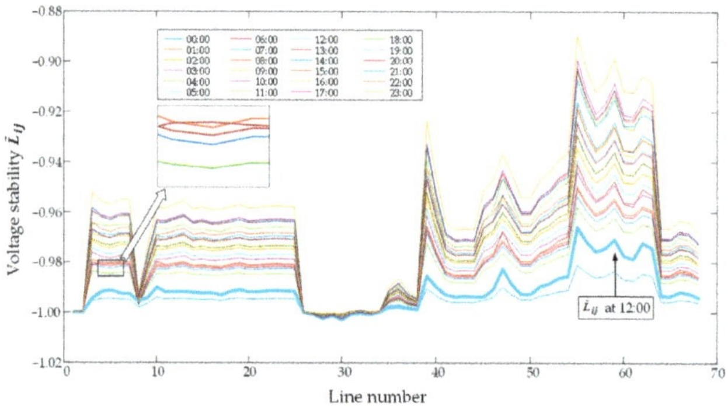

Figure 6. $\widetilde{L}_{ij}$ change diagram of reconstructed system under L3 line fault.

In Figure 6, the different curves represent the static voltage stability index of each line in the distribution network at various times throughout a day. The topology of the network remains unchanged during the period shown in the figure, which explains the general similarity in the trends of the curves. However, due to random fluctuations in the output of DGs, the curves at different times exhibit differences. For example, the variation in Line 5 from 5:00 to 8:00, marked in the figure, highlights the effect of DG output fluctuations on voltage stability.

Throughout the day, the output of DGs and load variations cause the voltage stability index to fluctuate. During long-term operation, the value of $\widetilde{L}_{ij}$ consistently meets the requirements (less than 0). Specifically, at 12:00, the value of $\widetilde{L}_{ij}$ is relatively small, indicating that the load pressure is lower at this time. This can be considered a special case during long-term system operation. Using the power value of the source and load at this time for reconfiguration may result in a relatively large deviation.

These observations highlight the ability of the proposed algorithm to maintain voltage stability under varying conditions, ensuring that the system can recover efficiently from faults while keeping voltage fluctuations within acceptable limits.

In the proposed post-recovery optimal reconfiguration algorithm, the value of weight settings may affect the optimization result. Using the 69-bus system in Case 3 as an example, the results with different weight settings of voltage stability and load shedding are as follows.

As shown in Table 7, the impact of different weight settings on the optimal reconfiguration is minimal, as there is a certain positive relationship between the static voltage stability index and the load shedding amount: the more stable the system, the smaller the load shedding.

It is important to note that α should remain at a relatively high value to adequately account for the variability in DG output and load. Experiments with different values of α, including 80% and 90%, showed that the outcomes, such as load shedding and voltage stability, remained largely unchanged. This suggests that as long as α is sufficiently large, the specific value does not substantially impact the final results.

Table 7. The results with different weight settings of voltage stability and load shedding.

ω1/ω2	Reconfiguration Operation	Worst $\tilde{L}_{ij}$	Loss of Load (MWh)
0.1/0.9	L72 and L73 are closed	−0.9479	1.2070
0.4/0.6	L72 and L73 are closed	−0.9479	1.2070
0.9/0.1	L72 and L73 are closed	−0.9479	1.2070

5.4. Analysis of Comparison with Other Methods

5.4.1. Comparison of Reconfiguration Results of Different Methods

The conventional mathematical optimization method, ordinary decision trees, and AdaBoost-enhanced decision trees were used to calculate the fault cases of the 69-bus system, in order to verify the correctness and effectiveness of the proposed method. The weak classifier generation method with improved decision trees adopted the same CART algorithm as the ordinary decision trees method. All methods used the Gini index as the node splitting criterion; the relevant training parameters were set in the same way, and the fault sample set used for training was also the same. The conventional mathematical optimization method was solved using the Gurobi 10.0.2 solver. The reconfiguration results of the three methods are shown in Table 8.

Table 8. Comparison of the reconfiguration results of the different algorithms.

Case Number	Reconfiguration Results of Conventional Mathematical Optimization	Reconfiguration Results of Ordinary Decision Trees	Reconfiguration Results of AdaBoost-Enhanced Decision Trees
1	L69 is closed	L69 is closed	L69 is closed
2	L71 is closed	L71 is closed	L71 is closed
3	L69 and L72 are closed	L70 and L73 are closed	L69 and L72 are closed
4	L69 and L72 are closed	L71 and L72 are closed	L69 and L72 are closed
5	L73 is closed	L73 is closed	L73 is closed
6	L72 is closed	L72 is closed	L72 is closed

By comparing the results of the above three methods, it can be observed that the improved AdaBoost decision tree algorithm is consistent with the reconfiguration results obtained from conventional mathematical optimization, while the ordinary decision tree algorithm exhibits noticeable deviations. For example, in Case 4, the AdaBoost-enhanced decision tree method produces the same optimal switching strategy as the mathematical solver, ensuring almost full load recovery. In contrast, the ordinary decision tree method yields a suboptimal configuration that causes a large portion of the network to disconnect from the main power source, resulting in lower load restoration and potential voltage instability.

In Case 3, while all three methods produce feasible solutions that allow the system to operate under acceptable conditions, only the mathematical optimization method and the AdaBoost-enhanced decision tree method generate optimal configurations. The ordinary decision tree algorithm, due to its limited learning capacity and the absence of adaptive weighting for decision nodes, can only provide a suboptimal switching configuration, which leads to higher load losses and reduced overall performance.

5.4.2. Comparison of Load Recovery Speed of Different Methods

In the process of load recovery, the speed of load recovery depends on the calculation speed of the different methods. The 69-bus system was chosen for comparing the computation speed of the different algorithms, due to its larger scale. In addition, a new fault case was added (Case 7), in which L3 and L8 line are faulty.

The conventional mathematical optimization algorithm and the AdaBoost-enhanced decision tree algorithm were applied to calculate the results for the different fault cases of the 69-bus system. The conventional mathematical optimization method was solved using the Gurobi 10.0.2 solver. The computation times for both methods under identical fault cases are presented in Table 9.

Table 9. Comparison of the solving time of the different algorithms.

Case Number	Solving Time of Conventional Mathematical Optimization (s)	Solving Time of AdaBoost-Enhanced Decision Trees (s)
1	25.6861	0.2426
2	23.5707	0.2815
3	27.7653	0.2328
4	24.3391	0.3042
5	22.9184	0.2436
6	25.5102	0.2761
7	Non-convergence	0.2591

By comparing the solving times of the two algorithms in Table 5, it is evident that the AdaBoost-enhanced decision tree algorithm offers significantly faster computation. In Cases 1–6, the solving time of the conventional mathematical optimization method ranges from approximately 22 to 28 s, while the AdaBoost-enhanced decision tree algorithm completes the calculation in around 0.23 to 0.30 s, representing a reduction of about 90% in solving time. The AdaBoost-enhanced decision tree algorithm can rapidly retrieve power restoration solutions by learning from historical fault scenarios and system operating conditions, thereby avoiding the time-consuming process of solving complex mixed-integer nonlinear programming problems.

In practical engineering applications, minimizing the outage duration is critical, and fault reconfiguration decisions are typically required within a few seconds. The proposed fast-solving algorithm is therefore more aligned with real-time operational requirements.

Notably, in Case 7, where the faulty line is special, the conventional mathematical optimization approach fails to converge, due to increased model complexity. In contrast, the AdaBoost-enhanced decision tree algorithm still produces a feasible solution in this case. Although the generated solution requires further validation through power flow analysis, it nonetheless provides a viable reconfiguration scheme. This demonstrates the practical advantage of the proposed algorithm in handling complex or atypical fault scenarios.

5.4.3. Comparison of Algorithm Methods, Considering Different Factors

In the proposed two-stage fault reconfiguration strategy in this paper, the fault ride-through characteristics and output uncertainty of DGs are taken into account, allowing for the achievement of a higher load restoration ratio and better system stability. It is necessary to compare the proposed method with traditional methods that do not consider these factors, so as to demonstrate the advantages of the proposed method.

(1) The factor of DG fault ride-through

Case 5 and Case 6 of the 69-bus system and Case 4 of 11-bus system were used to analyze the impact of DG fault ride-through on the algorithm results. The load recovery results with and without considering fault ride-through are shown in Figure 7

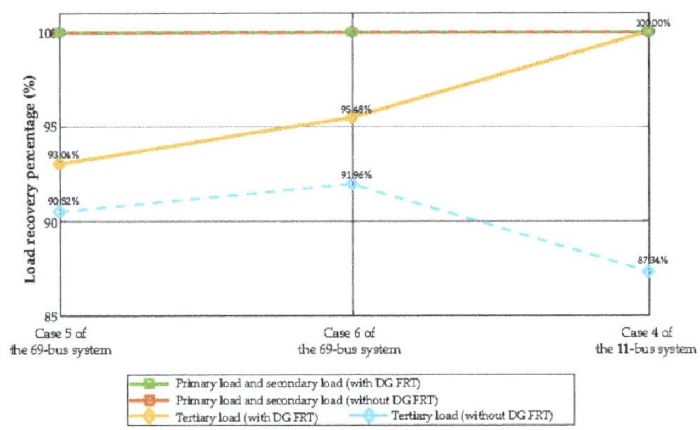

Figure 7. Comparison of load recovery results with/without considering FRT.

In Figure 7, the load recovery percentage of the method that does not consider the DG FRT capability is lower than that of the proposed method, which incorporates this factor. The case study results show that the proposed strategy achieves a higher load recovery rate, particularly in fault scenarios where DG disconnection is likely to occur.

For example, in Case 4 of the 11-bus system, the proposed method successfully restores all three levels of load. In contrast, the method without considering DG FRT restores only 87.34% of the tertiary load, resulting in a reduction of approximately 13%. Similarly, in the 69-bus system, Case 5 and Case 6 show that the tertiary load recovery rate in the method without considering DG FRT is about 4% lower than that of the proposed method.

It is worth noting that the 11-bus system has relatively limited resources and fewer available nodes for reconfiguration. Therefore, the observed gap in load recovery rate is more significant, which highlights the advantage of the proposed method in resource-constrained systems. These results demonstrate that by leveraging the FRT capability of DGs, the proposed strategy can significantly enhance the load recovery capability, particularly in smaller-scale or less redundant distribution networks.

(2) The factor of the uncertainty of DGs and loads

To demonstrate the necessity of considering the factor of the uncertainty of DGs and loads, a comparative calculation was performed for the 69-bus system without considering DG uncertainty (by assuming it to be a constant value for the calculation). In the optimal calculation without considering uncertainty, the power of source and load was set to a constant value. To avoid the influence of specific value choices, the power of source and load values at four typical times were selected for the calculation: 4:00, 10:00, 12:00, and 16:00. The worst case was chosen for comparison. The results are shown in Table 10.

Table 10. Comparison of optimal reconfiguration results with and without considering uncertainty.

Case Number	Considering Uncertainty		Without Considering Uncertainty	
	Worst $\tilde{L}_{ij}$	Loss of Load (MWh)	Worst $\tilde{L}_{ij}$	Loss of Load (MWh)
1	−0.8898	1.4664	−0.8898	1.4664
2	−0.9264	0.4602	−0.9264	0.4712
3	−0.9479	1.2070	−0.8425	1.5457
4	−0.9251	0.6349	−0.9251	0.7266

In Case 2, 3, and 4, when the uncertainty of DGs and loads is not considered, the load loss is higher. This suggests that fluctuations in DG power may lead to a larger net load power (the difference between the load and DG power), causing load loss. Therefore, the optimal reconfiguration strategy must account for the uncertainty of DGs and loads.

As for the static voltage stability index, in Case 3, the $\widetilde{L}_{ij}$ with consideration of uncertainty is smaller, indicating better system stability. Additionally, the largest deviation from the results occurs at 12:00. At 12:00, the DG output is highest, and the supply pressure is relatively low, so the scheme given at this time may not be optimal for other time periods.

In summary, considering uncertainty in the optimal reconfiguration can reduce load loss and improve the stability of the system in post-recovery long-term operation.

In terms of load loss, the results of Case 2, 3, and 4 show that the method considering uncertainty achieves lower load loss compared to the method that does not. Specifically, in Case 3, the load loss is reduced by approximately 22% when uncertainty in DG output and load demand is taken into account. Incorporating uncertainty into the optimization process enables the strategy to better anticipate and adapt to such variations, resulting in more effective load restoration.

As for the static voltage stability index $\widetilde{L}_{ij}$, it is also improved when uncertainty is considered. In Case 3, the index increases by about 12.5%, suggesting that the system operates under more stable conditions during post-recovery operation. It is also observed that the largest deviation in the voltage stability index occurs at 12:00, when the DG output is at its peak. At 12:00, the supply pressure is relatively low, and the reconfiguration scheme selected for that moment may not be optimal across other time periods, due to the dynamic nature of DG output and load.

In summary, considering uncertainty in DGs and loads during fault reconfiguration not only reduces load loss, but also enhances post-recovery voltage stability. This demonstrates the necessity of incorporating uncertainty into the reconfiguration strategy to ensure the long-term reliable operation of the distribution system.

5.5. Analysis of the Practicality of the Proposed Method

The aforementioned comparisons between the proposed method and other approaches have demonstrated its advantages in terms of computational speed, accuracy, load restoration, and system stability. In addition, the proposed two-stage reconfiguration strategy exhibits strong engineering applicability in real-world scenarios.

Taking the power supply company of a certain city in China as an example, we investigated the current fault reconfiguration process used by the operator, and identified several practical challenges they face during real-time fault restoration. These are summarized as follows:

1. Limited adaptability of current reconfiguration schemes

In existing practice, maintenance staff predefine several reconfiguration schemes during system planning. When a fault occurs, one of these fixed schemes is selected after manual verification, regardless of real-time system conditions. This rigid approach limits flexibility and responsiveness. The method proposed in this paper can automatically generate reconfiguration strategies based on real-time operational data, and considers multiple optimization objectives, thus offering dynamic adaptability.

2. Slow computation and infeasibility under complexity

Under certain scenarios with high DG penetration and complex network topology, existing methods often fail to provide timely or feasible solutions, especially with limited on-site computational capacity. Our proposed AdaBoost-enhanced decision tree algorithm significantly improves computational efficiency while ensuring optimality, enabling rapid decision-making in field environments.

3. Neglect of DG support capabilities

The existing strategy does not take into account the FRT capabilities of DGs. As DGs become more prevalent, this omission leads to underutilization of available resources. Our model explicitly considers the FRT capabilities of DGs, allowing them to actively support the system during fault recovery.

4. Lack of consideration of post-fault operational risks

The current approach focuses only on the immediate restoration of important loads, without evaluating the potential risks that may arise during subsequent operation. In contrast, our second-stage reconfiguration model considers the uncertainties in both DG output and load demand using fuzzy-based modeling, and ensures post-recovery voltage stability, contributing to long-term secure system operation.

6. Conclusions

This paper proposes a two-stage fault reconfiguration strategy for distribution networks with high penetration of DGs, considering the fault ride-through capability and output uncertainty of DGs. The results from the case studies demonstrate the following:

1. The proposed rapid power restoration reconfiguration model takes into account different types of DG behaviors during fault recovery, including the fault ride-through capability of DGs. After a fault occurs, it can effectively leverage the DGs' support capability for distribution networks and improve the load recovery percentage, thereby enhancing the restoration of lost load.
2. In the rapid power restoration strategy, the AdaBoost-enhanced improved decision tree algorithm offers faster computation speeds than traditional optimization methods, enhancing its ability to meet fault response requirements in practical engineering applications.
3. In the post-recovery optimal reconfiguration strategy, the reconfiguration model simultaneously considers load restoration and system stability as optimization objectives, and accounts for the uncertainty of DGs' output and load power, in order to enhance the long-term operational stability of the system.

The strategy proposed in this paper focuses primarily on fault reconfiguration within the distribution networks. It does not yet consider the issue of island operation after the failure of external power sources. The DG grid-connected control strategy does not analyze factors such as voltage–frequency control. Fault reconfiguration is typically based on static models, while distributed generators exhibit strong dynamic characteristics. The adaptability of more complex uncertainty models to the reconfiguration problem remains to be further validated. Future research will address and expand upon these aspects.

Author Contributions: Conceptualization, Y.H. and J.L.; methodology, Y.L.; software, J.L.; validation, Y.H., J.L. and Z.W.; formal analysis, F.S.; investigation, X.X.; resources, Y.F.; data curation, X.Z.; writing—original draft preparation, Y.H.; writing—review and editing, Y.L. All authors have read and agreed to the published version of the manuscript.

Funding: This research was funded by the State Grid Zhejiang Electric Power Company Science and Technology Project "Research on Key Technologies of Distributed Smart Distribution Network Operation Control for the New Power System" (Project No. 5211HZ240008).

Data Availability Statement: The original contributions presented in the study are included in the article, and further inquiries can be directed to the corresponding author.

Conflicts of Interest: Authors Yuwei He, Yanjun Li, Jian Liu, Xiang Xiang, Fang Sheng, Xinyu Zhu, and Yunpeng Fang were employed by the Hangzhou Power Supply Company. The remaining authors declare that the research was conducted in the absence of any commercial or financial relationships that could be construed as a potential conflicts of interest. The funders had no role in the design of the study; in the collection, analyses, or interpretation of data; in the writing of the manuscript; or in the decision to publish the results.

Appendix A

Appendix A.1

The constraints of the first stage (fast power restoration) of the strategy model are as follows:

(1) Power flow constraints

$$\begin{cases} P_{s,j} + \sum\limits_{i\in\Omega_{j,f}} \alpha_{ij}(P_{ij} - r_{ij}I_{ij}^2) = P_{load,j} + \sum\limits_{k\in\Omega_{j,z}} \alpha_{jk}P_{jk} \\ Q_{s,j} + \sum\limits_{i\in\Omega_{j,f}} \alpha_{ij}(Q_{ij} - x_{ij}I_{ij}^2) = Q_{load,j} + \sum\limits_{k\in\Omega_{j,z}} \alpha_{jk}Q_{jk} \end{cases} \tag{A1}$$

$$U_j^2 = U_i^2 - 2(r_{ij}P_{ij} + x_{ij}Q_{ij}) + (r_{ij}^2 + x_{ij}^2)I_{ij}^2 \tag{A2}$$

$$(U_i I_{ij})^2 = P_{ij}^2 + Q_{ij}^2 \tag{A3}$$

$$\begin{cases} P_{s,j} = P_{m,j} + P_{DG,j} + P_{ess,d,j} - P_{ess,c,j} \\ Q_{s,j} = Q_{m,j} + Q_{DG,j} \end{cases} \tag{A4}$$

where α_{ij} is the line switch status variable; $P_{s,j}$ and $Q_{s,j}$, and $P_{load,j}$ and $Q_{load,j}$, are the active and reactive power of the power source and load at node j, respectively; $P_{m,j}$, $P_{DG,j}$, $P_{ess,d,j}$, and $P_{ess,c,j}$ are the active power outputs of the main grid, DG, energy storage discharge, and energy storage charge at node j, respectively; $Q_{m,j}$ and $Q_{DG,j}$ represent the reactive power outputs of the main grid and DG at node i; and $\Omega_{j,f}$ and $\Omega_{j,z}$ are the parent and child node sets of node j.

The above power flow constraints form a non-convex model. The 0–1 variable α_{ij} in Equation (A1) is linearized using the Big-M method. Additionally, variables U_i^{sqr} and I_{ij}^{sqr} are introduced, and second-order cone relaxation (as shown in Equation (A5)) is applied to transform Equations (A1) to (A3) into a mixed-integer second-order cone programming model.

$$\begin{cases} U_j^{sqr} = U_j^2,\ I_{ij}^{sqr} = I_{ij}^2 \\ -\alpha_{ij}M \le P_{ij} \le \alpha_{ij}M \\ -\alpha_{ij}M \le Q_{ij} \le \alpha_{ij}M \\ -\alpha_{ij}M \le I_{ij}^{sqr} \le \alpha_{ij}M \end{cases} \tag{A5}$$

$$\begin{cases} P_{s,j} + \sum\limits_{i\in\Omega_{j,f}} (P_{ij} - r_{ij}I_{ij}^{sqr}) = P_{load,j} + \sum\limits_{k\in\Omega_{j,z}} P_{jk} \\ Q_{s,j} + \sum\limits_{i\in\Omega_{j,f}} (Q_{ij} - x_{ij}I_{ij}^{sqr}) = Q_{load,j} + \sum\limits_{k\in\Omega_{j,z}} Q_{jk} \end{cases} \tag{A6}$$

$$\begin{cases} U_j^{sqr} \le M(1-\alpha_{ij}) + U_i^{sqr} + (r_{ij}^2 + x_{ij}^2)I_{ij}^{sqr} - 2(r_{ij}P_{ij} + x_{ij}Q_{ij}) \\ U_j^{sqr} \ge -M(1-\alpha_{ij}) + U_i^{sqr} + (r_{ij}^2 + x_{ij}^2)I_{ij}^{sqr} - 2(r_{ij}P_{ij} + x_{ij}Q_{ij}) \end{cases} \tag{A7}$$

$$\left\| \begin{matrix} 2P_{ij} \\ 2Q_{ij} \\ I_{ij}^{sqr} - U_i^{sqr} \end{matrix} \right\| \le I_{ij}^{sqr} + U_i^{sqr} \tag{A8}$$

(2) Line operation safety constraints

$$\begin{cases} 0 \leq I_{ij}^{\text{sqr}} \leq I_{ij,\max}^2 \\ 0 \leq P_{ij} \leq P_{ij,\max} \\ 0 \leq Q_{ij} \leq Q_{ij,\max} \end{cases} \quad \text{(A9)}$$

where $I_{ij,\max}$, $P_{ij,\max}$, and $Q_{ij,\max}$ are the maximum allowable current, active power, and reactive power that can flow through the line ij, respectively.

(3) Node voltage constraints

For nodes without DG connections, the node voltage must remain within the allowable range. For nodes with DG connections, the voltage can operate in a low-voltage condition.

$$\begin{cases} U_{\text{ref}} = 1 \\ U_{i,\min}^2 \leq U_{i,\text{other}}^{\text{sqr}} \leq U_{i,\max}^2 \\ U_{i,\text{DG}}^{\text{sqr}} \leq U_{i,\max}^2 \end{cases} \quad \text{(A10)}$$

where U_{ref} is the reference node voltage magnitude; and $U_{i,\min}$ and $U_{i,\max}$ are the minimum and maximum allowable voltage magnitudes at node i, respectively.

(4) Energy storage operation constraints

$$\begin{cases} C_{i,\min}^{\text{soc}} \leq C_i^{\text{soc}} + \frac{\eta_{\text{c},i} P_{\text{ess,c},i} \Delta T - P_{\text{ess,d},i} \Delta T / \eta_{\text{d},i}}{E_{\text{ess}}} \leq C_{i,\max}^{\text{soc}} \\ 0 \leq P_{\text{ess,c},i} \leq u_{\text{c},i} P_{\text{ess,c},i,\max} \\ 0 \leq P_{\text{ess,d},i} \leq u_{\text{d},i} P_{\text{ess,d},i,\max} \\ u_{\text{c},i} + u_{\text{d},i} = 1 \end{cases} \quad \text{(A11)}$$

where C_i^{soc} denotes the state of charge of the energy storage at node i; $C_{i,\min}^{\text{soc}}$ and $C_{i,\max}^{\text{soc}}$ are the minimum and maximum states of charge at node i, respectively; $\eta_{\text{c},i}$ and $\eta_{\text{d},i}$ are the charging and discharging efficiencies at node i; $P_{\text{ess,c},i}$ and $P_{\text{ess,d},i}$ are the charging and discharging power at node i; $P_{\text{ess,c},i,\max}$ and $P_{\text{ess,d},i,\max}$ represent the maximum charging and discharging power at node i; and $u_{\text{c},i}$ and $u_{\text{d},i}$ represent the charging and discharging states at node i.

Appendix A.2

Compared to the first stage, the constraints of the second stage (post-recovery optimal reconfiguration) of the strategy model take into account multiple time periods. The constraints are as follows:

(1) Power flow constraints

$$\begin{cases} U_{j,t}^{\text{sqr}} = U_{j,t}^2,\ I_{ij,t}^{\text{sqr}} = I_{ij,t}^2 \\ -\alpha_{ij} M \leq P_{ij,t} \leq \alpha_{ij} M \\ -\alpha_{ij} M \leq Q_{ij,t} \leq \alpha_{ij} M \\ -\alpha_{ij} M \leq I_{ij,t}^{\text{sqr}} \leq \alpha_{ij} M \end{cases} \quad \text{(A12)}$$

$$\begin{cases} P_{\text{s},j,t} + \sum\limits_{i \in \Omega_{j,f}} (P_{ij,t} - r_{ij} I_{ij,t}^{\text{sqr}}) = P_{\text{load},j,t} + \sum\limits_{k \in \Omega_{j,z}} P_{jk,t} \\ Q_{\text{s},j,t} + \sum\limits_{i \in \Omega_{j,f}} (Q_{ij,t} - x_{ij} I_{ij,t}^{\text{sqr}}) = Q_{\text{load},j,t} + \sum\limits_{k \in \Omega_{j,z}} Q_{jk,t} \end{cases} \quad \text{(A13)}$$

$$\begin{cases} U_{j,t}^{\text{sqr}} \leq M(1 - \alpha_{ij}) + U_{i,t}^{\text{sqr}} + (r_{ij}^2 + x_{ij}^2) I_{ij,t}^{\text{sqr}} - 2(\text{r}_{ij} P_{ij,t} + x_{ij} Q_{ij,t}) \\ U_{j,t}^{\text{sqr}} \geq -M(1 - \alpha_{ij}) + U_{i,t}^{\text{sqr}} + (r_{ij}^2 + x_{ij}^2) I_{ij,t}^{\text{sqr}} - 2(\text{r}_{ij} P_{ij,t} + x_{ij} Q_{ij,t}) \end{cases} \quad \text{(A14)}$$

$$\left\| \begin{matrix} 2P_{ij,t} \\ 2Q_{ij,t} \\ I_{ij,t}^{\text{sqr}} - U_{i,t}^{\text{sqr}} \end{matrix} \right\| \leq I_{ij,t}^{\text{sqr}} + U_{i,t}^{\text{sqr}} \tag{A15}$$

$$\begin{cases} P_{\text{s},j,t} = P_{\text{m},j,t} + P_{\text{DG},j,t} + P_{\text{ess,d},j,t} - P_{\text{ess,c},j,t} \\ Q_{\text{s},j,t} = Q_{\text{m},j,t} + Q_{\text{DG},j,t} \end{cases} \tag{A16}$$

$P_{\text{s},j,t}$ and $Q_{\text{s},j,t}$, and $P_{\text{load},j,t}$ and $Q_{\text{load},j,t}$, are the active and reactive power of the power source and load at node j at time t, respectively; $P_{\text{m},j,t}$, $P_{\text{DG},j,t}$, $P_{\text{ess,d},j,t}$, and $P_{\text{ess,c},j,t}$ are the active power outputs of the main grid, DG, energy storage discharge, and energy storage charge at node j at time t, respectively; $P_{ij,t}$ and $Q_{ij,t}$ represent the active and reactive power flowing from node i to node j at time t; $I_{ij,t}$ represents the current flowing through line ij at time t; and $U_{i,t}$ denotes the voltage magnitude at node j at time t.

(2) Line operation safety constraints

$$\begin{cases} 0 \leq I_{ij,t}^{\text{sqr}} \leq I_{ij,\max}^2 \\ 0 \leq P_{ij,t} \leq P_{ij,\max} \\ 0 \leq Q_{ij,t} \leq Q_{ij,\max} \end{cases} \tag{A17}$$

where $I_{ij,\max}$, $P_{ij,\max}$, and $Q_{ij,\max}$ are the maximum allowable current, active power, and reactive power that can flow through the line ij, respectively.

(3) Node voltage constraints

$$\begin{cases} U_{\text{ref}} = 1 \\ U_{i,\min}^2 \leq U_{i,t}^{\text{sqr}} \leq U_{i,\max}^2 \end{cases} \tag{A18}$$

where U_{ref} is the reference node voltage magnitude; and $U_{i,\min}$ and $U_{i,\max}$ are the minimum and maximum allowable voltage magnitudes at node i, respectively.

(4) Energy storage operation constraints

$$\begin{cases} C_{i,t}^{\text{soc}} = C_{i,t-1}^{\text{soc}} + \frac{\eta_{\text{c},i} P_{\text{ess,c},i,t} \Delta T - P_{\text{ess,d},i,t} \Delta T / \eta_{\text{d},i}}{E_{\text{ess}}} \\ 0 \leq P_{\text{ess,c},i,t} \leq u_{\text{c},i,t} P_{\text{ess,c},i,\max} \\ 0 \leq P_{\text{ess,d},i,t} \leq u_{\text{d},i,t} P_{\text{ess,d},i,\max} \\ u_{\text{c},i,t} + u_{\text{d},i,t} = 1 \\ C_{i,\min}^{\text{soc}} \leq C_{i,t}^{\text{soc}} \leq C_{i,\max}^{\text{soc}} \end{cases} \tag{A19}$$

where $C_{i,t}^{\text{soc}}$ denotes the state of charge of the energy storage at node i at time t; $P_{\text{ess,c},i,t}$ and $P_{\text{ess,d},i,t}$ are the charging and discharging power at node i at time t; and $u_{\text{c},i,t}$ and $u_{\text{d},i,t}$ represent the charging and discharging states at node i at time t.

Appendix B

Appendix B.1. IEEE 69-Bus Distribution Network System with DGs

Table A1 lists the access nodes and parameters of the distributed power sources. In addition, the maximum allowable short-circuit current $I_{\max}$ of DGs is 1.2 times the rated current.

Table A2 lists the energy storage access nodes and parameters.

The total load of the system is 3 802 kW + j2 695 kVar, and the load weights of each node are shown in Table A3.

Table A1. Parameters of DGs in the IEEE 69-bus system.

Equipment	Connected Node	Rated Active Power (kW)	Equipment	Connected Node	Rated Active Power (kW)
DG1	5	800	DG5	42	200
DG2	9	400	DG6	48	100
DG3	19	250	DG7	63	800
DG4	32	250			

Table A2. Parameters of energy storage devices in the IEEE 69-bus system.

Equipment	Connected Node	Capacity (MWh)	Upper Limit of Charging and Discharging Power (MW)	Charging and Discharging Efficiency
Energy storage 1	4	1.2	0.45/0.45	90%/90%
Energy storage 2	51	1	0.3/0.3	90%/90%

Table A3. Parameters of loads in the IEEE 69-bus system.

Type of Load	Connected Nodes	Weight Factor
Primary load	6, 8, 21, 51	100
Secondary load	12, 17, 33, 42, 49, 64	10
Tertiary load	The rest of the nodes	1

Figure A1 shows the power prediction curves of DGs and loads in one day, where the ordinate refers to the DG output power as a percentage of the rated power, and the load demand power as a percentage of the maximum demand. The fuzzy triangle membership parameters of DGs, $k_{i,1}^t$ are $k_{i,2}^t$, 0.92 and 1.08, respectively, and the fuzzy triangle membership parameters of loads, $k_{i,1}^t$ $k_{i,2}^t$, are 0.95 and 1.05, respectively.

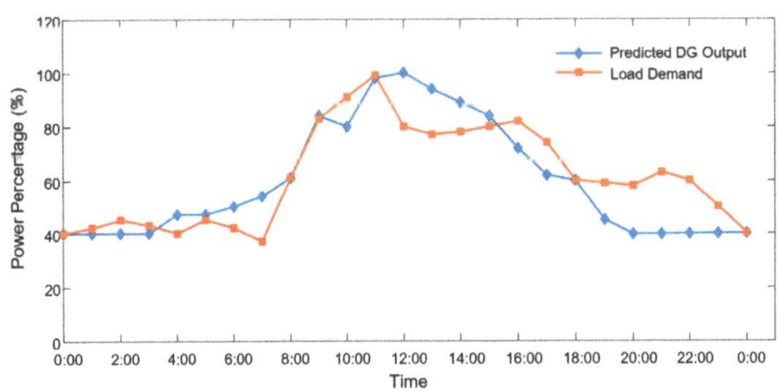

Figure A1. DG and load power forecasting curves.

Appendix B.2. 11-Bus Distribution Network System with DGs

The system voltage level is 10 kV, and the power forecast of DGs and loads in one day is the same as that of system 1.

Table A4 lists the access nodes and parameters of the distributed power sources. In addition, the maximum allowable short-circuit current Imax of DGs is 1.2 times the rated current.

Table A4. Parameters of DGs in the 11-bus system.

Equipment	Connected Node	Rated Active Power (MW)
DG1	4	4
DG2	7	10.6
DG3	8	5

Table A5 lists the load weights of each node.

Table A5. Parameters of loads in the 11-bus system.

Equipment	Connected Node	Power (MW)	Weight Factor
P1	2	2.14 MW + j0.35 MVar	100
P2	6	1.89 MW + j0.96 MVar	100
P3	7	4.29 MW + j1.34 MVar	100
P4	8	5.93 MW + j4.27 MVar	10
P5	9	6.34 MW + j4.71 MVar	1
P6	10	4.16 MW + j1.35 MVar	1
P7	11	7.85 MW + j2.23 MVar	1

References

1. Morales-Munoz, A.; Freijedo, F.D.; Pugliese, S.; Liserre, M. Selective Virtual Impedance for Overcurrent Limitation in Grid-Forming Inverters Under Asymmetrical Faults. In Proceedings of the 2024 IEEE 15th International Symposium on Power Electronics for Distributed Generation Systems (PEDG), Luxembourg, 23–26 June 2024; pp. 1–5. [CrossRef]
2. Liyanage, C.; Nutkani, I.; Meegahapola, L. A Comparative Analysis of Prominent Virtual Synchronous Generator Strategies Under Different Network Conditions. *IEEE Open Access J. Power Energy* **2024**, *11*, 178–195. [CrossRef]
3. Shobug, M.A.; Chowdhury, N.A.; Hossain, M.A.; Sanjari, M.J.; Lu, J.; Yang, F. Virtual Inertia Control for Power Electronics-Integrated Power Systems: Challenges and Prospects. *Energies* **2024**, *17*, 2737. [CrossRef]
4. Li, J.; Mo, H.; Sun, Q.; Wei, W.; Yin, K. Distributed Optimal Scheduling for Virtual Power Plant with High Penetration of Renewable Energy. *Int. J. Electr. Power Energy Syst.* **2024**, *160*, 110103. [CrossRef]
5. Taheri, S.I.; Salles, M.B.C.; Costa, E.C.M. Optimal Cost Management of Distributed Generation Units and Microgrids for Virtual Power Plant Scheduling. *IEEE Access* **2020**, *8*, 208449–208461. [CrossRef]
6. ALAhmad, A.K.; Verayiah, R.; Shareef, H. Long-Term Optimal Planning for Renewable Based Distributed Generators and Battery Energy Storage Systems toward Enhancement of Green Energy Penetration. *J. Energy Storage* **2024**, *90*, 111868. [CrossRef]
7. Nsaif, Y.M.; Lipu, M.S.H.; Ayob, A.; Yusof, Y.; Hussain, A. Fault Detection and Protection Schemes for Distributed Generation Integrated to Distribution Network: Challenges and Suggestions. *IEEE Access* **2021**, *9*, 142693–142717. [CrossRef]
8. Karimi, A.; Nayeripour, M.; Abbasi, A.R. Coordination in Islanded Microgrids: Integration of Distributed Generation, Energy Storage System, and Load Shedding Using a New Decentralized Control Architecture. *J. Energy Storage* **2024**, *98*, 113199. [CrossRef]
9. Behbahani, M.R.; Jalilian, A.; Bahmanyar, A.; Ernst, D. Comprehensive Review on Static and Dynamic Distribution Network Reconfiguration Methodologies. *IEEE Access* **2024**, *12*, 9510–9525. [CrossRef]
10. Zidan, A.; Khairalla, M.; Abdrabou, A.M.; Khalifa, T.; Shaban, K.; Abdrabou, A.; El Shatshat, R.; Gaouda, A.M. Fault Detection, Isolation, and Service Restoration in Distribution Systems: State-of-the-Art and Future Trends. *IEEE Trans. Smart Grid* **2017**, *8*, 2170–2185. [CrossRef]
11. Shen, F.; Wu, Q.; Xue, Y. Review of Service Restoration for Distribution Networks. *J. Mod. Power Syst. Clean Energy* **2020**, *8*, 1–14. [CrossRef]
12. Zarei, S.F.; Parniani, M. A Comprehensive Digital Protection Scheme for Low-Voltage Microgrids with Inverter-Based and Conventional Distributed Generations. *IEEE Trans. Power Deliv.* **2017**, *32*, 441–452. [CrossRef]
13. Musarrat, M.N.; Fekih, A.; Islam, M.R. An Improved Fault Ride Through Scheme and Control Strategy for DFIG-Based Wind Energy Systems. *IEEE Trans. Appl. Supercond.* **2021**, *31*, 1–6. [CrossRef]
14. Bagherzadeh, L.; Shayeghi, H.; Pirouzi, S.; Shafie-khah, M.; Catalão, J.P.S. Coordinated Flexible Energy and Self-Healing Management According to the Multi-Agent System-Based Restoration Scheme in Active Distribution Network. *IET Renew. Power Gener.* **2021**, *15*, 1765–1777. [CrossRef]

15. Ye, Z.; Chen, C.; Chen, B.; Wu, K. Resilient Service Restoration for Unbalanced Distribution Systems With Distributed Energy Resources by Leveraging Mobile Generators. *IEEE Trans. Ind. Inform.* **2021**, *17*, 1386–1396. [CrossRef]
16. Shi, X.; Zhang, H.; Wei, C.; Li, Z.; Chen, S. Fault Modeling of IIDG Considering Inverter's Detailed Characteristics. *IEEE Access* **2020**, *8*, 183401–183410. [CrossRef]
17. Cao, Y.; Liu, W.; Zhang, Y.; Zhang, H. An Improved Fault Recovery Strategy for Active Distribution Network Considering LVRT Capability of DG. In Proceedings of the 8th Renewable Power Generation Conference (RPG 2019), Shanghai, China, 24–25 October 2019; pp. 1–7.
18. Li, C.; Xi, Y.; Lu, Y.; Liu, N.; Chen, L.; Ju, L.; Tao, Y. Resilient Outage Recovery of a Distribution System: Co-Optimizing Mobile Power Sources with Network Structure. *Prot. Control Mod. Power Syst.* **2022**, *7*, 32. [CrossRef]
19. Lin, C.; Chen, C.; Bie, Z.; Li, G. Post-disaster Load Restoration Method for Urban Distribution Network with Dynamic Uncertainty and Frequency-Voltage Control. *Autom. Electr. Power Syst.* **2022**, *46*, 56–64.
20. Shi, X.; Ke, Q.; Lei, J.; Yuan, Z. Fault Reconfiguration of Distribution Networks with Soft Open Points Considering Uncertainties of Photovoltaic Outputs and Loads. *J. Phys. Conf. Ser.* **2020**, *1578*, 012205. [CrossRef]
21. Shen, Y.; Wang, G.; Zhu, J. Resilience Improvement Model of Distribution Network Based on Two-Stage Robust Optimization. *Electr. Power Syst. Res.* **2023**, *223*, 109559. [CrossRef]
22. Kanojia, S.S.; Suthar, B.N. Voltage Stability Index: A Review Based on Analytical Method, Formulation and Comparison in Renewable Dominated Power System. *IJAPE* **2024**, *13*, 508. [CrossRef]
23. Moradi, M.H.; Abedini, M. A Combination of Genetic Algorithm and Particle Swarm Optimization for Optimal Distributed Generation Location and Sizing in Distribution Systems with Fuzzy Optimal Theory. *Int. J. Green Energy* **2012**, *9*, 641–660. [CrossRef]
24. Bi, C.Q.; Chen, J.J.; Wang, Y.X.; Feng, L. Fuzzy Credibility Chance-Constrained Multi-Objective Optimization for Multiple Transactions of Electricity–Gas–Carbon under Uncertainty. *Electr. Power Syst. Res.* **2025**, *238*, 111089. [CrossRef]
25. Xiao, G.; Liu, H.; Nabatalizadeh, J. Optimal Scheduling and Energy Management of a Multi-Energy Microgrid with Electric Vehicles Incorporating Decision Making Approach and Demand Response. *Sci. Rep.* **2025**, *15*, 5075. [CrossRef] [PubMed]
26. Fan, G.; Peng, C.; Wang, X.; Wu, P.; Yang, Y.; Sun, H. Optimal Scheduling of Integrated Energy System Considering Renewable Energy Uncertainties Based on Distributionally Robust Adaptive MPC. *Renew. Energy* **2024**, *226*, 120457. [CrossRef]
27. Arjomandi-Nezhad, A.; Fotuhi-Firuzabad, M.; Moeini-Aghtaie, M.; Safdarian, A.; Dehghanian, P.; Wang, F. Modeling and Optimizing Recovery Strategies for Power Distribution System Resilience. *IEEE Syst. J.* **2021**, *15*, 4725–4734. [CrossRef]
28. Kong, F.; Guo, T.; Zhang, X. Design of Active Fault Diagnosis and Repair System for Active Distribution Networks Based on Decision Tree Algorithm. In Proceedings of the 6th International Conference on Information Technologies and Electrical Engineering, Hunan, China, 3–5 November 2023; Association for Computing Machinery: New York, NY, USA, 2024; pp. 759–764.
29. Wang, Q.; Meng, L. Distribution Network Fault Reconfiguration with Distributed Generation Based on Ant Colony Algorithm. *Adv. Mater. Res.* **2013**, *732–733*, 1328–1333. [CrossRef]
30. Rebollal, D.; Carpintero-Rentería, M.; Santos-Martín, D.; Chinchilla, M. Microgrid and Distributed Energy Resources Standards and Guidelines Review: Grid Connection and Operation Technical Requirements. *Energies* **2021**, *14*, 523. [CrossRef]
31. *IEEE Std 1547-2018 (Revision of IEEE Std 1547-2003)*; IEEE Standard for Interconnection and Interoperability of Distributed Energy Resources with Associated Electric Power Systems Interfaces. IEEE: New York, NY, USA, 2018; pp. 1–138. [CrossRef]
32. Li, H.; Deng, C.; Zhang, Z.; Liang, Y.; Wang, G. An Adaptive Fault-Component-Based Current Differential Protection Scheme for Distribution Networks with Inverter-Based Distributed Generators. *Int. J. Electr. Power Energy Syst.* **2021**, *128*, 106719. [CrossRef]
33. Xu, C.; Dong, S.; Zhu, J. Description Method of Radial Constraints for Distribution Network Based on Disconnection Condition of Power Supply Loop. *Autom. Electr. Power Syst.* **2019**, *43*, 82–94.
34. Ding, W.; Chen, Q.; Dong, Y.; Shao, N. Fault Diagnosis Method of Intelligent Substation Protection System Based on Gradient Boosting Decision Tree. *Appl. Sci.* **2022**, *12*, 8989. [CrossRef]
35. Shan, W.; Li, D.; Liu, S.; Song, M.; Xiao, S.; Zhang, H. A Random Feature Mapping Method Based on the AdaBoost Algorithm and Results Fusion for Enhancing Classification Performance. *Expert Syst. Appl.* **2024**, *256*, 124902. [CrossRef]
36. Han, Y.; Xu, K.; Qin, J. Based on the CART Decision Tree Model of Prediction and Classification of Ancient Glass-Related Properties. *Highlights Sci. Eng. Technol.* **2023**, *42*, 18–27. [CrossRef]
37. Hu, H.; Siala, M.; Hebrard, E.; Huguet, M.-J. Learning Optimal Decision Trees with MaxSAT and Its Integration in AdaBoost. In Proceedings of the Twenty-Ninth International Joint Conference on Artificial Intelligence, International Joint Conferences on Artificial Intelligence Organization, Yokohama, Japan, 11–17 July 2020; pp. 1170–1176.
38. Chen, S.; Shen, B.; Wang, X.; Yoo, S.-J. A Strong Machine Learning Classifier and Decision Stumps Based Hybrid AdaBoost Classification Algorithm for Cognitive Radios. *Sensors* **2019**, *19*, 5077. [CrossRef] [PubMed]

39. Aboshady, F.M.; Pisica, I.; Zobaa, A.F.; Taylor, G.A.; Ceylan, O.; Ozdemir, A. Reactive Power Control of PV Inverters in Active Distribution Grids With High PV Penetration. *IEEE Access* **2023**, *11*, 81477–81496. [CrossRef]
40. Bhutta, M.S.; Sarfraz, M.; Ivascu, L.; Li, H.; Rasool, G.; ul Abidin Jaffri, Z.; Farooq, U.; Ali Shaikh, J.; Nazir, M.S. Voltage Stability Index Using New Single-Port Equivalent Based on Component Peculiarity and Sensitivity Persistence. *Processes* **2021**, *9*, 1849. [CrossRef]

electronics

MDPI

Article

A Novel Modified Delta-Connected CHB Multilevel Inverter with Improved Line–Line Voltage Levels

Abdullah M. Noman

Electrical Engineering Department, College of Engineering, Prince Sattam Bin Abdulaziz University, Al-Kharj 16278, Saudi Arabia; a.noman@psau.edu.sa

Abstract: Numerous cascaded inverter configurations have been developed to generate higher voltage levels, thereby improving performance and lowering costs. Comparing conventional delta-connected cascaded H-bridge (CHB) multilevel inverters to star-connected CHB multilevel inverters reveals a disadvantage. In conventional delta-connected CHB multilevel inverters, more switches are unavoidably needed to achieve the same line-to-line grid voltage, since more H-bridges cascaded in series are required than in a star-connected CHB. This paper presents a modified topology based on the delta-connected CHB multilevel configuration to provide the same number of line-to-line voltage levels as a star-connected CHB, using an equivalent number of switches. The number of switches in the proposed multilevel inverter is decreased compared to conventional delta-connected CHB MLIs at the same voltage levels. The mathematical modeling of the proposed topology and the simulation results using a fixed load and a PV-grid connection are provided to validate the efficacy and dependability of the proposed topology. To validate the usefulness of the proposed configuration, it was practically implemented in the laboratory. Data acquisition and generation of gating signals to fire the switches were implemented using a MicroLab-Box real-time controller. The prototype was examined under a resistive–inductive load and tested under different modulation indices. To demonstrate the effectiveness and the functionality of the topology, the experimental results are also provided.

Keywords: multilevel inverter; cascaded H bridge MLI; delta connected CHB; star connected CHB

1. Introduction

Multilevel inverter (MLI) topologies have proven to be a feasible solution for medium- to high-voltage high-power applications. The basic idea behind an MLI is to create a voltage waveform using a staircase pattern by combining numerous power semiconductor switches with different direct-current (DC) sources of lower voltage to convert power [1]. With a three-level converter, the word "multilevel" was first introduced [2]. Afterwards, numerous MLI topologies have been proposed [3–10]. MLIs offer several appealing characteristics, including reduced voltage stress on switches, decreased total harmonic distortion (THD) of voltage and current, enhanced power quality, and minimal power dissipation on switches while in the ON state, reduced filter size, ability to reduce the switching frequency, and hence the reduction in switching losses, etc. [11–16]. The merits of MLIs make them attractive for medium-voltage and high-voltage high-power applications. Generally, classical MLIs can be classified according to the connected DC sources into single- or separated-DC-source MLIs, as shown in Figure 1.

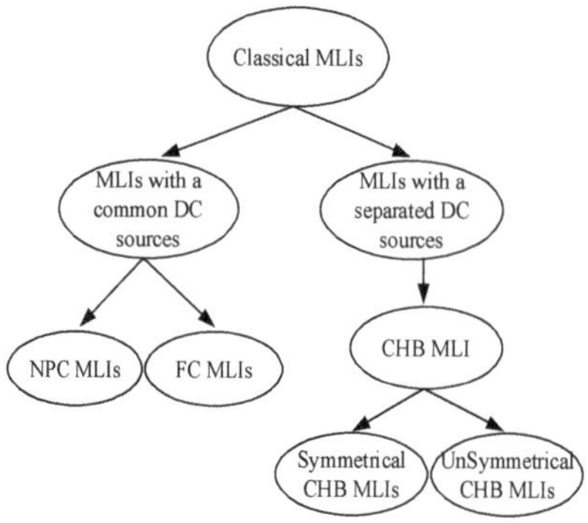

Figure 1. Asymmetric cascaded H-bridge topology.

The three-level neutral point multilevel inverter (NPC MLI) were initially introduced in 1981. After that, several MLIs based on the NPC have been proposed [14,15,17–20]. A three-level NPC MLI inverter is shown in Figure 2.

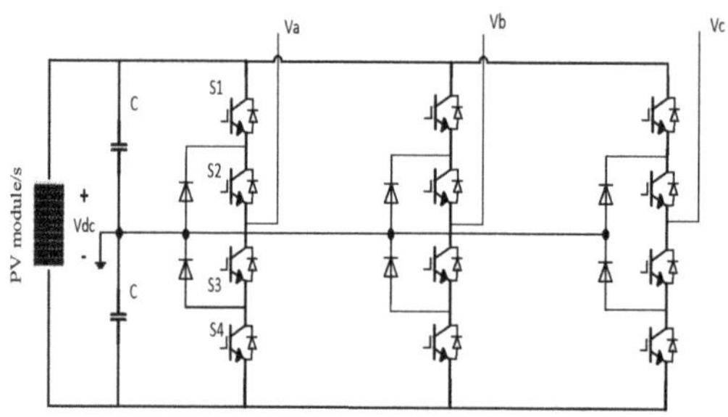

Figure 2. Three-phase three-level NPC MLI.

The NPC topology offers various merits, including the fact that voltage stress can be distributed between the series switches, and that current fluctuations are minimized. However, the NPC topology suffers from voltage drops on the clamped diodes, and the topology is bulky at high voltage levels.

Another topology called the flying capacitor multilevel inverter (FC MLI), in which the diodes of the NPC are replaced by capacitors, is designed in such a manner to reduce the disadvantages of NPC inverters [21,22]. A three-level flying capacitor MLI is shown in Figure 3. However, increasing the number of voltage levels with a flying capacitor MLI requires increasing the number of capacitors. In addition, the voltages of these capacitors need to be carefully controlled; otherwise, the output voltage is unbalanced [23].

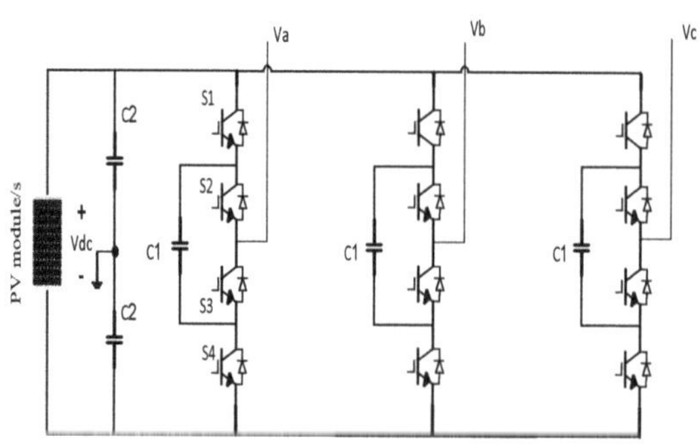

Figure 3. Three-phase three-level FC MLI.

As shown in Figures 2 and 3, NPC and FC MLIs require a single DC source [24]. CHB MLIs, however, require several separated DC sources. Each DC source is connected to one H-bridge cell. The series connection of H-bridge cells serves as the foundation for CHB MLIs. CHB MLIs require fewer components at the same voltage levels than other traditional MLI topologies. Photovoltaic (PV) modules, fuel cells, batteries, etc., can serve as DC sources [25]. When the magnitude of the DC sources is the same, CHB is referred to as symmetric. In a similar vein, the CHB is referred to as asymmetric when the DC sources have different magnitudes. The advantages of CHB MLIs are as follows:

- The CHB MLI topology has the fewest components among the alternative topologies to produce the same voltage levels.
- Extreme modularity.
- Low cost.
- High inverter efficiency.
- Easy to construct and control.

Cascaded MLI features have provided an appealing solution for various applications, including standalone systems, static VAR compensations, and grid-connected PV applications [6,26–31].

By using a unipolar modulation approach, each H-bridge can provide three output voltage levels [28,32]. Several CHB MLIs have been introduced in the literature to increase the number of generated voltage levels, such as five levels [33], seven levels [34], and more [35,36].

Numerous alternative cascaded MLI topologies have been documented in the literature with the aim of increasing the level count, thereby enhancing system performance and diminishing system expenses [15]. Asymmetric CHB MLI topologies with differing DC source magnitudes can generate more levels with fewer switches.

As mentioned above, cascaded H-bridge (CHB) MLIs comprise a series connection of several H-bridge cells.

The addition of cascaded MLI capabilities has rendered numerous applications [6,16, 28–31,37–42]. There are two possible connections for CHB configurations: delta and star. Applications vary for each connection. A star-connected and a delta-connected CHB MLIs are shown in Figures 4 and 5, respectively.

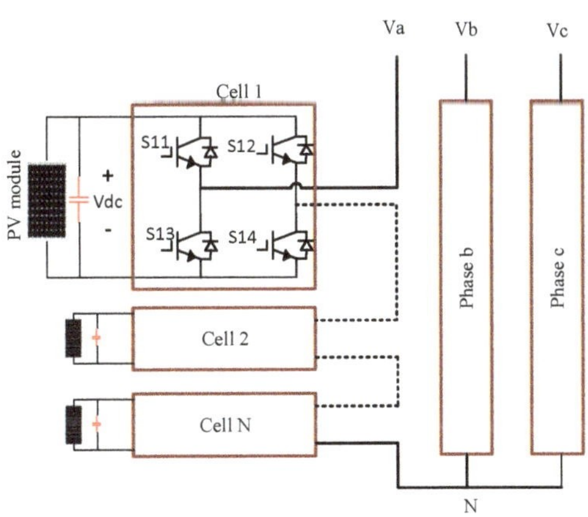

Figure 4. Star-connected CHB MLI.

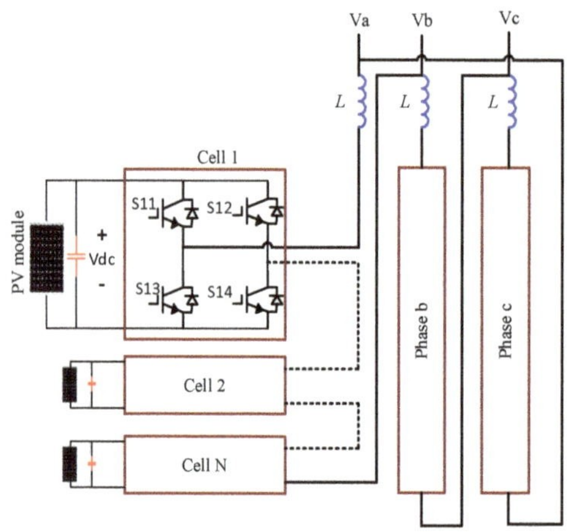

Figure 5. Delta-connected CHB MLI.

The authors in [39] proposed a STATCOM based on a star-connected CHB to suppress the grid voltage unbalance. In this paper, the limitation of star-connected CHB MLIs in compensating the unbalanced grid voltage is analyzed. The analysis is confirmed by simulation and experiments.

The authors in [40] investigated the use of three-phase star-connected CHB MLIs for active power filter applications. A simulation of a nine-level CHB MLI-based SAPF system was conducted in MATLAB/Simulink, demonstrating its efficacy in alleviating power quality problems stemming from non-linear loads.

The author in [38] proposed a star-connected three-phase seven-level CHB MLI. Each H-bridge is connected to a multistring PV, and each string is connected to one DC–DC converter. One DC–DC converter is assigned for each PV string. The DC–DC converter is responsible for the MPPT of the PV string and amplifying the DC voltage, if necessary. This connection method has the same issues of not correctly tracking the MPP of the PV modules, and its efficiency is reduced. Moreover, if many PV modules are coupled to a single DC–DC converter, the DC–DC converter's power rating will be high, raising the cost of the system. In [43], various modulation approaches were used to improve the power quality in a PV MLI. Harmonic suppression strategies were used in the inverter structure. To eliminate harmonics, appropriate switching states were applied to the inverter switches. The authors in [44] proposed a synchronization approach based on phase-shifted pulse width modulation (PSPWM) to control a star-connected CHB MLI. In [45], different modulation techniques based on specific sequential switching were presented and applied to CHB MLIs. The authors in [46] developed a control technique based on space vector modulation to improve the THD and the grid current of a star-connected CHB MLI. A star-connected CHB was employed for PV applications, and DC–DC converters were used for the maximum power point tracking.

Delta-connected CHB MLIs are gaining popularity as a viable alternative to star-connected CHB MLIs for PV applications. [47–50]. Individual phase control is simple to create with delta-connected CHB MLIs, which can be utilized as static synchronous compensation (STATCOM). STATCOM applications necessitate a careful consideration of handling negative-sequence reactive power [51–54].

STATCOM enhances supply quality and offers quick dynamic reactive power support [55,56]. Additionally, it is employed in large-scale industrial applications for compensating unbalanced and nonlinear loads [31]. One of the challenges that power converters face is to supply balanced grid currents even under unbalanced loads or unbalanced generation, especially with renewable energy resources [57]. Certain loads, such as traction drives and arc furnaces, can be sources of unbalanced current [31]. On the other hand, voltage unbalance can be caused due to the presence of line-to-ground faults or unbalanced loads. These unbalanced currents or voltages require that the grid-connected inverter supply negative-sequence current or voltage to balance the current or the voltage at the point of common coupling. One solution is to use STATCOM. STATCOM using MLIs has received more attention to suppress these unbalanced voltages and currents [58–60]. The enhancement of power system stability, unbalanced load adjustment, power factor improvement, line voltage control, active power filters (APF), and low-voltage ride-through (LVRT) are examples of common uses for STATCOM [55]. One interesting converter for STATCOM applications is the CHB converter topology. Because of its modular design, transformerless connection, and smaller output filter, it was anticipated to be the ideal option for MV applications [50,61]. The authors in [62] proposed delta-connected CHB MLIs as an alternative to star-connected CHB MLIs for STATCOM under imbalanced situations. Delta-connected CHB MLIs enable the compensation of imbalanced loads while simultaneously adjusting for reactive power [63]. The authors in [64] compared the two types of CHB MLIs (star-connected and delta-connected) and concluded that delta-connected CHB MLIs are preferable to star-connected CHB MLIs, especially for compensating unbalanced currents, since delta-connected CHB MLIs require less zero-sequence current injection than the zero-sequence voltage injection in star-connected CHB MLIs to compensate for the unbalanced conditions. Moreover, delta-connected CHB MLIs motivate researchers to further investigate and analyze delta-connected CHB MLIs for LVRT purposes [62,63]. The above literature review explains the motivations to use delta-connected CHB MLIs. However, in order to achieve the line-to-line grid voltage under balanced PV generation, the delta connection needs more bridges cascaded in series than the star connection, which unavoidably increases the converter's size. In addition, the minimization of circulating current inside the delta configuration necessitates the implementation of a solution, such as adding magnetic elements. As a result, delta-connected H-bridges are more expensive than star-connected H-bridges for generating the same line-to-line voltages. To synthesize line-to-line voltage, one phase leg needs a bridge number that is $\sqrt{3}$ times greater [49,64]. Minimizing the current that flows inside the delta connection must also be addressed.

This study suggested modifying the delta-connected CHB configuration in order to increase the number of line–line voltage levels to match the number of levels produced by the star-connected CHB. The number of required H-bridges in the proposed modified MLI is the same as that required by the star-connected CHB to produce the same line–line voltage levels. In addition, the proposed configuration eliminates the need for adding extra inductors since the solution can solve both the number of voltage levels and the circulating current. The proposed topology was mathematically analyzed and modeled inside the SIMULINK environment to evaluate its accuracy and efficacy. Moreover, the proposed MLI was experimentally implemented in the laboratory. The constructed inverter was tested under a resistive load at different modulation indices. The simulation and experimental results are provided to prove the advantages of the proposed topology.

The paper is organized as follows. Section 2 introduces the description of the proposed topology, while Section 3 shows the analysis of the proposed topology. The simulation results are explained in Section 4, and the experimental results are shown in Section 5. Section 6 describes the conclusions.

2. Description of the Proposed Topology

The proposed configuration is a modified configuration for traditional delta-connected CHB MLIs. Figure 6a shows the star-connected CHB topology, while Figure 6b shows a traditional delta-connected CHB MLI. As shown in this figure, the number of the line-to-line voltage levels generated from the traditional delta-connected CHB (Figure 6b) is the same as the line–neutral voltage levels generated from the star-connected CHB (Figure 6a).

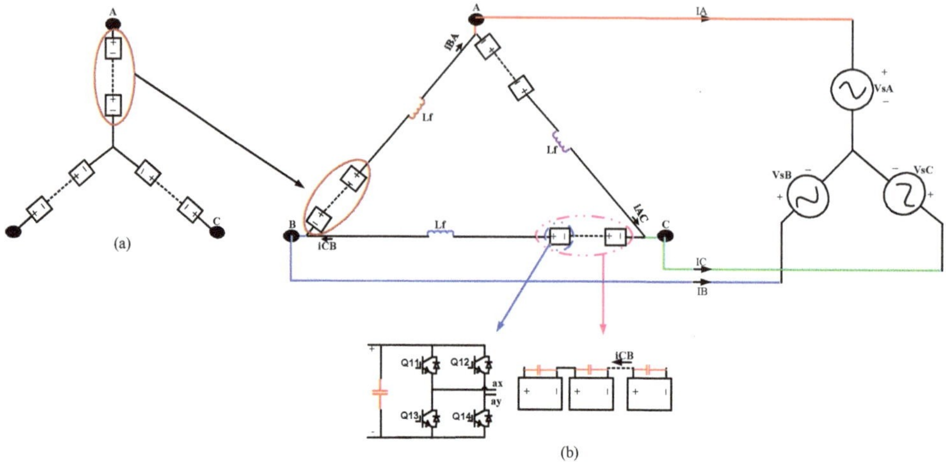

Figure 6. CHB MLI connections: (**a**) star-connected, (**b**) conventional delta-connected.

Thus, comparing delta-connected CHB MLIs to star-connected CHB, it is observed that the former require a number of H-bridges that is $\sqrt{3}$ times more at the same line-to-line voltage. Consequently, this results in an increase in both the size and cost of the delta-connected CHB. Furthermore, at the same line-to-line voltage levels, the delta-connected CHB exhibits increased switching losses and conduction losses, resulting in a reduced inverter efficiency compared to the star-connected CHB operation. Hence, this work proposes a modified delta-connected CHB to produce line-to-line voltage levels equivalent to those of a star-connected CHB, while utilizing the same number of H-bridges. The proposed topology is shown in Figure 7. As can be seen in this figure, the line-to-line voltage levels are synthesized using three coupled transformers rather than the inductors used in traditional delta-connected CHB MLIs. With the help of these coupled transformers, the line-to-line voltage levels of the proposed topology are the same as the line-to-line voltage levels of the star-connected CHB topology.

Interface inductors are unnecessary when using the proposed topology for grid connection applications. Alternatively, the coupled transformers in the proposed configuration can also amplify the produced voltage without requiring an extra voltage amplification device at the point of connection with the utility.

The transformer leakage inductance and core saturation behavior may have an effect on the inverter's dynamic response and harmonic performance. The winding resistance may cause an output voltage drop. On the other hand, the leakage inductance has little effect on the THD of the generated voltage. Therefore, these transformers should be characterized by high magnetization inductance, a ferrite core to avoid core saturation, and low winding resistance and leakage inductance.

As a summary, this topology generates higher voltage levels by reducing the number of required switches compared with the conventional delta-connected CHB multilevel inverter topology. By leveraging a reduced switch count, it achieves a more compact and simplified design. Furthermore, it effectively mitigates voltage and current distortion. The scalable nature of this design enables seamless expansion to accommodate additional voltage levels. The concept's validity has been substantiated through the simulation and experimental results.

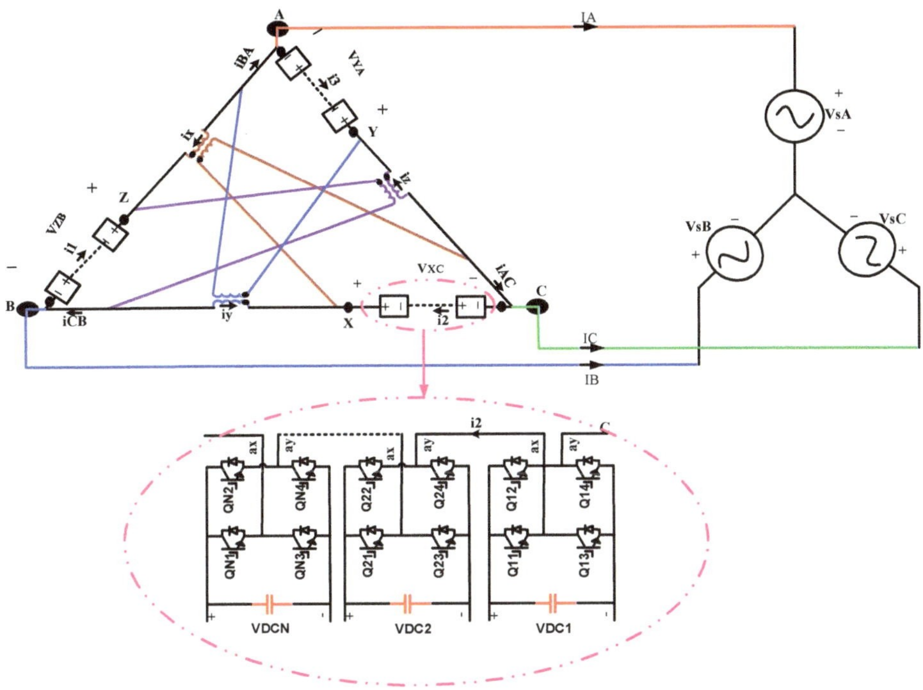

Figure 7. The proposed modified delta-connected CHB topology.

The proposed topology offers several key benefits compared to the conventional delta-connected CHB MLIs, including:

- Achieving more voltage levels with fewer switches than the traditional delta-connected multilevel inverter topology.
- Enhanced voltage quality and current waveform.
- Reduced voltage stress on switches compared to the conventional delta-connected CHB configurations.
  - ✓ Stresses on the switches are as follows:

$$\frac{\sqrt{2}V}{N\sqrt{3}} \tag{1}$$

    where V is the root mean square (RMS) of the line–line grid voltage and N is the number of cascaded H-bridge cells in each phase.

  - ✓ On the other hand, voltage stresses on the switches of the conventional delta-connected CHB are as follows:

$$\frac{\sqrt{2}V}{N} \tag{2}$$

- The proposed topology ensures uniform current distribution across all switches, with each switch subjected to the same RMS grid current (I); the current stress is $\sqrt{2}$I.

In the case of the conventional delta-connected CHB, and in order to increase the output voltage, a grid-side transformer in the conventional delta-connected CHB with a higher voltage ratio may be used. However, the conventional delta-connected CHB is a bulky topology compared to the modified proposed topology. That is, it is required to add grid-side transformers and three current-limiting inductors inside the conventional delta-connected CHB to reduce the circulating current. On the other hand, the proposed topology can eliminate the use of extra grid-side transformers as the intermediate transformers of the proposed topology act to amplify the voltage level (if needed) and, at the same time, eliminate the use of extra intermediate inductors. Moreover, one of the advantages of the higher voltage levels is the reduction in the THD of the grid voltage and grid currents. This indicates that the use of grid-side transformers in the conventional delta-connected CHB while reducing the cascaded H-bridges loses this advantage.

Table 1 provides a comparative evaluation of the proposed topology versus the conventional MLI architectures to generate seven line-to-live voltage levels. The proposed topologies exhibit a reduced switch count compared to the conventional NPC and FC MLIs. The proposed MLI is similar to star-connected CHB MLIs in the number of main switches, number of separated DC sources, and DC link capacitors.

Table 1. Comparison with some prior state-of-the-art MLI topologies.

Topology	NPC	FC	CHB (Star)	CHB (Delta)	Proposed Topology
Main switches	24	24	24	48	24
Clamping diodes	36	0	0	0	0
Flying capacitors	0	18	0	0	0
Dc link capacitors	4	4	6	12	6
Intermediate inductors	0	0	0	3	0
Intermediate transformers	0	0	0	0	3
DC sources	1	1	6	12	6
Modularity	No	No	Yes	Yes	Yes
Inherent voltage balancing capability	No	No	Yes	Yes	Yes
Required grid-side transformer (if necessary)	Yes	Yes	Yes	Yes	No

In addition, compared to the conventional delta-connected CHB MLIs, the proposed topology requires a lower number of components to generate the same line–line voltage levels. Moreover, the proposed topology uses three intermediate transformers to accomplish its function. On the other hand, the proposed topology does not require additional line frequency transformers at the grid side as the other topologies do, in which grid-side transformers can be used for voltage amplification and/or for isolation purposes if needed.

In NPC and FC MLIs, a single DC source is used, and multilevel voltages are generated by charging separated DC link capacitors, which may cause a voltage drift if the DC link voltage of the capacitors is not well-maintained. In the proposed topology, the unbalanced voltage problem is eliminated since separated DC sources are used, and, therefore, the proposed topology exhibits inherent voltage balancing capabilities. Moreover, the proposed topology exhibits high modularity compared to other state-of-the-art topologies.

3. Analysis of the Proposed Topology

3.1. Analysis of Voltage Relationships

Applying Kirchhoff's voltage law (KVL) for the proposed inverter shown in Figure 7 yields the following:

$$v_{sA} - v_{AZ} - v_{ZB} - v_{sB} = 0$$

$$v_{sB} - v_{BX} - v_{XC} - v_{sC} = 0 \quad (3)$$
$$v_{sC} - v_{CY} - v_{YA} - v_{sA} = 0$$

where v_{sA}, v_{sB}, and v_{sC} are the instantaneous phase voltages of the grid and the voltages v_{ZB}, v_{XC}, andv_{YA} are voltages of the CHB cells of phases A, B, and C, respectively.

Considering that the coupled transformers' turn ratios are $\left(\frac{N_1}{N_2}\right)$, Equation (3) can be written as follows:

$$v_{AB} = \frac{N_1}{N_2} v_{CX} + v_{ZB}$$
$$v_{BC} = \frac{N_1}{N_2} v_{AY} + v_{XC} \quad (4)$$
$$v_{CA} = \frac{N_1}{N_2} v_{BZ} + v_{YA},$$

where v_{AB}, v_{AB}, and v_{AB} are the generated instantaneous line–line voltages from the proposed topology and N_1 and N_2 are the turn ratios of the primary and secondary sides of each transformer in the proposed topology.

The root mean square (RMS) values of the voltages across each CHB arm in the proposed topology can be given as follows:

$$V_{ZB} = \frac{N m_a V_{dc}}{\sqrt{2}} \angle 0^o$$
$$V_{XC} = \frac{N m_a V_{dc}}{\sqrt{2}} \angle -120^o \quad (5)$$
$$V_{YA} = \frac{N m_a V_{dc}}{\sqrt{2}} \angle +120^o$$

where V_{dc} is the DC voltage connected to the H-bridge cell, N is the number of cascaded H-bridge cells, and m_a is the modulation index.

Assuming that $\left(\frac{N_1}{N_2} = 1\right)$ and substituting Equation (5) into Equation (4), we obtain the following:

$$v_{AB} = \sqrt{3} N m_a V_{dc} \, sin(\omega t + 30^o)$$
$$v_{BC} = \sqrt{3} N m_a V_{dc} \, sin(\omega t - 90^o) \quad (6)$$
$$v_{CA} = \sqrt{3} N m_a V_{dc} \, sin(\omega t + 150^o)$$

Equation (6) shows that the proposed topology produces the same line-to-line voltage as the star-connected CHB.

3.2. Analysis of Currents Relationships

On the other hand, the currents flowing through each CHB arm in each phase can be defined as follows:

$$\begin{bmatrix} i_1 \\ i_2 \\ i_3 \end{bmatrix} = \begin{bmatrix} i_z - i_x \\ i_x - i_y \\ i_y - i_z \end{bmatrix}, \quad (7)$$

where the currents in the primary side of the coupled transformers T_x, T_y and T_z are represented by i_x, i_y, and i_z, respectively. The currents i_1, i_2, and i_3 are the currents flowing inside the CHB cells of phases AB, BC, and CA, respectively.

Considering that the coupled transformers' magnetizing inductances are significant, the following is true:

$$i_x + i_y + i_z = 0 \quad (8)$$

Based on Equations (5) and (6), the current flowing through each transformer is as follows:

$$\begin{bmatrix} i_x \\ i_y \\ i_z \end{bmatrix} = \frac{\sqrt{2}}{\sqrt{3}} I \begin{bmatrix} sin(\omega t{+}150^o) \\ sin(\omega t{+}90^o) \\ sin(\omega t{-}30^o) \end{bmatrix} \tag{9}$$

Considering that the coupled transformers' turn ratios are $\left(\frac{N_1}{N_2}\right)$, the output terminal currents can be defined as follows:

$$\begin{bmatrix} i_A \\ i_B \\ i_C \end{bmatrix} = \begin{bmatrix} -i_3 - i_x + \frac{N_1}{N_2} i_y \\ -i_1 - i_y + \frac{N_1}{N_2} i_z \\ -i_2 - i_z + \frac{N_1}{N_2} i_x \end{bmatrix}, \tag{10}$$

where i_A, i_B, and i_C are the line currents of phases A, B, and C, respectively.

Considering that $\left(\frac{N_1}{N_2} = 1\right)$, substituting Equations (9) and (10) into Equation (7), the three-phase grid currents can be defined as follows:

$$\begin{bmatrix} i_A \\ i_B \\ i_C \end{bmatrix} = \begin{bmatrix} i_1 \\ i_2 \\ i_3 \end{bmatrix} = \sqrt{2} I \begin{bmatrix} sin(\omega t) \\ sin(\omega t){-}120^o \\ sin(\omega t{+}120^o) \end{bmatrix}, \tag{11}$$

where I is the RMS value of the output current. Equation (11) proves that the output terminal currents and the currents flowing through each CHB arm in the proposed topology are identical.

Moreover, the currents i_{AC}, i_{CB}, and i_{BA} can be defined as follows:

$$\begin{bmatrix} i_{AC} \\ i_{CB} \\ i_{BA} \end{bmatrix} = \begin{bmatrix} i_2 + i_C \\ i_1 + i_B \\ i_3 + i_A \end{bmatrix} = - \begin{bmatrix} i_1 \\ i_3 \\ i_2 \end{bmatrix} = \sqrt{2} I \begin{bmatrix} \sin(\omega t{-}180^o) \\ \sin(\omega t{-}60^o) \\ \sin(\omega t{+}60^o) \end{bmatrix} \tag{12}$$

The phasor diagrams of the voltage and current quantities for the above relationships of the proposed topology are shown in Figure 8.

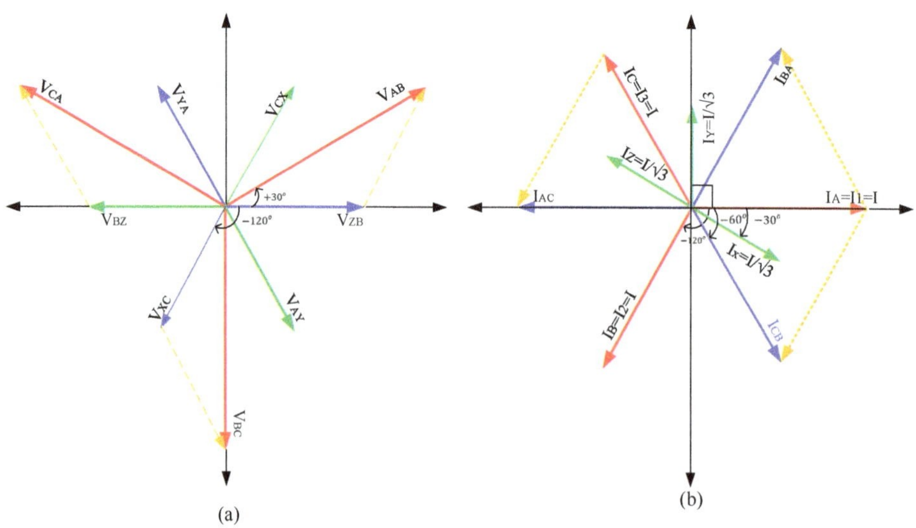

Figure 8. The phasor diagrams of the (**a**) voltages and (**b**) currents of the proposed topology.

3.3. Analysis of Power Relationships

From another point of view, the total apparent power of the N H-bridge cells of each phase meets the formulae below:

$$\begin{aligned}(S_{cells\_A}) &= V_{ZB} I_{1rms} = \frac{N m_a V_{dc}}{\sqrt{2}} I \\ (S_{cells\_B}) &= V_{XC} I_{2rms} = \frac{N m_a V_{dc}}{\sqrt{2}} I \\ (S_{cells\_C}) &= V_{YA} I_{3rms} = \frac{N m_a V_{dc}}{\sqrt{2}} I,\end{aligned} \quad (13)$$

where S_{cells_A}, S_{cells_B}, and S_{cells_C} are the apparent powers of the N H-bridge cells of phases A, B, and C, respectively, and I_{1rms}, I_{2rms}, and I_{3rms} are the currents flowing through the CHB cells of phases AB, BC, and CA, respectively.

Additionally, the apparent power of each transformer can be expressed as follows:

$$\begin{aligned}(S_{T_x}) &= V_{XC} I_{xrms} = \frac{N m_a V_{dc}}{\sqrt{6}} I \\ \left(S_{T_y}\right) &= V_{YA} I_{yrms} = \frac{N m_a V_{dc}}{\sqrt{6}} I \\ (S_{T_z}) &= V_{ZB} I_{zrms} = \frac{N m_a V_{dc}}{\sqrt{6}} I,\end{aligned} \quad (14)$$

where S_{T_x}, S_{T_y}, S_{T_z} are the apparent powers of the coupled transformers of phases AB, BC, and CA, respectively.

Equation (14) demonstrates that the apparent power of each coupled transformer is $1/\sqrt{3}$ of the N H-bridge cells of one phase.

From another point of view, in the case where grid-side transformers are used for the conventional delta-connected CHB, the power rating of each grid-side transformer used with the conventional delta-connected CHB is as follows:

$$S_{Tr/conv.delta} = V_{SA} I_A = \frac{N m_a V_{dc}}{\sqrt{2}} I \quad (15)$$

where $S_{Tr/conv.delta}$ is the apparent power rating of the grid-side transformer connected to each phase in the conventional delta-connected CHB, V_{SA} is the grid phase voltage, and I_A is the line grid current.

Comparing Equations (14) and (15) reveals that the ratings of the transformers used in the proposed topology are less that those of the grid-side transformers that may be used in the delta-connected CHB under the same DC link voltage and the same modulation index.

Moreover, the total apparent power (S_{TOT}) of the proposed topology can be expressed as follows:

$$S_{TOT} = \sqrt{3} V_{AB} I_{Arms} = 3 \frac{N m_a V_{dc}}{\sqrt{2}} I \quad (16)$$

3.4. Loss Analysis

The proposed topology's semiconductor device losses fall into three categories: conduction losses (caused by resistance and voltage drop), switching losses (stemming from non-ideal switch operation), and blocking losses (caused by leakage current in the off-state of the IGBT, can normally be neglected). Therefore, semiconductor losses can be expressed as follows [65,66]:

$$P_{SemiCond} = P_{Cond} + P_{Sw} \quad (17)$$

where $P_{SemiCond}$ is the semiconductor switch loss, P_{Cond} is the conduction loss, and P_{Sw} is the switching loss.

The conduction losses can be expressed as follows:

$$P_{Cond} = P_{Cond\_Trans} + P_{Cond\_Diode} \tag{18}$$

where P_{Cond_Trans} is the conduction loss in the transistor and P_{Cond_Diode} is the conduction loss in the antiparallel diode.

The conduction losses in Equation (18) can be expressed as follows [65,66]:

$$P_{Cond\_Trans} = V_{Trans} \times i(t) + R_{Trans} \times i^{2\beta}(t) \tag{19}$$

The conduction losses in the antiparallel diode can be expressed as follows [1,2]:

$$P_{Cond\_Diode} = V_{Diode} \times i(t) + R_{Diode} \times i^{2}(t), \tag{20}$$

where V_{Trans} and V_{Diode} signify the forward voltage drops of the transistor and the diode, $R_{Trans.}$ and R_{Diode} represent their resistances, β is a transistor-dependent constant, and $i(t)$ is the current flowing through each semiconductor switch.

$$P_{Sw} = f \times \sum_{J=1}^{N\_switches} \left[\sum_{i=1}^{N_{ON,J}} Enr._{ON,Ji} + \sum_{i=1}^{N_{OFF,J}} Enr._{OFF,Ji} \right], \tag{21}$$

where $N_{ON,J}$ and $N_{OFF,J}$ are the numbers of turns on and off of switch J during the fundamental frequency, $Enr._{ON,Ji}$ is the energy loss of switch J during the ith turn on, and $Enr._{OFF,Ji}$ is the energy loss of switch J during the ith turn off.

The energy losses during the switch's ON and OFF states are expressed as follows [65,66]:

$$\begin{aligned} Enr\_{}_{ON,J} &= \tfrac{1}{6} V_{Switch,\ J}\ I' t_{on} \\ and,\ \ Enr\_{}_{OFF,J} &= \tfrac{1}{6} V_{Switch,\ J}\ I\ t_{off} \end{aligned} \tag{22}$$

where $Enr\_{}_{ON,J}$ is the turn-on energy loss of switch J, $Enr\_{}_{OFF,J}$ is the turn-off energy loss of switch J, I represents the current through the switch before turning off, I' represents the current through the switch after turning on, and $V_{Switch,J}$ is the voltage of the switch during the off state.

On the other hand, the efficiency of the inverter can be expressed as follows [66]:

$$Efficiency = \frac{P_{out}}{P_{In}} = \frac{P_{In} - P_{losses}}{P_{In}} \tag{23}$$

where P_{In} is the input power, P_{Out} is the output power from the proposed topology, and P_{losses} is the total power loss in the proposed topology.

The power loss P_{losses} can be expressed as follows [66]:

$$P_{losses} = P_{SemiCond} + P_{Trs\_CU} + P_{Trs\_Core} \tag{24}$$

where P_{Trs_CU} is the copper loss of the transformers while P_{Trs_Core} is the transformers' core loss, which is the hysteresis and eddy current loss. The transformer copper loss can be expressed as follows:

$$P_{Trs\_CU} = I_{X1}{}^{2}R_{X1} + I_{X2}{}^{2}R_{X2} + I_{Y1}{}^{2}R_{Y1} + I_{Y2}{}^{2}R_{Y2} + I_{Z1}{}^{2}R_{Z1} + I_{Z2}{}^{2}R_{Z2}, \tag{25}$$

where I_{X1} and I_{X2} represent the current flowing through the primary and the secondary windings of the transformer x, respectively, and R_{X1} and R_{X2} are the resistances of the primary and the secondary windings of transformer x, respectively. The other quantities are for transformers y and z.

As revealed from Equation (9), the RMS values of I_{X1}, I_{Y1}, and I_{Z1} are the same $(= \frac{I}{\sqrt{3}})$. Substituting this into Equation (25), we obtain the following:

$$P_{Trs.CU} = \frac{I^2}{3}\left((R_{X1} + R_{Y1} + R_{Z1}) + \left(\frac{N_1}{N_2}\right)^2 (R_{X2} + R_{Y2} + R_{Z2})\right) \tag{26}$$

From another point of view, efficiency can be calculated as follows:

$$Efficiency = \frac{\sum_{i=1}^{N} \int (V_{dc,i}\, I_{dc,\ i})dt}{\int (v_{sA}(t)\, i_A(t)\, dt + \int (v_{sB}(t)\, i_B(t)\, dt + \int (v_{sC}(t)\, i_C(t)\, dt}, \tag{27}$$

where $V_{dc,i}$ is the DC voltage of each H-bridge cell, $I_{dc,i}$ is the input DC current flowing to each H-bridge cell, $v_{sA}(t)$, $v_{sA}(t)$, and $v_{sA}(t)$ are the instantaneous phase voltages of the grid for phases A, B, and C, respectively, and $i_A(t)$, $i_B(t)$, and $i_C(t)$ are the instantaneous phase currents of the grid.

4. Simulation Results

The simulation model of the proposed modified delta-connected CHB topology was built in SIMULINK software version 2024b. Figure 9 illustrates the utilization of the unipolar phase-shifted pulse-width modulation (PSPWM) to fire the switches in the suggested MLI. The diagram shows that the reference voltages V_{refa}, V_{refb}, and V_{refc} are compared with the phase-shifted carrier waveforms to generate the necessary pulses for the switches in the proposed MLI. The triangular waveform of the H-bridge cell (N) is phase-shifted by 180°/N compared to the triangular waveform of the H-bridge cell (N − 1).

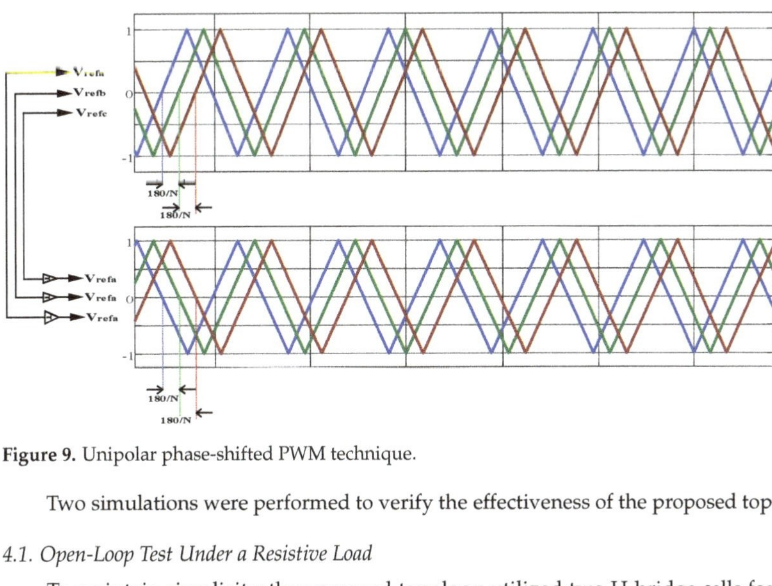

Figure 9. Unipolar phase-shifted PWM technique.

Two simulations were performed to verify the effectiveness of the proposed topology.

4.1. Open-Loop Test Under a Resistive Load

To maintain simplicity, the proposed topology utilized two H-bridge cells for each phase. Each H-bridge cell's input was coupled to a 40-Volt DC power source. The system parameters used for the simulation are shown in Table 2.

Table 2. Simulation system's parameters.

Load resistance and inductance (for the R and L testing)	48 Ω, 77 mH
DC input voltage (for the R and L testing)	40 V
RMS grid voltage	120 V
Grid interface inductance	2.5 mH
Fundamental frequency	60 Hz
Switching frequency, f_s	900 Hz

Line voltages were measured between the inverter's terminals A, B, and C. The modulation index was 0.95. The generated three-phase instantaneous voltages across the primary side of the transformers are shown in Figure 10. As shown, the number of voltage levels generated across each transformer winding is five. On the other hand, the generated voltages v_{ZB}, v_{XC}, and v_{YA} are seen in Figure 11. Since only two H-bridge cells were utilized per phase, the number of voltage levels generated was five, as illustrated in this figure. Moreover, the generated three-phase line-to-line voltage from the proposed inverter is seen in Figure 12. As depicted in this figure, nine levels were produced. From another point of view, only five levels were produced from the traditional delta-connected CHB MLIs when the same number of H-bridges was utilized (two H-bridges). That is, the number of voltage levels generated by the proposed MLI is higher (nine levels) than that generated by the traditional delta-connected MLIs for the same number of H-bridge cells. Figure 13 depicts three-phase line currents. As shown in this figure, the grid currents were not clean since the load was a purely resistive load. Adding an inductive element to the load acts as a filter and reduces these harmonics. The generated voltage levels were equivalent to the levels generated in case the star-connected CHB is used. In addition, the presence of three transformers helps in amplifying the voltage, if needed.

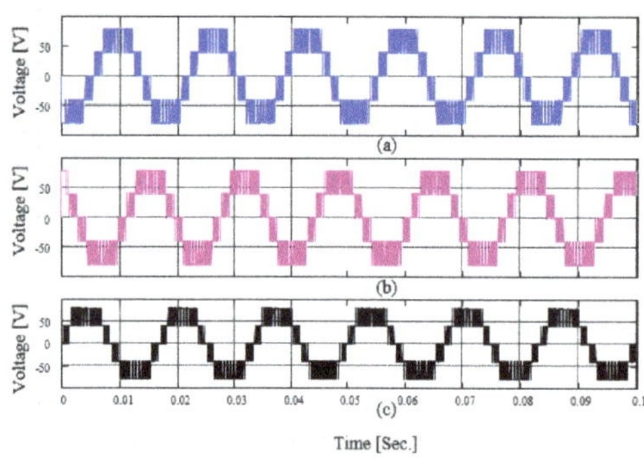

Figure 10. The generated voltages across (**a**) transformer x, (**b**) transformer y, and (**c**) transformer z.

Moreover, the proposed topology was tested while increasing the turn ratios of the intermediate transformers to amplify the output line–line voltages. Figure 14 shows the performance of the proposed topology when the turn ratios are (1.5:1) at 48 Ω, $V_{dc} = 40$ V, and $m_a = 1.15$. Figure 14a shows the generated three-phase voltages, while Figure 14b shows two voltage signals: the generated voltage v_{ZB} (the generated voltages from the CHB of phase AB) and the generated voltage across the primary side of transformer X (v_{Tx}). As seen in Figure 14b, the generated voltage across the primary side of the transformer X was higher due to a higher turn ratio and resulted in a higher amplitude of the generated line–line voltage.

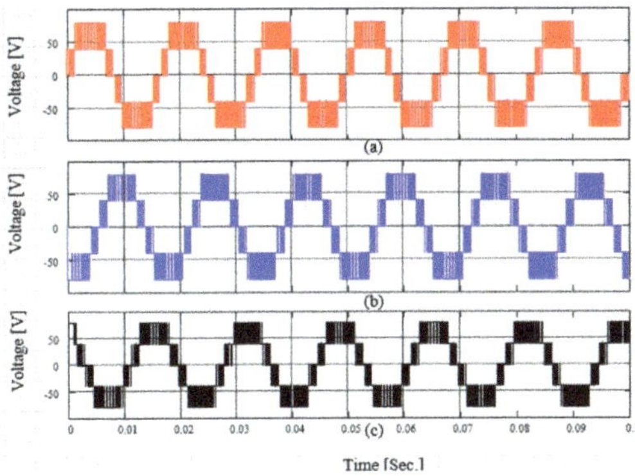

Figure 11. The generated voltages: (a) v_{ZB}, (b) v_{XC}, and (c) v_{YA}.

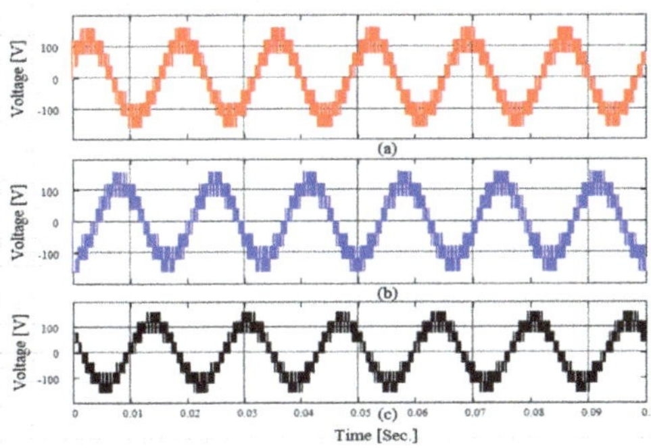

Figure 12. The generated three-phase line-to-line voltages from the proposed MLI under a resistive load (a–c).

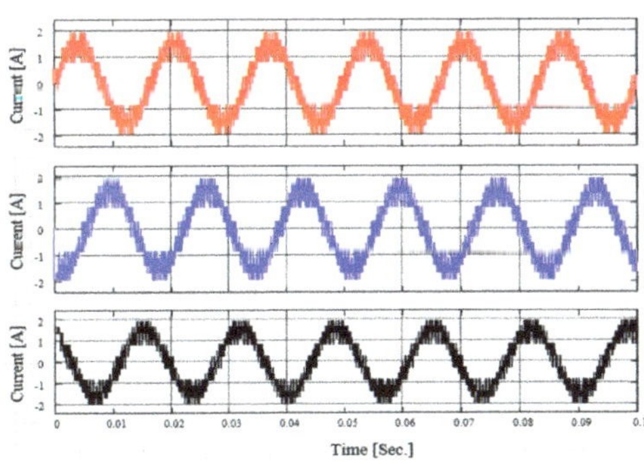

Figure 13. The three-phase grid currents from the proposed MLI under a resistive load.

On the other hand, the performance of the proposed topology was examined under varying modulation indices and under changing leakage inductances of the transformers. Figure 15 shows the THD of the generated voltage (v_{AB}) under varying modulation indices. As the modulation index increased, the THD decreased. Moreover, as the leakage inductance increased, the THD decreased.

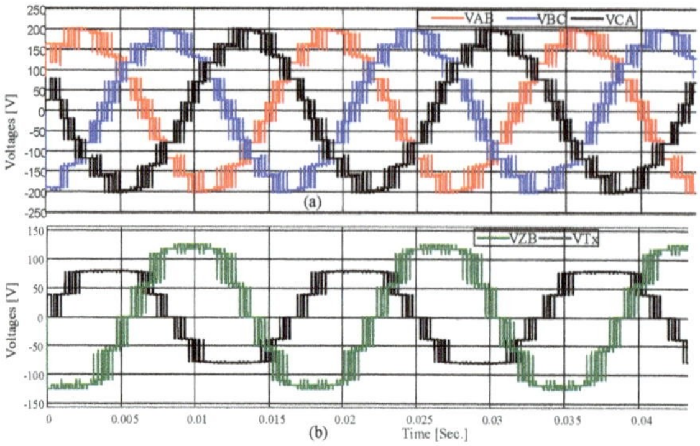

Figure 14. The performance of the proposed topology under the changing transformer's turn ratio: (**a**) three-phase line–line voltages; (**b**) voltages VZB and VTX.

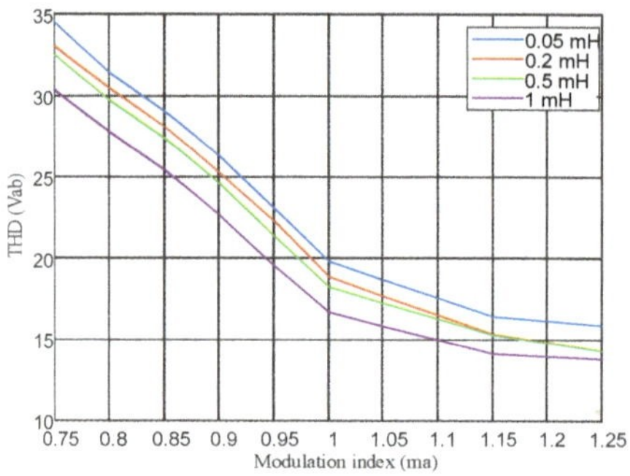

Figure 15. Effect of the leakage inductance on the THD of (v_{AB}) with varying modulation indices.

4.2. Open-Loop Test Under a Resistive–Inductive Load

The proposed MLI was also examined under a resistive–inductive load. Figure 16 shows the generated three-phase line-to-line voltages, while Figure 17 shows the load currents.

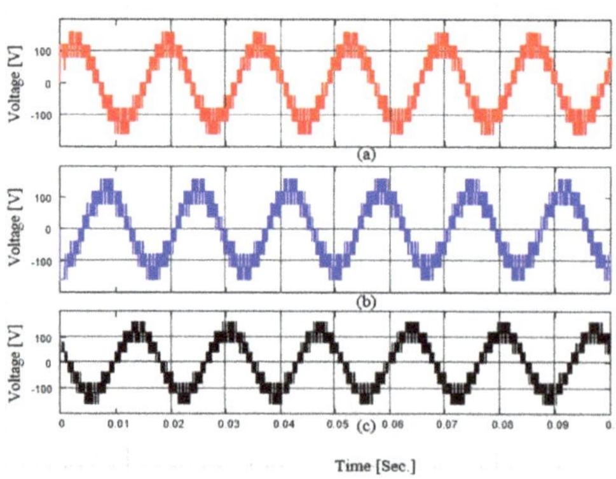

Figure 16. The generated three-phase line-to-line voltages from the proposed MLI under a resistive–inductive load (**a**–**c**).

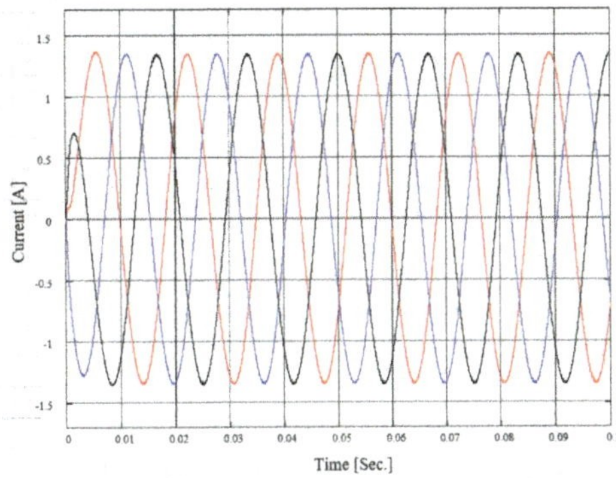

Figure 17. The three-phase grid currents from the proposed MLI under a resistive–inductive load.

4.3. PV–Grid Connection Application

To verify the effectiveness of the proposed topology, it was tested with PV–grid interconnection. For simplicity, two cascaded H-bridge cells were considered for each phase in the proposed topology. Similarly to the star-connected and conventional delta-connected cascaded H-bridge multilevel inverter topologies, each H-bridge cell should be powered by a separated DC source/PV panel. In the proposed topology, and for the PV–grid connection application, one PV module was used for each H-bridge unit. In order to solve the problem of shading of one PV panel, distributed maximum power point tracking (MPPT) is commonly used, in which each PV module is connected to only one DC–DC converter. Table 2 shows the system parameters, while the parameters of the PV module are shown in Table 3.

Table 3. KACST 245 PV module parameters.

Maximum power (Pmax)	245 W
Maximum power voltage (Vmax)	28.8 V
Maximum power current (Imax)	8.5 A
Open-circuit voltage (Voc)	31.5 V
Short-circuit current (Isc)	9.5 A

Figure 18 shows the control scheme. In this control scheme, the "perturb and observe" MPPT algorithm is used for extracting the maximum power point of the PV module using DC–DC isolated Cuk converters. The voltage-oriented technique is used to control the grid current. The outer loop controller is the DC link voltage controller which uses the proportional–integral controller (PI) to keep the DC link voltages of each phase equal so that the output voltage is maintained the same in all the phases in which the DC link voltages are controlled. The inner controller is the current controller in which two PI controllers are used to control the grid currents in the d- and q-axes. The current (Iq) is compared with zero to achieve the unity power factor. The resulting signals from the current controllers are converted into ABC, and then they are used to generate the required pulses that drive the IGBTs of the proposed topology.

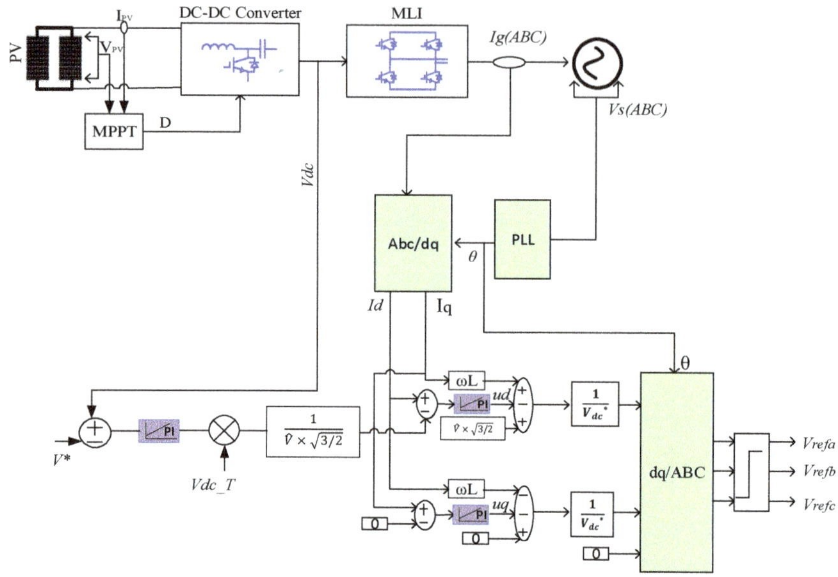

Figure 18. The control scheme. * Reference voltage, Vdc_T total DC link voltages.

The control scheme succeeded in keeping the DC link voltages at the reference voltages (100 V) as illustrated in Figure 19. In this figure, six DC link voltages are shown. DC link voltages V_{dc11} and V_{dc12} are the DC link voltages of the two H-bridge cells of phase AB. In addition, DC link voltages V_{dc21} and V_{dc22} are the DC link voltages of the two H-bridge cells of phase BC. Moreover, V_{dc31} and V_{dc32} are the DC link voltages of the two H-bridge cells of phase CA.

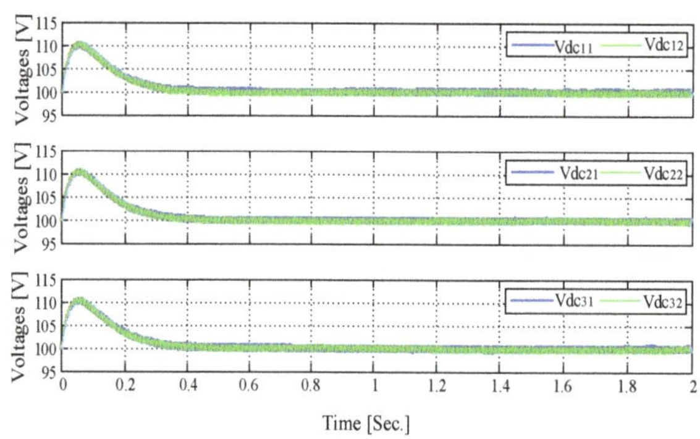

Figure 19. The simulated DC link voltages.

Figure 20a shows the generated three-phase line–line voltages. As seen in this figure, nine voltage levels were generated. In addition, Figure 20b shows the voltages generated by the CHB of phase AB (v_{ZB}) and the voltage generated across transformer T_x. As stated in the analysis, the grid currents coincided with the phase currents passing through each CHB, as revealed in Figure 21. The harmonic spectrum of the grid current i_A is revealed in Figure 22. The THD of the grid current i_A is 3.44%, which has an acceptable harmonic content of less than 5% as defined by the IEEE standard. The magnitude of the fifth harmonic order was approximately 1%. The magnitude of the other harmonic orders were less than 1%.

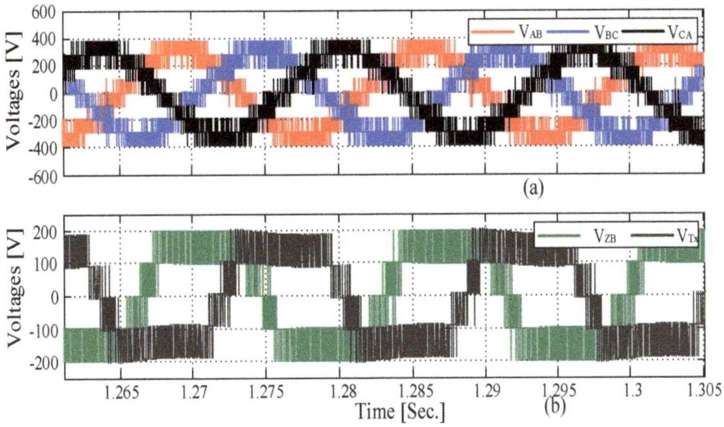

Figure 20. The generated voltages: (**a**) three-phase line–line voltages; (**b**) voltages V_{ZB} and V_{Tx}.

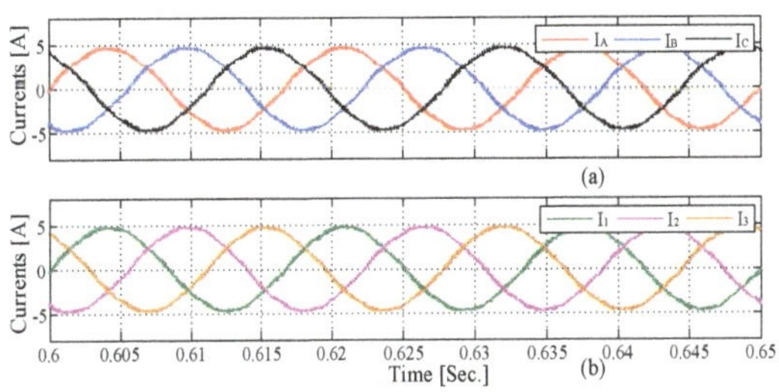

Figure 21. The generated currents: (**a**) three-phase grid currents; (**b**) the currents flowing inside the CHB cells.

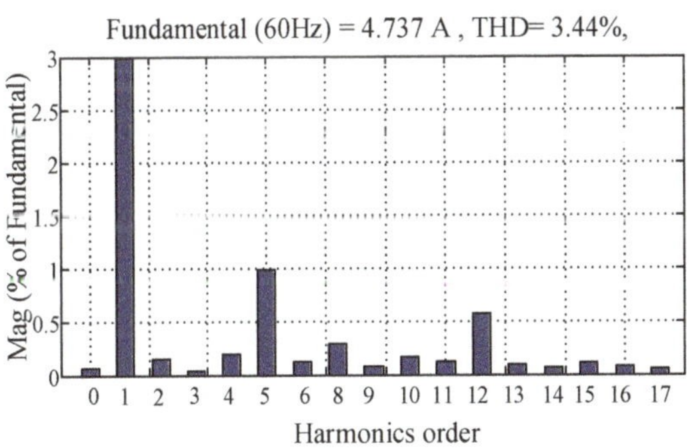

Figure 22. The THD of the grid current (i_A).

5. Experimental Results

The modified MLI's performance was validated through experimental tests conducted in a laboratory setting using various modulation indices and resistive–inductive loads. A DS1202 MicroLabBox data acquisition system was utilized to obtain data measurements and generate the required switching pulses. Figure 23 reveals the hardware setup of the proposed topology. For simplicity, and to prove the concept of the proposed MLI, two H-bridges were cascaded in each phase. One KACST 245 PV module was connected to the input of each H-bridge cell. All PV modules were identical, and they were installed in the same area. The specifications of the PV modules utilized in the experiment are presented in Table 3. The parameters used in the hardware setup are listed in Table 4.

Figure 23. Hardware arrangement of the proposed MLI. (**A**) Gate drives for the IGBTs; (**B**) level-shifting circuits; (**C**) voltage and current sensors; (**D**) IGBTs and DC link capacitors; (**E**) MicroLabBox; (**F**) coupled transformers; (**G**) photovoltaic modules.

Table 4. Experimental system's parameters.

Load resistance	48 Ω
Load inductance	77 mH
Inverter's switching frequency, f_s	900 Hz
Three coupled transformers	Part No. 810.1201 Each: 230 V, 1.5 KVA
IGBTs	Part No. SKM75GB12V
IGBT drives	Part No. SKYPER 32 R
PV modules	Part No. KACST 245

In order to validate the advantage of the proposed topology, it was tested under open loop control with a resistive–inductive load. The test involved connecting a resistive–inductive load with a series resistance of 48 Ω and an inductance of 77 mH to the A, B, and C terminals of the constructed inverter. Three modulation indices were employed to demonstrate the efficacy of the suggested topology. The performance of the proposed topology at a modulation index (m_a) of 0.85 is revealed in Figure 24. Figure 24a shows the line-to-line voltages (v_{AB} and v_{BC}) produced from the constructed MLI, the voltage v_{ZB}, and the voltage across the transformer X at $m_a = 0.85$. As shown in this figure, nine line-to-line voltage levels were generated from the modified delta-connected MLI. The number of voltage levels per arm (Green plot, v_{ZB}) was five, as shown in Figure 24.

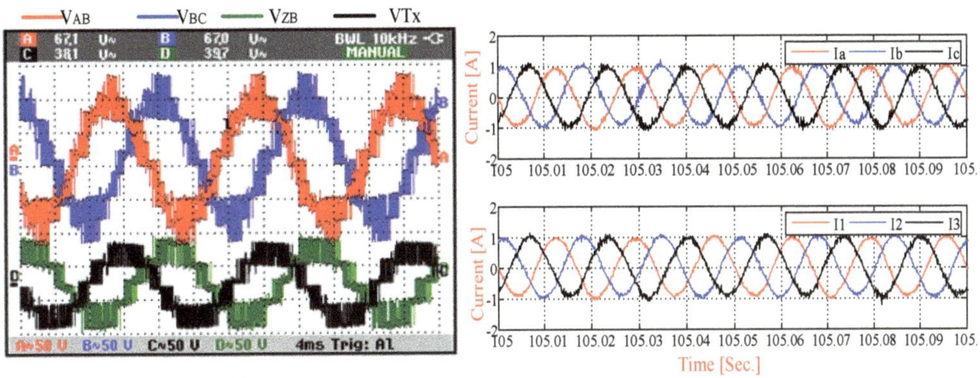

(a) (b)

Figure 24. The performance of the proposed MLI at modulation index m_a = 0.85. (**a**) The experimental generated voltages and (**b**) the experimental load and the CHB currents [A].

Moreover, the generated voltage across the coupled transformer (Black plot, v_{Tx}) is also shown in Figure 24a. On the other hand, the three-phase grid currents as well as the phase currents of each arm in the modified delta connected CHB are shown in Figure 24b. As shown in this figure, the line currents and the currents passing through the CHB cells coincided, as demonstrated in the analysis.

In order to demonstrate the effectiveness of the constructed MLI, an additional test was conducted at m_a = 0.95, as depicted in Figure 25.

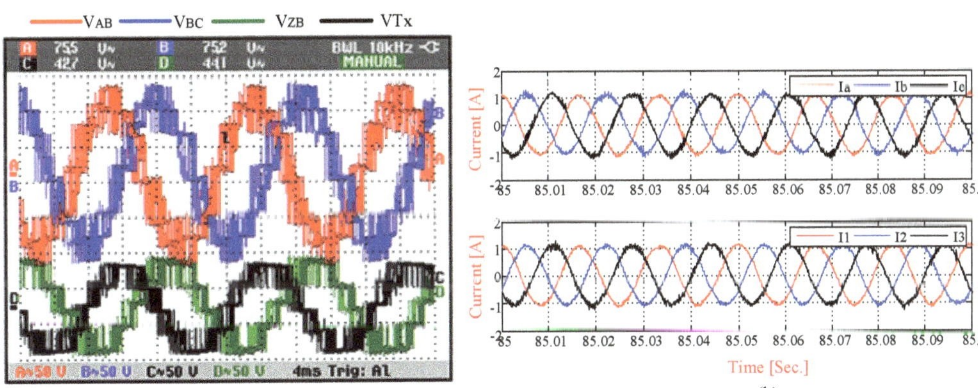

(a) (b)

Figure 25. The performance of the proposed MLI at modulation index m_a = 0.95. (**a**) The experimental generated voltages and (**b**) the experimental load and the CHB currents [A].

Moreover, Figure 26a shows the generated line-to-line voltages, the voltage per arm, and the voltage across the transformer at m_a = 1.25, while Figure 26b shows the currents.

As shown in these experimental results, there were nine generated line-to-line voltage levels, which is the advantage of the proposed modified delta-connected CHB MLI. Moreover, the line currents and the currents passing through the CHB cells in each arm coincided.

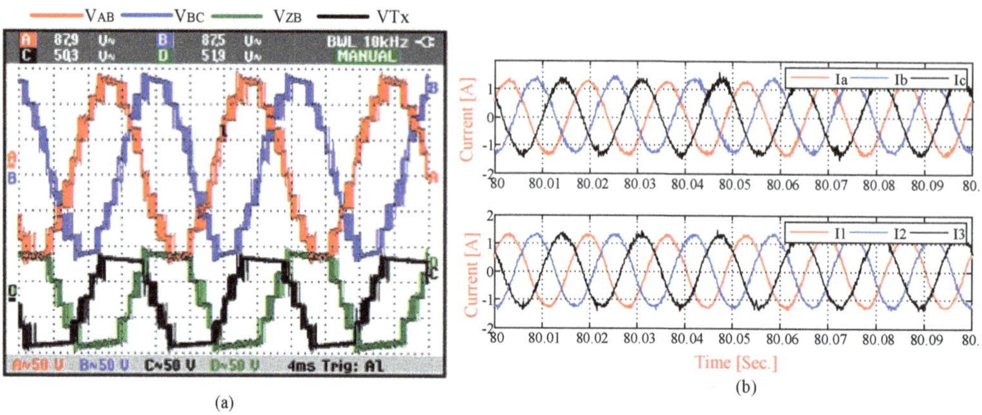

Figure 26. The performance of the proposed MLI at modulation index m_a = 1.25. (**a**) The experimental generated voltages and (**b**) the experimental load and the CHB currents [A].

6. Conclusions

The conventional delta-connected CHB MLIs necessitate a higher number of H-bridges connected in series compared to star-connected MLIs to attain the equivalent line-to-line grid voltage. Consequently, a greater number of switches is unavoidably needed. This paper proposes a modified delta-connected CHB MLI that generates the same number of line-to-line voltage levels as a star-connected CHB for the same number of switches. Three coupled transformers are used in the proposed topology. The rating of each one is $1/\sqrt{3}$ of the CHB cells of one phase. These transformers can have another advantage in case it is required to amplify the generated voltage, and it is not necessary to utilize transformers on the grid side. The functionality of the proposed topology is accomplished by simulating it using SIMULINK software. The simulation results showcased the capacity of the proposed topology to produce a higher number of levels. To verify the effective functioning of the suggested configuration, the configuration was constructed in the laboratory, and the produced switching signals were applied to a MicroLabBox data acquisition system, which in turn activated the IGBTs via gate drives. The proposed topology was experimentally tested under a resistive–inductive load with different modulation indices. The simulation and experimental findings are provided to demonstrate the feasibility of the proposed topology.

Funding: Prince Sattam bin Abdulaziz University, project number (PSAU/2024/01/32106).

Data Availability Statement: Data are contained within the article.

Acknowledgments: The authors extend their appreciation to Prince Sattam bin Abdulaziz University for funding this research work through the project number (PSAU/2024/01/32106).

Conflicts of Interest: The author declares no conflicts of interest.

References

1. Gupta, K.K.; Ranjan, A.; Bhatnagar, P.; Sahu, L.K.; Jain, S. Multilevel inverter topologies with reduced device count: A review. *IEEE Trans. Power Electron.* **2015**, *31*, 135–151. [CrossRef]
2. Nabae, A.; Takahashi, I.; Akagi, H. A new neutral-point-clamped PWM inverter. *IEEE Trans. Ind. Appl.* **1981**, *5*, 518–523. [CrossRef]
3. Rodriguez, J.; Lai, J.S.; Peng, F.Z. Multilevel inverters: A survey of topologies, controls, and applications. *IEEE Trans. Ind. Electron.* **2002**, *49*, 724–738. [CrossRef]

4. Colak, I.; Kabalci, E.; Bayindir, R. Review of multilevel voltage source inverter topologies and control schemes. *Energy Convers. Manag.* **2011**, *52*, 1114–1128. [CrossRef]
5. Kincic, S.; Chandra, A.; Babic, S. Multilevel inverter and its limitations when applied as statcom. In Proceedings of the 9th Mediterranean conference on control and automation, Dubrovnik, Croatia, 27–29 June 2001.
6. Daher, S.; Schmid, J.; Antunes, F.L.M. Multilevel inverter topologies for stand-alone PV systems. *IEEE Trans. Ind. Electron.* **2008**, *55*, 2703–2712. [CrossRef]
7. Jakkula, S.; Jayaram, N.; Pulavarthi, S.V.K.; Shankar, Y.R.; Rajesh, J. A generalized high gain multilevel inverter for small scale solar photovoltaic applications. *IEEE Access* **2022**, *10*, 25175–25189. [CrossRef]
8. Memon, A.J.; Mahar, M.A.; Larik, A.S.; Shaikh, M.M. A comprehensive review of reduced device count multilevel inverters for pv systems. *Energies* **2023**, *16*, 5638. [CrossRef]
9. Noman, A.M.; Al-Shamma'A, A.A.; Asef, P.; Alkuhayli, A. Hybrid cascaded MLI development for PV-grid connection applications. *IET Power Electron.* **2023**, *16*, 1717–1731. [CrossRef]
10. Noman, A.M.; Alkuhayli, A.; Al-Shamma'a, A.A.; Addoweesh, K.E. Hybrid MLI Topology Using Open-End Windings for Active Power Filter Applications. *Energies* **2022**, *15*, 6434. [CrossRef]
11. Yousefizad, S.; Azimi, E.; Nasiri-Zarandi, R.; Hafezi, H. A cascaded multilevel inverter based on new basic units. *Int. J. Electron.* **2022**, *109*, 2158–2177. [CrossRef]
12. Seifi, A.; Hosseinpour, M.; Hosseini, S.H. Applications. A novel bidirectional modular multilevel inverter utilizing diode-based bidirectional unit. *J. Circuit Theory Appl.* **2023**, *51*, 3226–3245. [CrossRef]
13. Balal, A.; Dinkhah, S.; Shahabi, F.; Herrera, M.; Chuang, Y.L. A review on multilevel inverter topologies. *Emerg. Sci. J.* **2022**, *6*, 185–200. [CrossRef]
14. Venkataramanaiah, J.; Suresh, Y.; Panda, A.K. A review on symmetric, asymmetric, hybrid and single DC sources based multilevel inverter topologies. *Renew. Sustain. Energy Rev.* **2017**, *76*, 788–812. [CrossRef]
15. Malinowski, M.; Gopakumar, K.; Rodriguez, J.; Perez, M.A. A survey on cascaded multilevel inverters. *IEEE Trans. Ind. Electron.* **2010**, *57*, 2197–2206. [CrossRef]
16. Tolbert, L.; Peng, F. Multilevel converters as a utility interface for renewable energy systems. In Proceedings of the IEEE Power Engineering Society Summer Meeting, Seattle, WA, USA, 16–20 July 2000.
17. Mohan, H.M.; Dash, S.K.; Ram, S.K.; Caesarendra, W. Performance assessment of three-phase PV tied NPC multilevel inverter based UPQC. In Proceedings of the International Conference on Intelligent Controller and Computing for Smart Power (ICICCSP), Hyderabad, India, 21–23 July 2022.
18. Kurdkandi, N.V.; Marangalu, M.G.; Hemmati, T.; Mehrizi-Sani, A.; Rahimpour, S.; Babaei, E. Five-level NPC based grid-tied inverter with voltage boosting capability and elimnated leakage current. In Proceedings of the 13th Power Electronics, Drive Systems, and Technologies Conference (PEDSTC), Tehran, Iran, 1–3 February 2022.
19. Salim, C.H. Electronica; Automatica; Series APF based on Five and Seven-level NPC Inverters using Modified PQ Method. *Electroteh. Electron. Autom. (EEA)* **2023**, *71*, 22–29.
20. Latran, M.B.; Teke, A. Investigation of multilevel multifunctional grid connected inverter topologies and control strategies used in photovoltaic systems. *Renew. Sustain. Energy Rev.* **2015**, *42*, 361–376. [CrossRef]
21. Cheng, H.; Zhao, Z.; Wang, C. A Novel Unidirectional Three-Phase Multilevel Rectifier Composed of Star-Connected Three Single-Phase Topology Based on Five-Level Flying Capacitor DC–DC Converter. *IEEE Trans. Ind. Electron.* **2022**, *70*, 5493–5503. [CrossRef]
22. Kampitsis, G.; Batzelis, E.I.; Mitcheson, P.D.; Pal, B.C. A clamping-circuit-based voltage measurement system for high-frequency flying capacitor multilevel inverters. *IEEE Trans. Power Electron.* **2022**, *37*, 12301–12315. [CrossRef]
23. Bouamrane, O.; Khalili, T.; Tyass, I.; Rafik, M.; Raihani, A.; Bahati, L.; Benhala, B. Flying capacitors multilevel inverter: Architecture, control and active balancing. In Proceedings of the International Conference on Energy and Green Computing, Meknes, Morocco, 9–10 December 2021.
24. Vishvakarma, R.P.; Singh, S.P.; Shukla, T.N. Multilevel inverters and its control strategies: A comprehensive review. In Proceedings of the 2nd International Conference on Power, Control and Embedded Systems, Allahabad, India, 17–19 December 2012.
25. Rohner, S.; Bernet, S.; Hiller, M.; Sommer, R. Modulation, losses, and semiconductor requirements of modular multilevel converters. *IEEE Trans. Ind. Electron.* **2009**, *57*, 2633–2642. [CrossRef]
26. Pamujula, M.; Ohja, A.; Kulkarni, R.D.; Swarnkar, P. Cascaded 'H'bridge based multilevel inverter topologies: A review. In Proceedings of the International Conference for Emerging Technology (INCET), Belgaum, India, 5–7 June 2020.
27. Dhanamjayulu, C.; Girijaprasanna, T. Experimental Implementation of Cascaded H-Bridge Multilevel Inverter with an Improved Reliability for Solar PV Applications. *Int. Trans. Electr. Energy Syst.* **2023**, *2023*, 8794874. [CrossRef]
28. Cao, Y.; Tolbert, L.M. 11-Level cascaded H-bridge grid-tied inverter interface with solar panels. In Proceedings of the Applied Power Electronics Conference and Exposition (APEC), Palm Springs, CA, USA, 21–25 February 2010.

29. Xiao, B.; Hang, L.; Mei, J.; Riley, C.; Tolbert, L.M.; Ozpineci, B. Modular Cascaded H-Bridge Multilevel PV Inverter with Distributed MPPT for Grid-Connected Applications. *IEEE Trans. Ind. Appl.* **2015**, *51*, 1722–1731. [CrossRef]
30. Choi, H.; Zhao, W.; Ciobotaru, M.; Agelidis, V.G. Large-scale PV system based on the multiphase isolated dc/dc converter. In Proceedings of the 3rd IEEE International Symposium on Power Electronics for Distributed Generation Systems (PEDG), Aalborg, Denmark, 25–28 June 2012.
31. Rivera, S.; Wu, B.; Kouro, S.; Wang, H.; Zhang, D. Cascaded H-bridge multilevel converter topology and three-phase balance control for large scale photovoltaic systems. In Proceedings of the 3rd IEEE International Symposium on Power Electronics for Distributed Generation Systems (PEDG), Aalborg, Denmark, 25–28 June 2012.
32. Ozdemir, S.; Altin, N.; Sefa, I. Single stage three level grid interactive MPPT inverter for PV systems. *Energy Convers. Manag.* **2014**, *80*, 561–572. [CrossRef]
33. Vinayaka, B.C.; Prasad, S.N. Modeling and design of five level cascaded h-bridge multilevel inverter with DC/DC boost converter. *Int. J. Eng. Res. Appl.* **2014**, *4*, 50–55.
34. Sivapriya, A.; Kalaiarasi, N. A review on cascaded h-bridge and modular multilevel converter: Topologies, modulation technique and comparative analysis. In *Advanced Power Electronics Converters for Future Renewable Energy Systems*, 1st ed.; Taylor & Francis Group: London, UK, 2023; pp. 195–222.
35. Nagarjuna, A.; Kumar, D.; Reddy, B.; Udaykiran, M. Fifteen level cascaded H-bridge multilevel inverter fed induction motor. *Int. J. Innov. Technol. Explor. Eng.* **2019**, *8*, 640–645. [CrossRef]
36. Rivera, S.; Kouro, S.; Wu, B.; Leon, J.I.; Rodriguez, J.; Franquelo, L.G. Cascaded H-bridge multilevel converter multistring topology for large scale photovoltaic systems. In Proceedings of the IEEE International Symposium on Industrial Electronics (ISIE) 2011, Gdansk, Poland, 27–30 June 2011.
37. Zhao, W.; Choi, H.; Konstantinou, G.; Ciobotaru, M.; Agelidis, V.G. Cascaded H-bridge multilevel converter for large-scale PV grid-integration with isolated DC-DC stage. In Proceedings of the 3rd IEEE International Symposium on Power Electronics for Distributed Generation Systems (PEDG), Aalborg, Denmark, 25–28 June 2012.
38. Villanueva, E.; Correa, P.; Rodriguez, J.; Pacas, M. Control of a single-phase cascaded H-bridge multilevel inverter for grid-connected photovoltaic systems. *IEEE Trans. Ind. Electron.* **2009**, *56*, 4399–4406. [CrossRef]
39. Li, J.; He, Y.; Liu, J. Research on the Capability of Star-connected CHB-STATCOM to Compensate the Unbalanced Voltage of Power Grid. *IEEE Trans. Power Electron.* **2025**, *40*, 8577–8591. [CrossRef]
40. Tandekar, J.K.; Ojha, A.; Jain, S. Application of CHB-MLI as a Three-Phase Star-Connected Nine-Level Shunt Active Power Filter. In *Multilevel Converters*, 1st ed.; Ahmad, S., Bakhsh, F.I., Sanjeevikumar, P., Eds.; Wiley: Hoboken, NJ, USA, 2024; pp. 339–360.
41. Marquez, A.; Leon, J.I.; Vazquez, S.; Franquelo, L.G. Advanced control of a multilevel cascaded H-bridge converter for PV applications. In Proceedings of the 40th Annual Conference of the IEEE Industrial Electronics Society, IECON, Dallas, TX, USA, 29 October–1 November 2014.
42. Alexander, S. Development of solar photovoltaic inverter with reduced harmonic distortions suitable for Indian sub-continent. *Renew. Sustain. Energy Rev.* **2016**, *56*, 694–704. [CrossRef]
43. Sahoo, S.K.; Bhattacharya, T. Phase-shifted carrier-based synchronized sinusoidal PWM techniques for a cascaded H-bridge multilevel inverter. *IEEE Trans. Power Electron.* **2017**, *33*, 513–524. [CrossRef]
44. Govindaraju, C.; Baskaran, K. Efficient sequential switching hybrid-modulation techniques for cascaded multilevel inverters. *IEEE Trans. Power Electron.* **2011**, *26*, 1639–1648. [CrossRef]
45. Khemili, F.Z.; Bouhali, O.; Lefouili, M.; Chaib, L.; El-Fergany, A.A.; Agwa, A.M. Design of Cascaded Multilevel Inverter and Enhanced MPPT Method for Large-Scale Photovoltaic System Integration. *Sustainability* **2023**, *15*, 9633. [CrossRef]
46. Chang, W.-N.; Liao, C.-H. Design and Implementation of a STATCOM Based on a Multilevel FHB Converter with Delta-Connected Configuration for Unbalanced Load Compensation. *Energies* **2017**, *10*, 921. [CrossRef]
47. Yu, Y.; Konstantinou, G.; Townsend, C.D.; Aguilera, R.P.; Agelidis, V.G. Delta-Connected Cascaded H-Bridge Multilevel Converters for Large-Scale Photovoltaic Grid Integration. *IEEE Trans. Ind. Electron.* **2017**, *64*, 8877–8886. [CrossRef]
48. Yu, Y.; Konstantinou, G.; Townsend, C.D.; Aguilera, R.P.; Hredzak, B.; Agelidis, V.G. Delta-connected cascaded H-bridge multilevel photovoltaic converters. In Proceedings of the 41st Annual Conference of Industrial Electronics Society, Yokohama, Japan, 9–12 November 2015.
49. Wen, J.; Smedley, K.M. Synthesis of multilevel converters based on single-and/or three-phase converter building blocks. *IEEE Trans. Power Electron.* **2008**, *23*, 1247–1256. [CrossRef]
50. Peng, F.Z.; Wang, J. A universal STATCOM with delta-connected cascade multilevel inverter. In Proceedings of the 35th IEEE Power Electronics Specialists Conference, Aachen, Germany, 20–25 June 2004.
51. Zhixing, H.; Fujun, M.; An, L.; Qianming, X.; Yandong, C.; Huagen, X.; Guobin, J. Circulating current derivation and comprehensive compensation of cascaded STATCOM under asymmetrical voltage conditions. *IET Gener. Transm. Distrib.* **2016**, *10*, 2924–2932. [CrossRef]

52. Babu, N.S.; Al Hosani, K. A novel DC voltage control for a cascade H-bridge multilevel STATCOM. In Proceedings of the 41st Annual Conference of the IEEE Industrial Electronics Society IECON, Yokohama, Japan, 9–12 November 2015.
53. Jung, J.-J.; Lee, J.-H.; Sul, S.-K.; Son, G.T.; Chung, Y.-H. DC Capacitor Voltage Balancing Control for Delta-Connected Cascaded H-Bridge STATCOM Considering Unbalanced Grid and Load Conditions. *IEEE Trans. Power Electron.* **2018**, *33*, 4726–4735. [CrossRef]
54. Padiyar, K.R.; Kulkarni, A.M. Modeling and analysis of FACTS and HVDC controllers. In *Dynamics and Control of Electric Transmission and Microgrids*, 1st ed.; Padiyar, K.R., Kulkarni, A.M., Eds.; Wiley: Hoboken, NJ, USA, 2019.
55. Akagi, H. Classification, terminology, and application of the modular multilevel cascade converter (MMCC). *IEEE Trans. Power Electron.* **2011**, *26*, 3119–3130. [CrossRef]
56. Molinas, M.; Suul, J.A.; Undeland, T. Low voltage ride through of wind farms with cage generators: STATCOM versus SVC. *IEEE Trans. Power Electron.* **2008**, *23*, 1104–1117. [CrossRef]
57. Mohammadi, P.H.; Bina, M.T. A transformerless medium-voltage STATCOM topology based on extended modular multilevel converters. *IEEE Trans. Power Electron.* **2010**, *26*, 1534–1545.
58. Liang, Y.; Nwankpa, C. A new type of STATCOM based on cascading voltage-source inverters with phase-shifted unipolar SPWM. *IEEE Trans. Ind. Appl.* **1999**, *35*, 1118–1123. [CrossRef]
59. Lee, C.; Leung, J.; Hui, S.; Chung, H.-H. Circuit-level comparison of STATCOM technologies. *IEEE Trans. Power Electron.* **2003**, *18*, 1084–1092. [CrossRef]
60. Marzo, I.; Sanchez-Ruiz, A.; Barrena, J.A.; Abad, G.; Muguruza, I. Power balancing in cascaded H-bridge and modular multilevel converters under unbalanced operation: A review. *IEEE Access* **2021**, *9*, 110525–110543. [CrossRef]
61. Hameed, U.; Sadiq, H.; Khalid, H.A.; Khan, M.U.; Ali, M. Investigation of zero sequence injection method for balancing of multi-level Cascaded-H bridge inverter. In Proceedings of the IEEE International Conference on Computing, Electronic and Electrical Engineering, Quetta, Pakistan, 12–13 November 2018.
62. He, Z.; Ma, F.; Xu, Q.; Chen, Y.; Li, C.; Li, M.; Guerrero, J.M.; Luo, A. Reactive power strategy of cascaded delta-connected STATCOM under asymmetrical voltage conditions. *IEEE J. Emerg. Sel. Top. Power Electron.* **2017**, *5*, 784–795. [CrossRef]
63. Wu, P.-H.; Chen, H.-C.; Chang, Y.-T.; Cheng, P.-T. Delta-connected cascaded H-bridge converter application in unbalanced load compensation. *IEEE Trans. Ind. Appl.* **2016**, *53*, 1254–1262. [CrossRef]
64. Liu, H.; Tolbert, L.M.; Khomfoi, S.; Ozpineci, B.; Du, Z. Hybrid cascaded multilevel inverter with PWM control method. In Proceedings of the Power Electronics Specialists Conference, Rhodes, Greece, 15–19 June 2008.
65. Kangarlu, M.F.; Babaei, E. A generalized cascaded multilevel inverter using series connection of submultilevel inverters. *IEEE Trans. Power Electron.* **2012**, *28*, 625–636. [CrossRef]
66. Gandomi, A.A.; Saeidabadi, S.; Hosseini, S.H.; Babaei, E.; Sabahi, M. Transformer-based inverter with reduced number of switches for renewable energy applications. *IET Power Electron.* **2015**, *8*, 1875–1884. [CrossRef]

electronics

MDPI

Article

Robust Sensorless PMSM Control with Improved Back-EMF Observer and Adaptive Parameter Estimation

Ayyoub Zeghlache [1,*], Ali Djerioui [1], Hemza Mekki [1], Samir Zeghlache [2] and Mohamed Fouad Benkhoris [3]

1 Laboratoire de Génie Electrique, Department of Electrical Engineering, Faculty of Technology, University of M'sila, M'sila 28000, Algeria; ali.djerioui@univ-msila.dz (A.D.); hemza.mekki@univ-msila.dz (H.M.)
2 Laboratoire d'Analyse des Signaux et Systemes, University of M'sila, M'sila 28000, Algeria; samir.zeghlache@univ-msila.dz
3 IREENA Laboratory, Nantes University, 44600 Saint-Nazaire, France; mohamed-fouad.benkhoris@univ-nantes.fr
* Correspondence: ayyoub.zeghlache@univ-msila.dz

Abstract: This paper presents an enhanced sensorless control strategy for permanent magnet synchronous motors (PMSMs) by improving back-electromotive force (back-EMF) estimation and control robustness. An improved back-EMF extended state observer (ESO) is proposed, incorporating back-EMF differentiation to compensate for DC position error without requiring an increased observer bandwidth. Furthermore, an ESO-based quadrature phase-locked loop (QPLL) is developed to improve position tracking accuracy and enhance the robustness of the speed loop sliding mode controller (SMC) against unknown disturbances. To address parameter uncertainties in the back-EMF observer and current controller, a recursive least squares (RLSs) algorithm with an adaptive forgetting factor is introduced, providing a balance between adaptation speed and noise suppression. Simulation results validate the proposed approach, demonstrating improved estimation accuracy, disturbance rejection, and overall robustness in sensorless PMSM control.

Keywords: permanent magnet synchronous motor; sensorless control; extended state observer; recursive least squares; adaptive forgetting factor

1. Introduction

Modern electric cars are increasingly widespread as advancements in power electronics continue to grow [1]. Among various motor options, the permanent magnet synchronous motor (PMSM) stands out in electric car applications due to its superior efficiency, power factor, power density, and torque density [2]. These advantages come from its design, which relies on permanent magnets, considered one of the most remarkable innovations in modern engineering, with much still to be explored [3]. The emergence of field-oriented control (FOC) has addressed one of the biggest challenges in PMSM control by simplifying its strong coupling issue, making it as easy to manage as a DC motor. However, PMSMs still face challenges, particularly parameter variations caused by external factors, such as high temperatures, which can also damage sensors, as they are essential components in the control system [4–6].

Accurate speed measurement is a key component in implementing FOC [7,8]. Traditionally, sensors are used to measure speed, but they add cost, weight, and are prone to failure under harsh conditions [9]. To address these challenges, researchers have introduced sensorless control techniques that operate solely on a simple mathematical model of the PMSM [10–12]. In sensorless control, estimating the back electromotive force (back-EMF) is crucial because it indirectly provides speed information [13,14]. However, since no back-EMF is generated at a standstill, this approach has traditionally been effective only in medium- to high-speed ranges [15]. In recent studies, various model-based back-EMF strategies have emerged to extend the operational range into low-speed conditions [15,16].

An observer-based approach is commonly employed for back-EMF estimation in sensorless PMSM control, relying on a mathematical model of the motor. However, model mismatches can degrade estimation accuracy, making robust observers a preferred solution [17,18]. Sliding mode observer (SMO) and extended state observer (ESO) are widely recognized for their ability to handle uncertainties and disturbances with relatively simple designs [18–20]. Despite their advantages, SMO suffers from chattering due to the discontinuous switching function, often necessitating additional filtering stages to obtain an accurate back-EMF estimate. In contrast, ESO inherently acts as a low-pass filter, eliminating the need for extra filtering [14,20]. However, its low-pass filtering characteristic can lead to DC position estimation errors when the observer bandwidth is insufficient. Simply increasing the bandwidth is not a straightforward solution, as excessive bandwidth can amplify noise sensitivity [21]. To address these limitations, various studies have focused on enhancing both SMO and ESO to improve back-EMF estimation accuracy while mitigating their respective drawbacks [22,23]. The work in [20] introduced a quasi-resonant controller with reduced order into the ESO structure to minimize DC offset estimation errors and reduce the phase lag in back-EMF estimation. While this method enhances estimation precision, it also increases computational complexity and requires tuning multiple parameters, challenges that are commonly observed across various proposed approaches, along with insufficient consideration of parameter uncertainties.

Once the back-EMF signal is estimated, the next challenge is accurately extracting the rotor position and speed from it. While the arctangent function can directly compute rotor position from the estimated back-EMF, and its derivative theoretically yields speed, these methods are highly sensitive to high-frequency noise, leading to inaccuracies in practical implementations [24]. To overcome this limitation, phase-locked loop (PLL) techniques have been widely adopted, as they inherently suppress noise by leveraging a PI controller to estimate speed, using integration to determine rotor position. This approach avoids the noise amplification associated with differentiation, resulting in smoother and more reliable estimates [25]. Recent advancements in PLL techniques, such as quadrature PLL (QPLL), along with ESO-based methods, have further improved speed and position estimation accuracy by effectively rejecting unknown disturbances and enabling higher precision [26,27].

This paper proposes several improvements to enhance sensorless control performance. The back-EMF ESO design is improved by incorporating the differentiation of the back-EMF, which compensates for the DC position error without requiring a higher observer bandwidth compared to conventional ESOs. A QPLL based on ESO is developed to improve position tracking accuracy and reinforce the sliding mode control (SMC) of the speed loop controller, leading to greater robustness and faster convergence against unknown disturbances. To further enhance robustness against parameter uncertainties in the back-EMF ESO and the current controller, a recursive least squares (RLSs) method with an adaptive forgetting factor based on an exponential function is introduced. These advancements lead to a more robust and reliable sensorless control system.

The discussion in this paper progresses as follows: Section 2 describes the PMSM model, incorporating parameter uncertainties to account for real-world variations. Section 3 presents the back-EMF observer design, comparing the conventional ESO with the proposed method. Section 4 explores the ESO-based QPLL approach for extracting speed and position estimates. Section 5 outlines the control strategies, where SMC regulates speed, and predictive current control manages current. Section 6 discusses parameter estimation using the RLS algorithm with an adaptive forgetting factor. Section 7 assesses the proposed sensorless control strategy through simulation results. Finally, Section 8 summarizes the key findings and contributions.

2. Model of PMSM

This paper considers a surface-mounted permanent magnet synchronous motor with equal d-axis and q-axis inductances. To simplify the analysis, magnetic saturation, eddy currents, and hysteresis losses are neglected. The PMSM with parameter uncertainties can be modeled in the dq-frame as follows:

$$\begin{cases} \frac{\mathrm{d}i_d}{\mathrm{dt}} = -\frac{R_s}{L_s} i_d + \omega i_q + \frac{1}{L_s} v_d + g_d \\ \frac{\mathrm{d}i_q}{\mathrm{dt}} = -\frac{R_s}{L_s} i_q - \omega i_d - \frac{\psi_f \omega}{L_s} + \frac{1}{L_s} v_q + g_q \\ \frac{\mathrm{d}\omega}{\mathrm{dt}} = k_T i_q - \frac{f}{J}\omega - \frac{n_p}{J} T_L + g_\omega \end{cases} \tag{1}$$

where i_d and i_q denote the stator currents in the dq-frame, and v_d and v_q denote the corresponding stator voltages, R_s represents the stator resistance, and L_s represents the stator inductance; the torque constant is defined as $k_T = 1.5 n_p^2 \psi_f / J$, with n_p being the number of pole pairs, ψ_f the rotor permanent magnet flux linkage, J the moment of inertia, f the friction coefficient, ω the electrical angular velocity, and T_L the load torque; additionally, g_d, g_q, and g_ω are introduced to account for parameter uncertainties in the PMSM model and are modeled as follows:

$$\begin{cases} g_d = -\left(\frac{\Delta R_s}{L_s+\Delta L_s} - \frac{R_s \Delta L_s}{L_s(L_s+\Delta L_s)}\right) i_d - \frac{\Delta L_s}{L_s(L_s+\Delta L_s)} v_d \\ g_q = -\left(\frac{\Delta R_s}{L_s+\Delta L_s} - \frac{R_s \Delta L_s}{L_s(L_s+\Delta L_s)}\right) i_q + \frac{\Delta L_s \psi_f \omega}{L_s(L_s+\Delta L_s)} - \frac{\Delta L_s}{L_s(L_s+\Delta L_s)} v_d \\ g_\omega = -\frac{\Delta J}{J+\Delta J} k_T i_q - \left(\frac{\Delta f}{J+\Delta J} - \frac{f \Delta J}{J(J+\Delta J)}\right)\omega + \frac{\Delta J}{J(J+\Delta J)} n_p T_L \end{cases} \tag{2}$$

where ΔR_s and ΔL_s denote the deviations of the stator resistance and inductance from their nominal values, while ΔJ and Δf account for variations in the moment of inertia and friction coefficient from their assumed values.

3. Back-EMF Observer

The stationary frame model facilitates the extraction of the rotor's electrical position via the back-EMF observer. In this frame, the current model is represented as:

$$\begin{cases} \frac{\mathrm{d}i_\alpha}{\mathrm{dt}} = -\frac{R_s}{L_s} i_\alpha + \frac{1}{L_s} v_\alpha + E_\alpha \\ \frac{\mathrm{d}i_\beta}{\mathrm{dt}} = -\frac{R_s}{L_s} i_\beta + \frac{1}{L_s} v_\beta + E_\beta \end{cases} \tag{3}$$

where i_α and v_α represent the current and voltage along the α-axis, and i_β and v_β represent the current and voltage along the β-axis. The unknown signals in the model, denoted as E_α and E_β, are defined by:

$$\begin{cases} E_\alpha = -\frac{1}{L_s} e_\alpha \\ E_\beta = -\frac{1}{L_s} e_\beta \end{cases} \tag{4}$$

where e_α and e_β correspond to the back-EMF in the $\alpha\beta$-frame, defined as follows:

$$\begin{cases} e_\alpha = -\omega\psi_f \sin(\theta) \\ e_\beta = \omega\psi_f \cos(\theta) \end{cases} \tag{5}$$

where θ represents the electrical position of the rotor, it is clear that the back-EMF in (4) is independent of the current and voltage states, thereby making the stationary frame an ideal option for estimating the back-EMF signals.

3.1. Conventional ESO

The ESO estimates the unknown signals E_α and E_β by defining them as system states, with its structure depicted in Figure 1 and detailed as follows:

$$\begin{cases} \frac{\mathrm{d}\hat{i}_{\alpha\beta}}{\mathrm{dt}} = f_{\alpha\beta} + \hat{E}_{\alpha\beta} + h_1(i_{\alpha\beta} - \hat{i}_{\alpha\beta}) \\ f_{\alpha\beta} = -\frac{R_s}{L_s} i_{\alpha\beta} + \frac{1}{L_s} v_{\alpha\beta} \\ \frac{\mathrm{d}\hat{E}_{\alpha\beta}}{\mathrm{dt}} = h_2(i_{\alpha\beta} - \hat{i}_{\alpha\beta}) \end{cases} \tag{6}$$

where the subscript "$\alpha\beta$" indicates that the equation applies identically to both the α and β axes due to their symmetrical structure, and the hat symbol "^" denotes that the associated state variable is an estimate. Additionally, the gains h_1 and h_2 are defined as constant observer gains.

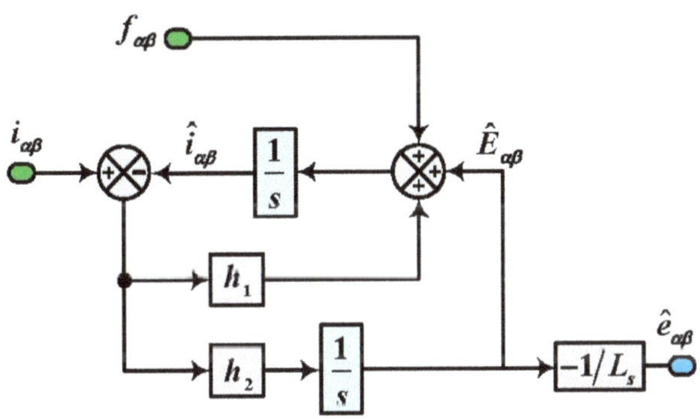

Figure 1. Structural diagram of the conventional back-EMF ESO.

By applying the Laplace transform to (3) and (6), we obtain a transfer function that relates the estimated back-EMF to its actual value, as detailed below:

$$G_{\alpha\beta}(s) = \frac{\hat{e}_{\alpha\beta}(s)}{e_{\alpha\beta}(s)} = \frac{h_2}{s^2 + h_1 s + h_2} \tag{7}$$

where s is defined as the Laplace operator.

As established by (7), the Routh–Hurwitz stability criterion requires h_1 and h_2 to be strictly positive. To satisfy these conditions, the observer bandwidth ω_o is chosen to be positive, ensuring the gains are parameterized as $h_1 = 2\omega_o$ and $h_2 = \omega_o^2$.

Figure 2 presents the frequency response of $G_{\alpha\beta}(s)$ for different observer bandwidths, demonstrating its low-pass filtering behavior. A higher observer bandwidth (ω_o) effectively reduces phase lag and amplitude attenuation, leading to a more accurate estimation of back-EMFs. However, this improvement increases the sensitivity of the observer to high-frequency noise, which may affect the robustness of the estimation.

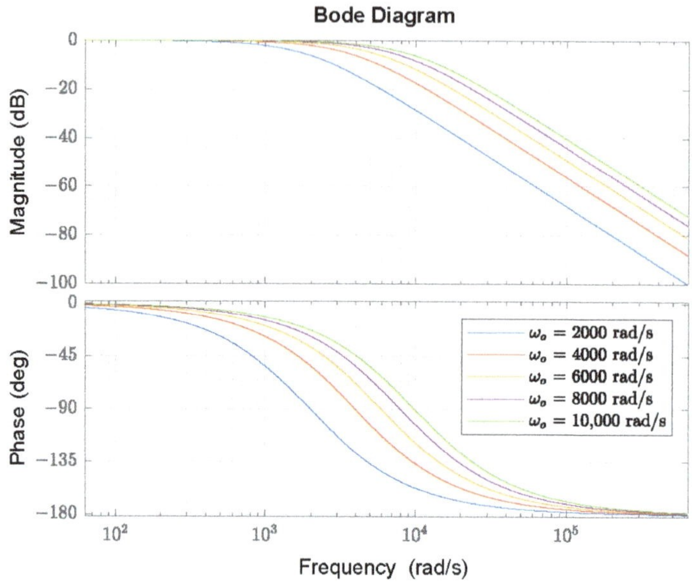

Figure 2. Bode diagram of the conventional back-EMF ESO.

3.2. Proposed ESO

The proposed ESO operates by tracking the frequency of the back-EMF, which equals the estimated electrical speed. Its inputs are the estimated electrical speed $\hat{\omega}$, the current $i_{\alpha\beta}$, and the known signal $f_{\alpha\beta}$, as shown in Figure 3. This design is expanded by incorporating the differentiation of the back-EMF to compensate for the DC position error and phase lag. The design of the proposed ESO is detailed as follows:

$$\begin{cases} \dfrac{d\hat{i}_{\alpha\beta}}{dt} = f_{\alpha\beta} + \hat{E}_{\alpha\beta} + h_1(i_{\alpha\beta} - \hat{i}_{\alpha\beta}) \\ \dfrac{d\hat{E}_{\alpha\beta}}{dt} = -\hat{\omega}^2\hat{z}_{\alpha\beta} + \hat{\partial}_{\alpha\beta} + h_2(i_{\alpha\beta} - \hat{i}_{\alpha\beta}) \\ \dfrac{d\hat{\partial}_{\alpha\beta}}{dt} = h_3(i_{\alpha\beta} - \hat{i}_{\alpha\beta}) \\ \dfrac{d\hat{z}_{\alpha\beta}}{dt} = \hat{E}_{\alpha\beta} \end{cases} \tag{8}$$

where h_1, h_2, and h_3 represent constant observer gains.

Similarly, applying the Laplace transform to (3) and (8) produces a transfer function that connects the estimated back-EMF to its actual value, as presented below.

$$H_{\alpha\beta}(s) = \frac{\hat{e}_{\alpha\beta}(s)}{e_{\alpha\beta}(s)} = \frac{h_2 s + h_3}{s^3 + h_1 s^2 + (\hat{\omega}^2 + h_2)s + h_1\hat{\omega}^2 + h_3} \tag{9}$$

To ensure the stability of the transfer function $H_{\alpha\beta}(s)$, the parameters must satisfy the following criterion derived from the Routh–Hurwitz stability analysis:

$$\begin{cases} h_1 > 0 \\ h_2 > \dfrac{h_3}{h_1} \\ h_3 > \hat{\omega}^2 \end{cases} \tag{10}$$

The observer gains h_1, h_2, and h_3 are tuned according to the selected ESO bandwidth and the estimated electrical rotor speed; their adjustments are implemented as follows:

$$\begin{bmatrix} h_1 & h_2 & h_3 \end{bmatrix} = \begin{bmatrix} 3\omega_o & 3\omega_o^2 - \hat{\omega}^2 & \omega_o^3 - 3\omega_o\hat{\omega}^2 \end{bmatrix} \tag{11}$$

where ω_o is the ESO bandwidth and must satisfy $\omega_o^3/(1+3\omega_o) \;\; >> \hat{\omega}^2$ to ensure observer stability.

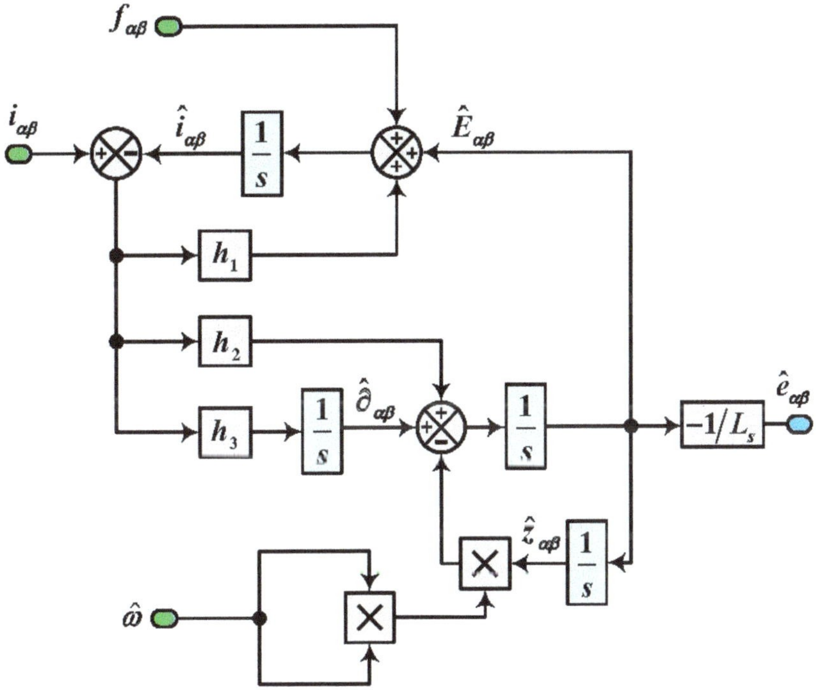

Figure 3. Structural diagram of the proposed back-EMF ESO.

Figure 4 presents the Bode diagram of $H_{\alpha\beta}(s)$ for back-EMF frequencies of 100, 200, 300, and 400 rad/s. The proposed ESO effectively minimizes DC error and phase lag across all tested frequencies with a fixed observer bandwidth of $\omega_o = 3000$ rad/s. This ensures accurate estimation without the need for higher bandwidth, improving robustness while maintaining performance.

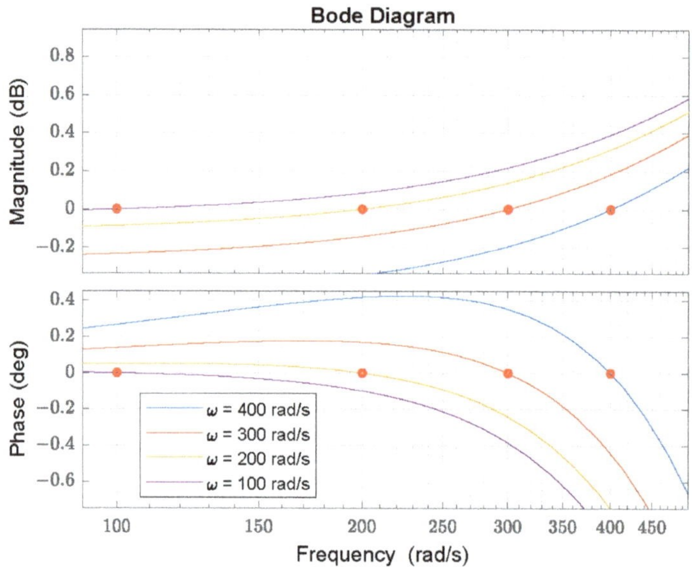

Figure 4. Bode diagram of the proposed back-EMF ESO.

4. QPLL Based on ESO

The direct method using the arc-tangent function introduces high-frequency noise in position estimation and may face computational challenges in extracting the estimated position and speed. To address these issues, the QPLL provides a robust and smooth approach for calculating the position estimation error from the estimated back-EMF, which can be formulated as follows:

$$\begin{aligned}\Delta E &= -\hat{e}_\alpha \cos\hat{\theta} - \hat{e}_\beta \sin\hat{\theta} \\ &= \hat{E}_{eq}(\sin\theta\cos\hat{\theta} - \cos\theta\sin\hat{\theta}) \\ &= \hat{E}_{eq}\sin(\theta-\hat{\theta}) \approx \hat{E}_{eq}(\theta-\hat{\theta}) \approx \hat{E}_{eq}\varepsilon_\theta\end{aligned} \tag{12}$$

where $\hat{E}_{eq} = \sqrt{\hat{e}_\alpha^2 + \hat{e}_\beta^2}$, and ε_θ is the position estimation error.

Conventional phase-locked loop (PLL) methods, including QPLL based on a PI controller, are commonly used to extract estimated speed and position from the position estimation error. However, in PMSM control, uncertainties and disturbances affect the estimated back-EMF, making it difficult for the PI controller to provide accurate estimations. A more effective approach is to replace the PI controller with an extended state observer (ESO) to account for these uncertainties and disturbances.

Based on the equation of motion presented in (1), by clearly identifying the known and unknown disturbances, the dynamics can be reformulated as:

$$\begin{cases}\dfrac{\mathrm{d}\theta}{\mathrm{dt}} = \omega \\ \dfrac{\mathrm{d}\omega}{\mathrm{dt}} = k_T i_q^* + d_\omega \\ d_\omega = k_T(i_q - i_q^*) - \dfrac{f}{J}\omega - \dfrac{n_p}{J}T_L + g_\omega\end{cases} \tag{13}$$

By treating the unknown disturbance d_ω as an extra state in the dynamics described by (13), we can derive the design of the third-order ESO as follows:

$$\begin{cases} \dfrac{d\hat{\theta}}{dt} = \hat{\omega} + \beta_1 \varepsilon_\theta \\ \dfrac{d\hat{\omega}}{dt} = k_T i_q^* + \hat{d}_\omega + \beta_2 \varepsilon_\theta \\ \dfrac{d\hat{d}_\omega}{dt} = \beta_3 \varepsilon_\theta \end{cases} \quad (14)$$

where β_1, β_2, and β_3 correspond to the observer gains. The gains are then adjusted based on the ESO bandwidth parameter σ, as indicated by the following expressions: $\beta_1 = 3\sigma$, $\beta_2 = 3\sigma^2$, and $\beta_3 = \sigma^3$.

5. Design of Speed and Current Controllers

5.1. Speed Controller Design

The sliding surface is defined as:

$$s = \omega^* - \hat{\omega} \quad (15)$$

The output speed control loop for generating the reference quadratic current can be designed as:

$$i_q^* = \frac{1}{k_T}\left[\frac{d\hat{\omega}^*}{dt} - \hat{d}_\omega + k_1|s|^{0.5}\text{sign}(s) + \int k_2\text{sign}(s)\right] \quad (16)$$

The unknown disturbance estimation error ($\tilde{d}_\omega = d_\omega - \hat{d}_\omega$) and the time derivative of the speed estimation error ($\dot{\tilde{\omega}} = \dot{\omega} - \dot{\hat{\omega}}$) are treated as a perturbation term, which is assumed to be globally bounded according to the following condition:

$$|\tilde{d}_\omega| + |\dot{\tilde{\omega}}| \leq \delta_\omega |s|^{0.5} \quad (17)$$

where δ_ω represents a positive bounding constant. As stated in [28], the system achieves strong global asymptotic stability if the gains fulfill the following conditions:

$$\begin{cases} k_1 > 2\delta_\omega \\ k_2 > k_1 \dfrac{5\delta_\omega k_1 + 4\delta_\omega}{2(k_1 - 2\delta_\omega)} \end{cases} \quad (18)$$

5.2. Current Controller Design

Predictive Current Control is chosen for the current controller design because it eliminates the need for parameter tuning, unlike conventional PI controllers, while ensuring optimal control performance.

To discretize the $\alpha\beta$-frame current model of a PMSM, the current derivative is approximated using the forward Euler method with a sampling time of T_s, as follows:

$$\frac{di_{\alpha\beta}}{dt} \approx \frac{i_{\alpha\beta}(k+1) - i_{\alpha\beta}(k)}{T_s} \quad (19)$$

By applying (19) and (3), the predicted current model in the stationary frame is derived as follows:

$$\begin{cases} i_\alpha(k+1) = \left(1 - \dfrac{R_s}{L_s}T_s\right) i_\alpha(k) + \dfrac{T_s}{L_s}\left(v_\alpha(k) - \hat{e}_\alpha(k)\right) \\ i_\beta(k+1) = \left(1 - \dfrac{R_s}{L_s}T_s\right) i_\beta(k) + \dfrac{T_s}{L_s}\left(v_\beta(k) - \hat{e}_\beta(k)\right) \end{cases} \quad (20)$$

The voltage components $v_\alpha(k)$ and $v_\beta(k)$ are computed from the switching pulses applied to each up-leg of the inverter (S_1, S_2, S_3) in conjunction with the DC-link voltage (v_{dc}). Concurrently, the estimated back-EMF $\hat{e}_\alpha(k)$ and $\hat{e}_\beta(k)$ are determined based on the estimated speed and position. The following equations detail how $v_\alpha(k)$, $v_\beta(k)$, $\hat{e}_\alpha(k)$, and $\hat{e}_\beta(k)$ are calculated.

$$\begin{cases} \hat{e}_\alpha(k) = -\hat{\omega}(k)\psi_f \sin(\hat{\theta}(k)) \\ \hat{e}_\beta(k) = \hat{\omega}(k)\psi_f \cos(\hat{\theta}(k)) \\ v_\alpha(k) = \frac{v_{dc}}{3}(2S_1 - S_2 - S_3) \\ v_\beta(k) = \frac{\sqrt{3}v_{dc}}{3}(S_2 - S_3) \end{cases} \tag{21}$$

A three-phase inverter comprises eight switching states. However, (0,0,0) and (1,1,1) generate the same zero-voltage vector, yielding seven unique vectors (six active and one zero). At each sampling period, the voltage vector minimizing a cost function is selected. This cost function is designed to track the torque current reference, optimize torque per ampere, and enforce current magnitude limits, as defined below:

$$\begin{aligned} g =& (i_\alpha^* - i_\alpha(k+1))^2 + \left(i_\beta^* - i_\beta(k+1)\right)^2 \\ &+ \begin{cases} 0 & \text{if } \sqrt{i_\alpha(k+1)^2 + i_\beta(k+1)^2} \le i_{\max} \\ \infty & \text{if } \sqrt{i_\alpha(k+1)^2 + i_\beta(k+1)^2} \ge i_{\max} \end{cases} \end{aligned} \tag{22}$$

where i_α^* and i_β^* are the $\alpha\beta$-frame current references, derived from the dq-frame current references via the inverse Park transformation. The quadrature-axis current reference (i_q^*) is generated by the speed controller, while the direct-axis current reference (i_d^*) is set to zero to achieve torque decoupling.

6. Recursive Least Squares

In the $\alpha\beta$ plane, the discrete model of the current equation for the PMSM can also be formulated as:

$$y = \Theta x \tag{23}$$

where y is the output vector, Θ is the matrix of parameters to be estimated, and x is the input vector. Their definitions are as follows:

$$x = \begin{bmatrix} i_\alpha(k-1) & i_\beta(k-1) & v_\alpha(k-1) & v_\beta(k-1) & r_\alpha(k-1) & r_\beta(k-1) \end{bmatrix}^T$$

$$\begin{bmatrix} r_\alpha(k-1) & r_\beta(k-1) \end{bmatrix}^T = \begin{bmatrix} \hat{\omega}(k-1)\sin(\hat{\theta}(k-1)) & -\hat{\omega}\sin(\hat{\theta}(k-1)) \end{bmatrix}^T$$

$$y = \begin{bmatrix} i_\alpha(k) & i_\beta(k) \end{bmatrix}^T$$

$$\Theta = \begin{bmatrix} \left(1 - \frac{R_s}{L_s}T_s\right) & 0 & \frac{T_s}{L_s} & 0 & \frac{T_s}{L_s}\psi_f & 0 \\ 0 & \left(1 - \frac{R_s}{L_s}T_s\right) & 0 & \frac{T_s}{L_s} & 0 & \frac{T_s}{L_s}\psi_f \end{bmatrix}$$

To estimate the parameter matrix Θ in real time, a recursive least squares (RLSs) method is employed [29], which can be represented as:

$$\begin{cases} e = y - \hat{y} \\ \hat{y} = \widehat{\mathbf{\Theta}}(k)x \\ \widehat{\mathbf{\Theta}}(k) = \widehat{\mathbf{\Theta}}(k-1) + (y - \widehat{\mathbf{\Theta}}(k-1)x)(x^T\mathbf{P}(k)) \\ \mathbf{P}(k) = (1/\lambda)\left\{\mathbf{P}(k-1) - \dfrac{\mathbf{P}(k-1)x}{(\lambda + x^T\mathbf{P}(k-1)x)}x^T\mathbf{P}(k-1)\right\} \end{cases} \tag{24}$$

where P denotes the covariance matrix, $\hat{y}$ is the estimate of y computed using the estimated parameter matrix $\widehat{\mathbf{\Theta}}$, and e represents the estimation error. The initial conditions are chosen to facilitate robust convergence: $P(0)$ is typically initialized as a sufficiently large positive scalar multiplied by the identity matrix (i.e., $P(0) = \delta I$ with $10^{10} > \delta > 10^4$), $\widehat{\mathbf{\Theta}}(0)$ is set as a sufficiently small positive or zero vector, and the forgetting factor λ is commonly set to 1.

In sensorless PMSM control, accurate estimation of parameters such as inductance is particularly challenging due to their dependence solely on operating currents, similar to nonlinear inductors in power electronics applications [30]. Due to the complexity, nonlinearity, and strong coupling of PMSMs, using a fixed forgetting factor of 1 is ineffective. In recursive least squares, the forgetting factor (λ) controls the weight of past data in the identification process. A lower λ prioritizes recent data, while a higher λ focuses more on older data. Therefore, selecting an optimal λ is crucial for balancing the rapid tracking of parameter variations with minimal excessive fluctuations.

To address these challenges, the conventional RLS algorithm was modified by incorporating an adaptive forgetting factor that adjusts dynamically based on the magnitude of the estimation error. In the modified algorithm, the forgetting factor is updated according to

$$\lambda = \lambda_{\min} + (\lambda_{\max} - \lambda_{\min})\exp\left(-\kappa\|e\|^2\right) \tag{25}$$

where $\lambda_{\min}$ is the minimum allowable forgetting factor (facilitating rapid adaptation), $\lambda_{\max}$ is the maximum allowable forgetting factor (ensuring robust steady-state performance), and κ is a design parameter that sets the sensitivity of the update to the squared error magnitude. A large estimation error makes the exponential term very small, so λ moves closer to $\lambda_{\min}$, allowing the algorithm to quickly forget old data and adapt to new changes. Conversely, when the error is small, the exponential term stays near 1, and λ remains close to $\lambda_{\max}$, ensuring that more past data are used, which helps to reduce noise.

7. Results Analysis

To evaluate the effectiveness of the proposed sensorless control strategy for PMSM, simulations were conducted in MATLAB (R2022b)/Simulink with a sampling time of 1×10^{-5} s, following the control structure shown in Figure 5, with motor parameters detailed in Table A1. Since back-EMF is absent at a standstill, the proposed method is not applicable at zero speed or during the brief startup phase. Sensorless control techniques relying on back-EMF estimation require the motor to be accelerated to a certain speed for accurate signal detection. In this study, the startup process followed the strategy outlined in [31], where a speed ramp-up frequency was applied while maintaining a q-axis current level dependent on load torque and regulating the d-axis current to zero. Once the motor reached the desired conditions, the proposed sensorless control was activated at 10 ms. To ensure a fair comparison, both the conventional and proposed back-EMF ESO bandwidths were set to ω_o = 50,000 rad/s, while the QPLL ESO bandwidth for speed and position estimation was configured to $\sigma = 500$ rad/s. To evaluate the robustness of the proposed sensorless control under parameter uncertainties, simulations were also performed with parameter deviations modeled as in (2), where the uncertainties were set to $\Delta R_s = 0.8R_s$, $\Delta L_s = -0.3L_s$, $\Delta J = 0.6J$, and $\Delta f = 0.6f$.

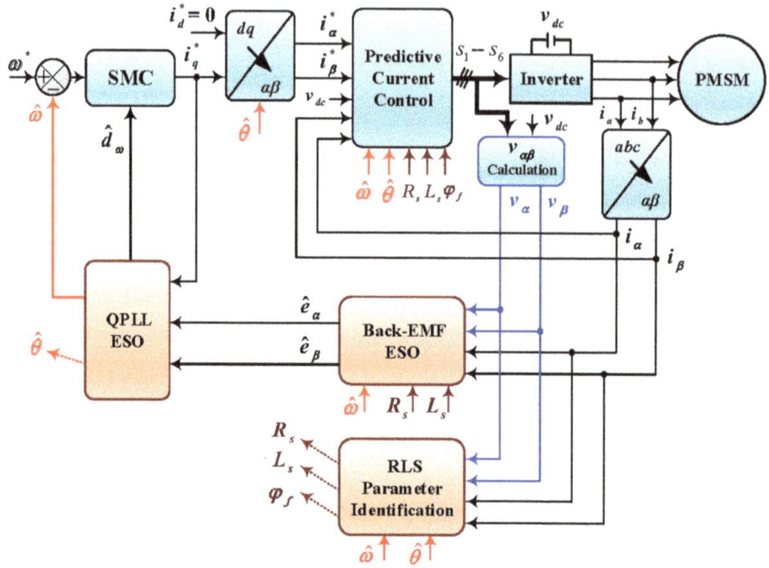

Figure 5. Sensorless control structure diagram.

Figure 6 presents the performance comparison between the conventional and proposed back-EMF ESO under varying speed conditions with a rated load torque of 28.4 Nm. The top plot illustrates the actual and estimated speed trajectories, where both observers closely follow the reference speed. The reference speed starts at 900 rpm, then drops to 500 rpm at 0.5 s, followed by 200 rpm at 1 s and 100 rpm at 1.5 s, and then returns to 900 rpm at 2 s. The middle plot depicts the speed estimation error, showing that the conventional ESO experiences noticeable oscillations during speed transitions, whereas the proposed ESO maintains near-zero error. The bottom plot compares the position estimation error, demonstrating that the proposed ESO effectively eliminates DC position error and phase lag across different frequency operations, ensuring more accurate estimation than the conventional ESO.

Figures 7 and 8 present the performance comparison of the sensorless control of PMSM with and without the use of RLS parameter identification under parameter uncertainties. As previously described, parameter uncertainties ($\Delta R_s = 0.8R_s$, $\Delta L_s = -0.3L_s$, $\Delta J = 0.6J$, and $\Delta f = 0.6f$) were applied at 1 s. The reference speed was set to 600 rpm, while the load torque was initially maintained at the rated value of 28.4 Nm and then reduced to 5% of the rated torque at 1.5 s. Figure 7 illustrates the actual speed and torque responses, where a noticeable deviation occurs after 1 s due to the applied uncertainties. The PMSM with RLS maintains a more stable speed and torque trajectory despite these variations, compared to the system without RLS. Figure 8 highlights the speed and position estimation errors, showing a significant increase in error after 1 s for the system without RLS, whereas the RLS-based approach effectively mitigates these errors, ensuring robustness and accuracy even in the presence of parameter uncertainties.

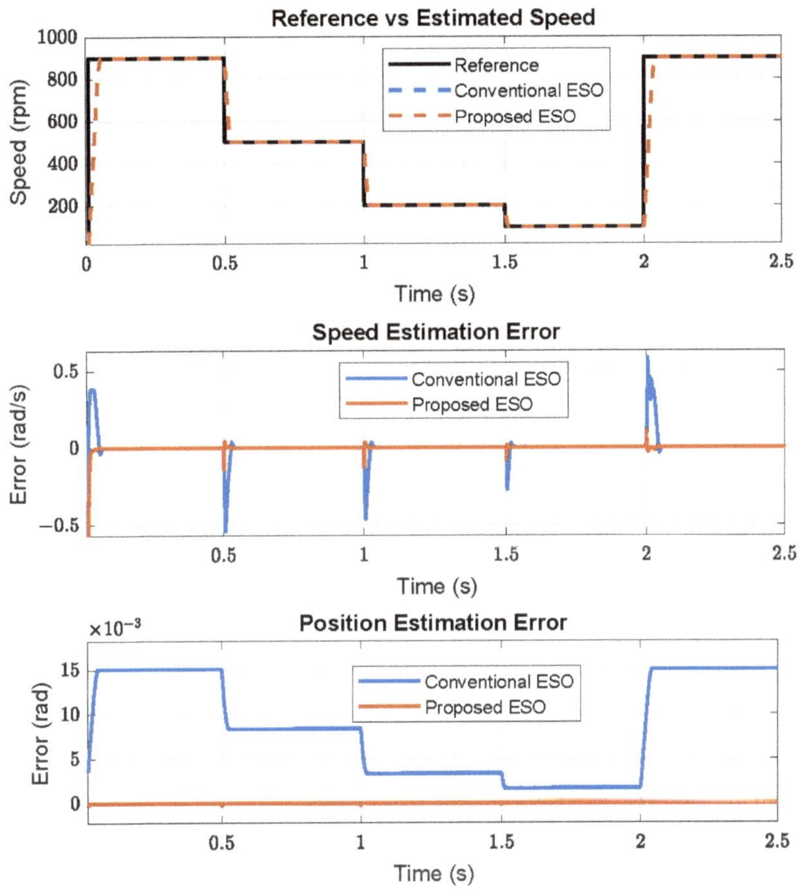

Figure 6. Comparison of back-EMF ESO performance.

Figure 9 shows the performance of the back-EMF in the α-axis for the PMSM under parameter uncertainties, comparing the cases without and with RLS parameter identification at the same speed reference of 600 rpm (40 Hz) with a rated load torque. In the top plot, parameter uncertainties were applied at 1 s, and the results show that the RLS-based approach achieves a smoother and more accurate back-EMF estimation despite these deviations. The middle plot presents the total harmonic distortion (THD) of the back-EMF without RLS, revealing a THD of 5.37% with prominent harmonic components at the 5th, 7th, 17th, and 19th orders. In contrast, the bottom plot demonstrates that the use of RLS significantly reduces the THD, effectively minimizing harmonic distortion and improving the overall quality of the back-EMF estimation under parameter uncertainty conditions. As shown in Table 1, this reduction in THD is not limited to the back-EMF but is also evident in the stator current, where the phase "a" current THD decreases from 8.36% to 2.77% with RLS. Additionally, Figure 10 further illustrates this effect by depicting the three-phase stator current waveforms for both cases, demonstrating that with RLS, the current waveforms remain smooth and unaffected by parameter uncertainties, closely resembling the ideal case without uncertainties.

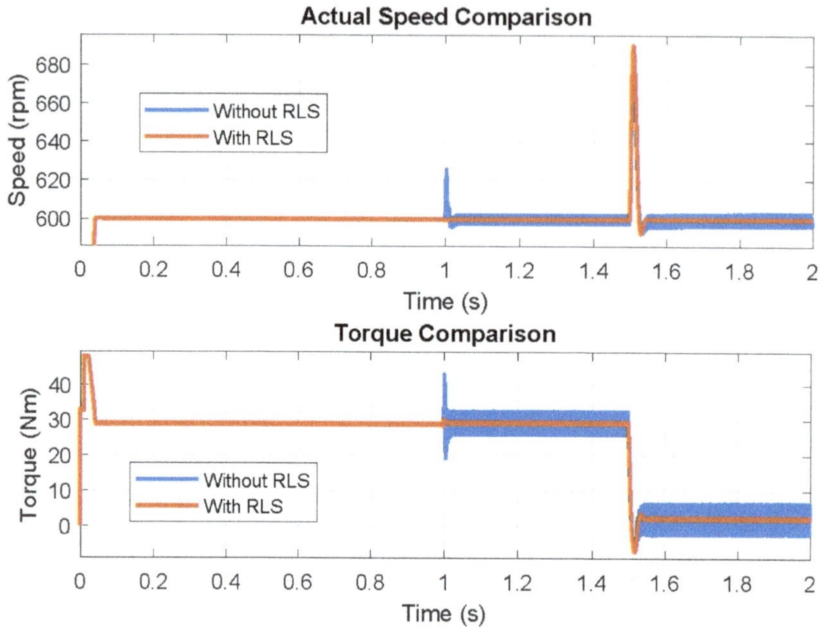

Figure 7. Speed and torque with and without RLS.

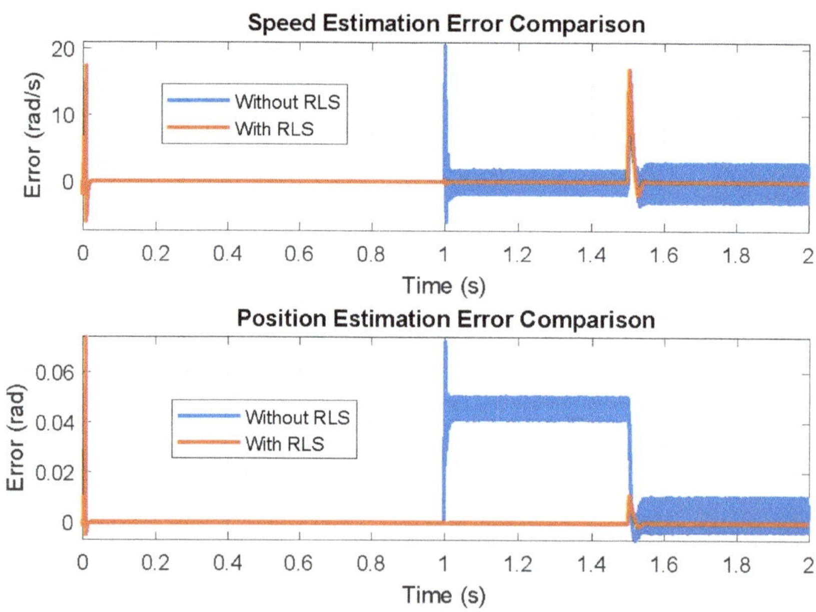

Figure 8. Speed and position errors with and without RLS.

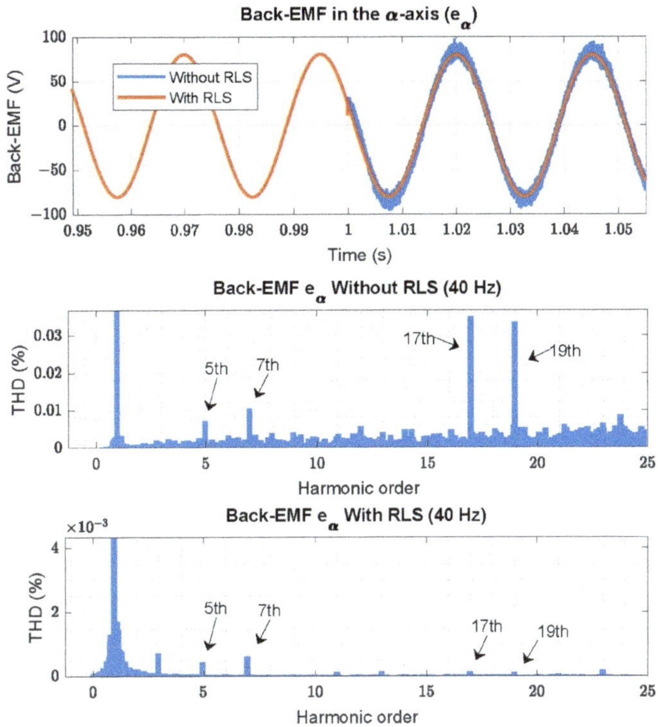

Figure 9. Back-EMF and THD with and without RLS.

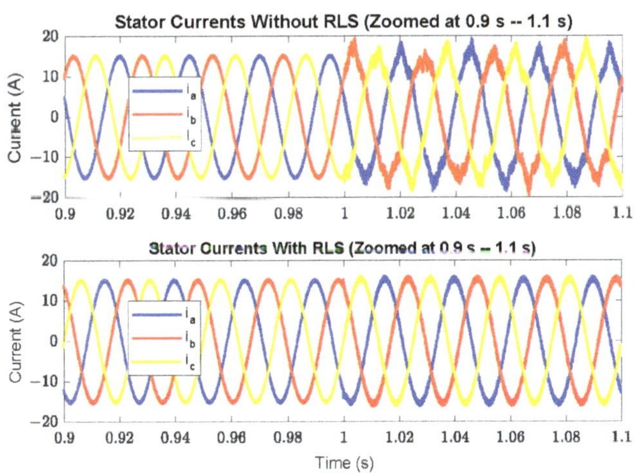

Figure 10. Stator current waveforms under parameter uncertainties with and without RLS.

Table 1. THD of back-EMF and stator current under parameter uncertainties.

THD Metric	Without RLS	With RLS
THD of Current (Phase a)	8.36%	2.77%
THD of Back-EMF (α-axis)	5.37%	0.03%

Figure 11 presents the estimation errors for RLS, speed, and position under both fixed and variable forgetting factor (λ) values when parameter uncertainties occur at 1 s. The results demonstrate that a fixed λ struggles to balance adaptation speed and noise suppression, leading to increased fluctuations in estimation errors. In contrast, the adaptive λ effectively reduces estimation errors, ensuring better tracking performance. Figure 12 illustrates the dynamic behavior of the adaptive λ, which varies between 0.05 and 1, allowing the algorithm to adjust in real time. This flexibility enhances the robustness of the estimation process by providing fast adaptation during transients while maintaining stability in steady state.

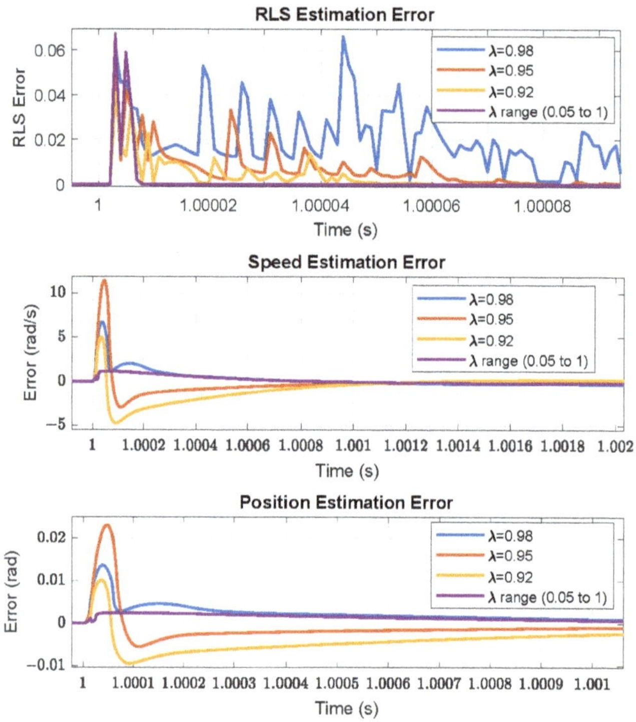

Figure 11. Different forgetting factor (λ) values.

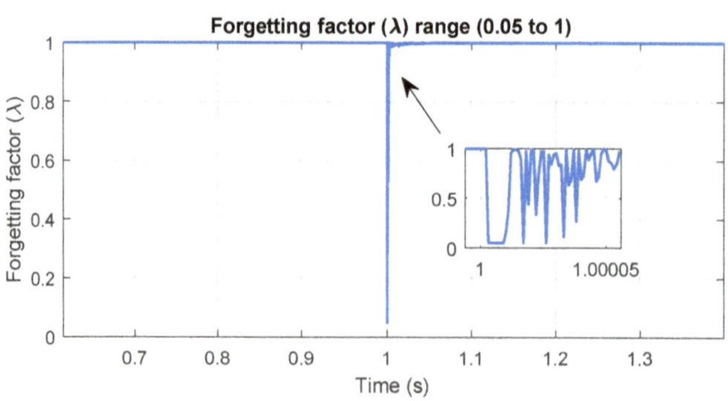

Figure 12. Forgetting factor (λ) range (0.05 to 1).

8. Conclusions

This paper presents a robust sensorless control strategy for PMSM based on a back-EMF extended state observer (ESO), which effectively eliminates DC position error compared to conventional ESO. With the addition of another ESO in the quadrature phase-locked loop (QPLL), not only can speed and position be accurately estimated, but it also enables the identification of unknown disturbances in the speed loop, allowing the sliding mode controller (SMC) to achieve faster convergence and improved disturbance rejection. Additionally, the proposed approach integrates a recursive least squares (RLSs) algorithm to enhance robustness against parameter uncertainties. A key improvement is the use of a variable forgetting factor instead of a fixed value, addressing the trade-off between adaptation speed and noise suppression. The results demonstrate that a fixed λ struggles to maintain this balance, leading to increased fluctuations in estimation errors, whereas the adaptive λ effectively reduces estimation errors, ensuring better tracking performance by enabling fast adaptation during transients while maintaining stability in the steady state. This flexibility significantly enhances the robustness of the estimation process.

Author Contributions: Methodology, A.D., H.M., M.F.B. and S.Z.; writing—original draft, A.Z., A.D. and H.M. All authors have read and agreed to the published version of the manuscript.

Funding: This research received no external funding.

Institutional Review Board Statement: Not applicable.

Informed Consent Statement: Not applicable.

Data Availability Statement: Data are contained within the article.

Conflicts of Interest: The authors declare no conflicts of interest.

Appendix A

Table A1. Motor parameters.

Parameter	Value
Rated Power	4400 W
Rated Current	16.5 A
Rated Torque	28.4 N·m
Number of Pole Pairs (n_p)	4
Stator Resistance (R_s)	0.25 Ω
Stator Inductance (L_s)	4.8 mH
Rotor Flux Linkage (ψ_f)	0.32 Wb
Moment of Inertia (J)	0.00774 Kg·m^2
Friction Coefficient (f)	0.0089 Nms

References

1. Tu, C.C.; Hung, C.L.; Hong, K.B.; Elangovan, S.; Yu, W.C.; Hsiao, Y.S.; Lin, W.C.; Kumar, R.; Huang, Z.H.; Hong, Y.H.; et al. Industry perspective on power electronics for electric vehicles. *Nat. Rev. Electr. Eng.* **2024**, *1*, 435–452. [CrossRef]
2. Chen, H.; Tang, J.; Liu, Y.; Jiang, B.; Boscaglia, L. Electromagnetic Performance Investigation of a Brushless Electrically Excited Synchronous Machine for Long-Distance Heavy-Duty Electric Vehicles. *IEEE Trans. Transp. Electrif.* **2025**, *11*, 225–235. [CrossRef]
3. Guo, Y.; Liu, L.; Ba, X.; Lu, H.; Lei, G.; Yin, W.; Zhu, J. Designing High-Power-Density Electric Motors for Electric Vehicles with Advanced Magnetic Materials. *World Electr. Veh. J.* **2023**, *14*, 114. [CrossRef]
4. Liu, T.; Zhao, Q.; Zhao, K.; Li, L.; Zhu, G. Sensorless model predictive control of permanent magnet synchronous motor based on hybrid parallel observer under parameter uncertainty. *IET Power Electron.* **2024**, *17*, 438–449. [CrossRef]

5. Akrami, M.; Jamshidpour, E.; Baghli, L.; Frick, V. Application of Low-Resolution Hall Position Sensor in Control and Position Estimation of PMSM—A Review. *Energies* **2024**, *17*, 4216. [CrossRef]
6. Huang, Y.; Zhao, M.; Zhang, J.; Lu, M. The Hall Sensors Fault-Tolerant for PMSM Based on Switching Sensorless Control with PI Parameters Optimization. *IEEE Access* **2022**, *10*, 114048–114059. [CrossRef]
7. Chi, W.C.; Cheng, M.Y. Implementation of a sliding-mode-based position sensorless drive for high-speed micro permanent-magnet synchronous motors. *ISA Trans.* **2014**, *53*, 444–453. [CrossRef]
8. Sahebjam, M.; Sharifian, M.B.; Feyzi, M.R.; Sabahi, M. Novel methodology for direct speed control of a permanent magnet synchronous motor with sensorless operation. *IET Electr. Power Appl.* **2021**, *15*, 728–741. [CrossRef]
9. Dong, S.; Zhou, M.; Wang, C.; Gou, L.; You, X. Improved Hybrid Sensorless Control for IPMSM in Full Speed Range at Low Switching Frequency. *IEEE Trans. Energy Convers.* **2024**, *40*, 123–135. [CrossRef]
10. Shi, L.; Lv, M.; Li, P. Sensorless Position Control in High-Speed Domain of PMSM Based on Improved Adaptive Sliding Mode Observer. *Processes* **2024**, *12*, 2581. [CrossRef]
11. R, S.; Singh, B. Sensorless Predictive Control of SPMSM-Driven Light EV Drive Using Modified Speed Adaptive Super Twisting Sliding Mode Observer with MAF-PLL. *IEEE J. Emerg. Sel. Top. Ind. Electron.* **2021**, *2*, 42–52. [CrossRef]
12. Zhou, Q.; Wang, Y.; Zhang, Y.; Zhang, L.; Shi, K. Sensorless control of permanent magnet synchronous motor based on improved sliding mode observer. *Trans. Inst. Meas. Control* **2025**, *47*, 194–205. [CrossRef]
13. Zeghlache, A.; Mekki, H.; Djerioui, A.; Benkhoris, M.F. Sensorless Control of BLDC Motor Based on ESO with an Active Harmonic Compensator. *J. Control Autom. Electr. Syst.* **2024**, *35*, 960–969. [CrossRef]
14. Wang, H.; Zhang, G.; Liu, X. Sensorless Control of Surface-Mount Permanent-Magnet Synchronous Motors Based on an Adaptive Super-Twisting Sliding Mode Observer. *Mathematics* **2024**, *12*, 2029. [CrossRef]
15. Tian, B.; Li, Y.; Hu, J.; Wang, G.; Molinas, M.; Zhang, G.; Wang, K.; Zhang, Z. A Wide Speed Range Sensorless Control for Three-Phase PMSMs Based on a High-Dynamic Back EMF Observer. *IEEE Trans. Transp. Electrif.* **2025**, *11*, 4336–4349. [CrossRef]
16. Harshit Mohan, M.K.P.; Dwivedi, S.K. Sensorless Control of Electric Drives—A Technological Review. *IETE Tech. Rev.* **2020**, *37*, 504–528. [CrossRef]
17. Jiang, N.; Cao, R.; Sun, W.; Chen, D.; Wang, K. MRAS-based Sensorless Control of PMSM Drives Using Extended State Observer in Shaftless Rim-driven Thruster System. *IEEE Trans. Transp. Electrif.* **2024**, 1. [CrossRef]
18. Zuo, Y.; Lai, C.; Iyer, K.L.V. A Review of Sliding Mode Observer Based Sensorless Control Methods for PMSM Drive. *IEEE Trans. Power Electron.* **2023**, *38*, 11352–11367. [CrossRef]
19. Jiang, F.; Sun, S.; Liu, A.; Xu, Y.; Li, Z.; Liu, X.; Yang, K. Robustness Improvement of Model-Based Sensorless SPMSM Drivers Based on an Adaptive Extended State Observer and an Enhanced Quadrature PLL. *IEEE Trans. Power Electron.* **2021**, *36*, 4802–4814. [CrossRef]
20. Chen, S.; Ding, W.; Hu, R.; Wu, X.; Shi, S. Sensorless Control of PMSM Drives Using Reduced Order Quasi Resonant-Based ESO and Newton–Raphson Method-Based PLL. *IEEE Trans. Power Electron.* **2023**, *38*, 229–244. [CrossRef]
21. Jiang, F.; Yang, F.; Sun, S.; Yang, K. Static-Errorless Rotor Position Estimation Method Based on Linear Extended State Observer for IPMSM Sensorless Drives. *Energies* **2022**, *15*, 1943. [CrossRef]
22. Sun, H.; Zhang, X.; Liu, X.; Su, H. Adaptive Robust Sensorless Control for PMSM Based on Improved Back EMF Observer and Extended State Observer. *IEEE Trans. Ind. Electron.* **2024**, *71*, 16635–16643. [CrossRef]
23. Sreejith, R.; Singh, B. Sensorless Predictive Current Control of PMSM EV Drive Using DSOGI-FLL Based Sliding Mode Observer. *IEEE Trans. Ind. Electron.* **2021**, *68*, 5537–5547. [CrossRef]
24. Li, X.; Cui, Y.; Wu, X. Sensorless Control of Surfaced-Mounted Permanent Magnet Synchronous Motor in a Wide-Speed Range. *Electronics* **2024**, *13*, 1131. [CrossRef]
25. Chen, S.; Ding, W.; Wu, X.; Hu, R.; Shi, S. Finite Position Set-Phase-Locked Loop with Low Computational Burden for Sensorless Control of PMSM Drives. *IEEE Trans. Ind. Electron.* **2023**, *70*, 9672–9676. [CrossRef]
26. Yu, Y.; Shao, Y.; Chai, F.; Cui, M. Static-Errorless Position Estimation for Sensorless PMSM Drives with Enhanced Robustness Against the Full-Frequency Domain Disturbance. *IEEE Trans. Power Electron.* **2022**, *37*, 5884–5897. [CrossRef]
27. Yu, K.; Li, S.; Zhu, W.; Wang, Z. Sensorless Control Scheme for PMSM Drive via Generalized Proportional Integral Observers and Kalman Filter. *IEEE Trans. Power Electron.* **2025**, *40*, 4020–4033. [CrossRef]
28. Moreno, J.A.; Osorio, M. A Lyapunov approach to second-order sliding mode controllers and observers. In Proceedings of the 2008 47th IEEE Conference on Decision and Control, Cancun, Mexico, 9–11 December 2008 ; pp. 2856–2861. [CrossRef]
29. Ichikawa, S.; Tomita, M.; Doki, S.; Okuma, S. Sensorless control of permanent-magnet synchronous motors using online parameter identification based on system identification theory. *IEEE Trans. Ind. Electron.* **2006**, *53*, 363–372. [CrossRef]

30. Yao, Z.; Lan, H.; He, X.; Deng, F.; Wang, C.; Lu, S.; Tang, Y. Nonlinear Inductor-Based Single Sensor Current Balancing Method for Interleaved DC–DC Converters. *IEEE Trans. Power Electron.* **2024**, *39*, 3996–4000. [CrossRef]
31. Wang, Z.; Lu, K.; Blaabjerg, F. A Simple Startup Strategy Based on Current Regulation for Back-EMF-Based Sensorless Control of PMSM. *IEEE Trans. Power Electron.* **2012**, *27*, 3817–3825. [CrossRef]

electronics

Article

Application of Voltage Optimization Strategy for Rotary Power Flow Controllers in Loop Closing of Distribution Networks

Wenqiang Xie *, Yubo Yuan, Xian Zheng, Hui Chen, Jian Liu and Chenyu Zhang

Power Science Research Institute of State Grid Jiangsu Electric Power Co., Ltd., Nanjing 210024, China
* Correspondence: xiewenqiang11@js.sgcc.com.cn

Abstract: To mitigate voltage limit issues in the operation of a novel electromagnetic voltage regulation device, this paper presents a flexible loop-closing control strategy with voltage optimization. The approach uses a two-stage path optimization: in the first stage, the voltage phase at the loop-closing point is adjusted to ensure smooth operation, while in the second stage, the voltage magnitude is optimized to prevent voltage limits and achieve seamless regulation. By integrating phase angle difference calculations with coordinated rotation angle control, the simulation results show that this strategy reduces loop-closing current by approximately 95.87% compared to direct loop closing, decreases voltage fluctuations by around 50.0% compared to traditional methods, and shortens operation time by 40.14%. This approach significantly enhances system stability and response speed, effectively addressing the issue of excessive loop-closing current caused by voltage deviations at distribution network tie switches.

Keywords: flexible loop-closing control; rotary power flow controller; two-stage path optimization; voltage optimization

1. Introduction

To enhance power supply reliability in distribution networks, open-loop operation is typically employed within a closed-loop design [1,2]. During equipment maintenance, loop-closing transfer operations are necessary to enable uninterrupted load transfer. Prior to loop-closing, a voltage difference exists at the loop-closing point, which diminishes to zero upon completing the operation. Due to the presence of various inertia elements in the system, the loop-closing process may cause oscillations that gradually decay until the system reaches a new steady state, simultaneously generating a current through the loop-closing switch, commonly referred to as loop current [3,4].

The voltage requirements at the loop-closing point are as follows: the phase sequence at both ends of the loop must be consistent; the maximum voltage magnitude difference between the two ends should not exceed 20%; and the phase difference between the voltages at the loop-closing point should not exceed 20°. However, with the increasing integration of distributed energy sources, the voltage magnitude and phase difference across the tie switch grow, resulting in higher loop currents. If the loop current surpasses the relay protection device's threshold, the loop-closing process may fail, jeopardizing grid safety and power supply reliability [5]. The loop current's magnitude and characteristics are crucial for assessing system stability and response [6]. Thus, proactive adjustments to the voltage magnitude and phase difference at the loop-closing point are essential to mitigate the loop current's impact on the system.

https://doi.org/10.3390/electronics14030630

One method to suppress loop current involves leveraging intelligent algorithms [7–9], which utilize existing grid equipment to enhance the radial distribution network's configuration, thereby improving loop-closing conditions and facilitating system recovery after faults. Reference [10] introduces a decentralized approach using a multi-agent system, factoring in customer priorities and distributed generators, while addressing operational constraints in both intact and restored feeder sections to achieve loop restoration. Meanwhile, Reference [11] proposes a multi-objective evolutionary algorithm for service restoration, emphasizing the use of remote-controlled switches and effectively managing distribution systems without simplifying the network topology. However, the limitation of these methods lies in the fact that when dealing with large-scale systems, the computational complexity increases significantly, and the feasibility and implementation difficulty of the algorithms also rise.

Another method involves using phase shifters and other adjustment devices to precisely regulate voltage magnitude and phase at the loop-closing point. For instance, in cases where a 30° phase difference caused by star and delta connections hinders loop-closing, Reference [12] proposes a flexible control device utilizing a thyristor-controlled phase-shifting transformer to manage significant voltage and phase discrepancies. Similarly, Reference [13] examines the operation of phase-shifting transformer devices, enhances their design to optimize voltage quality, and suggests a strategy for smooth regulation in scenarios with substantial phase differences. Although these methods can effectively adjust voltage and phase differences, they still face the challenge of not achieving smooth regulation. Furthermore, these methods inherently rely on power electronic devices, which results in relatively higher costs and increased maintenance complexity.

For multi-factor adjustments, power flow control capabilities are essential. Reference [14] proposes a loop-closing device that meets the demand for non-interruptible load transfer under extreme conditions and proposes adjustment schemes for both AC and DC sides, effectively reducing the impact of loop current. Reference [15] studies the application of static var generators in voltage regulation and non-interruptible loop-closing operations and proposes strategies for non-interruptible load transfer under voltage disturbances.

Although these control devices can effectively suppress loop current, power electronic devices are expensive and difficult to maintain, and some devices can only handle single loop-closing scenarios, which have certain limitations. The Rotary Power Flow Controller (RPFC), as a new type of electromagnetic flexible interconnection device, can continuously control the voltage magnitude and phase of the line and has broad application potential [16,17]. The RPFC consists of two rotary phase shifting transformers (RPSTs), with their primary windings connected in parallel to the power supply, and the secondary windings connected in series with the line. Driven by servo motors, the stator windings of the two RPSTs rotate by a certain angle, generating a series voltage with continuously adjustable amplitude and phase. Regarding the basic characteristics and modeling of RPFC, Reference [18] analyzes the power control characteristics of the RPFC, highlighting its advantages of low cost, high durability, and reliability, making it a promising option for applications. References [19,20] systematically studies the modeling and application of RPFC, focusing on its physical structure and mathematical models, and verifying its potential to enhance power system security and stability. Regarding the planning and optimization applications of RPFC in distribution networks, Reference [21] proposes a planning method combining RPFC and energy routers to address energy imbalance and voltage distribution challenges, validating its effectiveness through a two-layer optimization model. Reference [22] enhances RPFC's PQ decoupling control by introducing variable speed and fuzzy PI adjustments, resolving oscillation and overshoot issues. Regarding the applications of RPFC in reactive power compensation and voltage regulation, Reference [23] develops a steady-state mathematical model and dual closed-loop control strategy for RPFC, enabling bidirectional reactive power regulation and validating its dynamic performance. Reference [24] proposes an RPFC-based co-phase power supply compensation method for electrified railways, using a dual closed-loop control strategy and verifying its effectiveness through simulation. Reference [25] introduces an RPFC-based rolling optimization strategy for bidirectional voltage regulation and loss reduction, improving system stability and reliability. Regarding the applications of RPFC in loop-closing control, Reference [26] proposes a virtual power-based RPFC loop-closing control method to regulate voltage seamlessly at the loop-closing point, preventing system misjudgment and ensuring safe and stable operation. Reference [27] constructs a steady-state power decoupling control model for RPFC based on instantaneous reactive power theory. It addresses the coordination of rotary phase transformers with a two-stage speed control scheme, demonstrating effective performance for flexible loop-closing operations.

This study introduces a flexible loop-closing approach using a two-stage optimal path tailored for RPFC applications in distribution networks. The key contributions of this work are summarized as follows:

- A two-stage optimal path-based RPFC loop-closing control strategy is proposed, focusing on adjusting voltage phase and magnitude at the loop-closing point to prevent the voltage fluctuations commonly seen in traditional methods.
- A phase angle difference calculation and rotation angle coordination control method are proposed, dynamically adjusting the RPFC rotation angle setpoint, significantly improving the stability and response speed of the loop-closing operation.
- Comparative simulations confirm the practicality and reliability of the proposed method, demonstrating the considerable potential of RPFC in enabling flexible loop-closing operations within distribution networks.

The paper is structured as follows: Section 2 explains the fundamental working principle of RPFC. Section 3 outlines the two-stage optimal path-based RPFC loop-closing control strategy. Section 4 provides simulation results and analysis, while Section 5 summarizes the conclusions.

2. Basic Principle of RPFC

To prevent excessive loop-closing currents during the process, it is generally necessary to ensure that the voltage difference at the loop-closing point remains within a reasonable range. Figure 1 demonstrates a typical loop-closing scenario in a distribution network. The green switches in the figure indicate that the loop-closing current is small, and the loop can be directly closed. The red switches indicate that the loop-closing current is large, and the loop cannot be directly closed. The arrows represent the load or the next-level network. As shown, switches B1 and B3 are positioned at either end of the same feeder, resulting in a relatively small voltage difference at the loop-closing point. On the other hand, switches B2 and B4 connect feeders from different substations, leading to a more pronounced voltage difference. Additionally, switch B5 connects feeders originating from transformers with distinct connection groups, introducing a 30° phase shift and creating a larger voltage difference at the loop-closing point.

This paper employs RPFC to minimize the voltage difference at the loop-closing point, effectively reducing the loop-closing current. The RPFC is composed of two RPSTs, as illustrated in its main circuit in Figure 2. In the figure, the red and blue arrows represent the stator voltages of the two RPSTs, the black dashed arrows represent their corresponding set values, and the green arrow represents their combined voltage, which is the RPFC output voltage. In the diagram, $\boldsymbol{U}_1$ represents the sending-end voltage of the line where the RPFC is installed; $\boldsymbol{U}_2$ and $\boldsymbol{U}_3$ denote the voltages at the loop-closing point; $\Delta\boldsymbol{U}$ is the RPFC output voltage; $\boldsymbol{U}_{st1}$ and $\boldsymbol{U}_{st2}$ are the stator voltages of RPST1 and RPST2, respectively; k is the turn ratio of the stator and rotor windings of the RPST; a_1 and a_2 are the rotation angles of the two RPSTs; and QF_b is the tie switch; Z_{eq} is the equivalent impedance; j indicates the phase of the output voltage, and d refers to the angle between the output voltage and the stator voltage. The sending-end voltage $\boldsymbol{U}_1$ excites the RPFC to generate constant amplitude stator voltages $\boldsymbol{U}_{st1}$ and $\boldsymbol{U}_{st2}$, which are combined to produce an adjustable output voltage with flexible amplitude and phase control, thus enabling precise regulation of the left-end voltage at the loop-closing point.

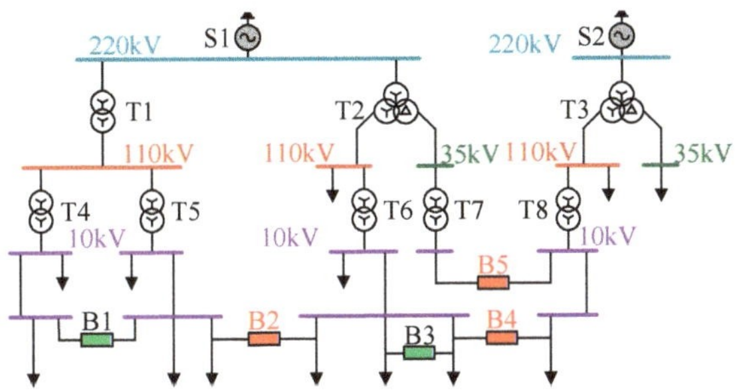

Figure 1. Schematic diagram of a typical distribution network.

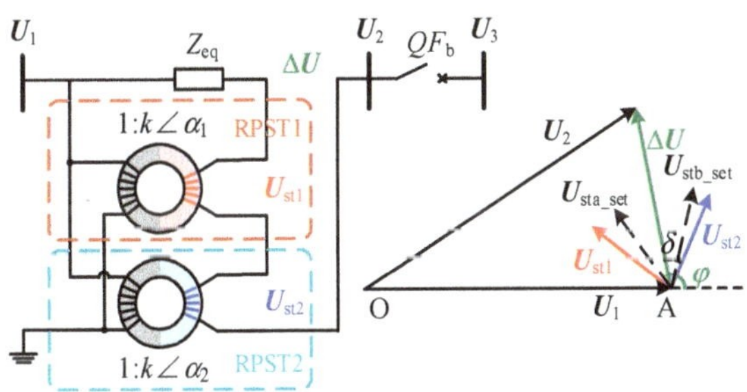

Figure 2. Schematic diagram of RPFC main circuit.

In addition, the power limitation of the RPFC is closely related to its capacity, similar to the working principle of a transformer. Due to the star and delta connection on the primary and secondary sides, the harmonic components in the RPFC output voltage are very small, resulting in low total harmonic distortion (THD). This gives the RPFC a significant advantage in power quality, effectively reducing harmonic pollution and ensuring stable system operation. Therefore, the power limitation of the RPFC is primarily determined by its design capacity, while providing efficient voltage regulation and low harmonic distortion.

The RPFC is modeled as a controlled voltage source D$\boldsymbol{U}$ in series with an equivalent impedance Z_{eq}. Under standard operating conditions, the stator voltages of the RPFC have equal magnitudes, denoted as $U_{st1} = U_{st2} = U_{st}$. Therefore, δ, between the stator voltage and the output voltage is given by:

$$\delta = \arccos\left(\frac{\Delta U}{2U_{st}}\right) \tag{1}$$

The RPFC calculates the output voltage amplitude and phase setpoints, ΔU_{set} and φ_{set}, based on the phasor relationship between $\boldsymbol{U}_1$ and $\boldsymbol{U}_3$. These output voltage setpoints are then converted into the corresponding rotation angle setpoints α_{a_set} and α_{b_set} for the two rotary phase shifters:

$$\begin{cases} \alpha_{a\_set} = \varphi_{set} + \delta_{set} \\ \alpha_{b\_set} = \varphi_{set} - \delta_{set} \end{cases} \tag{2}$$

3. RPFC Loop-Closing Control Strategy

3.1. Two-Stage Operation Path for Output Voltage

The operational path planning for the RPFC output voltage is illustrated in Figure 3, where b and g denote the voltage phases of $\boldsymbol{U}_2$ and $\boldsymbol{U}_3$, respectively. The planning process is divided into two stages. The primary goal of the first stage is to align the voltage magnitude and phase at the loop-closing point, ensuring $\beta = \gamma$. The constraint in this stage is to ensure that the magnitudes of $\boldsymbol{U}_2$ and $\boldsymbol{U}_1$ remain equal throughout the process, i.e., $U_2 = U_1$. Therefore, in the first stage, the output voltage is required to follow the arc path ADB, with the output voltage magnitude and phase setpoints as follows:

$$\begin{cases} \Delta U_{set1} = 2U_1 \sin\frac{\beta_{set}}{2} \\ \varphi_{set1} = 90 + \frac{\beta_{set}}{2} \end{cases} \tag{3}$$

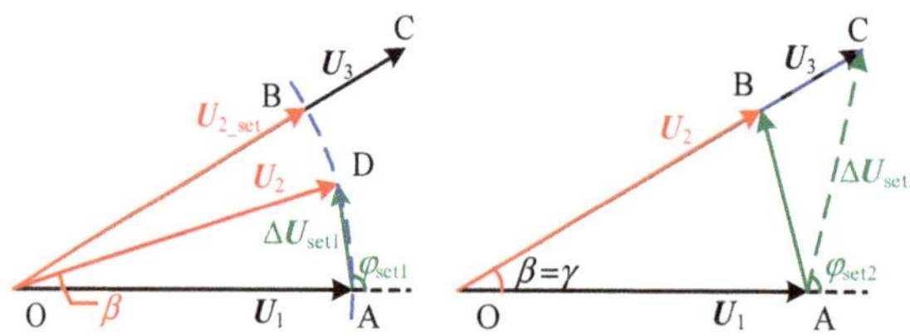

(**a**) **First stage** (**b**) **Second stage**

Figure 3. Phase diagram of two-stage operation path planning for RPFC output voltage.

The setpoint β_{set} for the RPFC, depending on different operational periods, can be divided into the following three processes:

$$\begin{cases} \beta_{set} = 0 & t < t_1 \\ \beta_{set} = k_r(t - t_1) & t_1 \le t < t_2 \\ \beta_{set} = \gamma & t \le t_2 \end{cases} \tag{4}$$

In this equation, t_1 and t_2 denote the initial moments of the first and second stages; k_r represents the rate of change in the phase setpoint β_{set}. When $t < t_1$, the RPFC is in its initial state, and the two stator voltages differ by 180°, resulting in zero output voltage, and the node voltage $\boldsymbol{U}_2 = \boldsymbol{U}_1$. When $\beta = \beta_{set} = \gamma$, the first stage of the RPFC operation concludes, and it transitions to the second stage.

The second stage aims to equalize the voltage magnitudes at both ends of the loop-closing point, ensuring $U_2 = U_3$. The constraint during this stage is to maintain consistent phase alignment, i.e., $\beta = \gamma$. Therefore, in the second stage, the RPFC output voltage is required to follow the path of BC, with the output voltage magnitude and phase setpoints expressed as follows:

$$\begin{cases} \Delta U_{set2} = \text{abs}(\boldsymbol{U}_3 - \boldsymbol{U}_1) \\ \varphi_{set2} = \arg(\boldsymbol{U}_3 - \boldsymbol{U}_1) \end{cases} \tag{5}$$

The phase setpoint remains unchanged, i.e., $\beta = \beta_{set} = \gamma$, while the magnitude setpoint is $U_{2\_set} = U_3$. When the voltage magnitudes satisfy $U_2 = U_{2\_set} = U_3$, the second stage of RPFC operation concludes, ensuring zero voltage difference at the loop-closing point.

3.2. Rotation Angle Coordination Control Method

Figure 4 illustrates the RPFC rotation angle coordination control process. In the figure, θ_{1a} denotes the initial angle between the stator voltage $\boldsymbol{U}_{st10}$ and $\boldsymbol{U}_{sta\_set}$, while θ_{2b} represents the initial angle between $\boldsymbol{U}_{st20}$ and $\boldsymbol{U}_{stb\_set}$. The control system compares the initial positions of the two stator voltages with their respective setpoints and selects the closest stator voltage setpoint as the target value. This approach helps to reduce the operation time and prevent voltage limit violations during the control process.

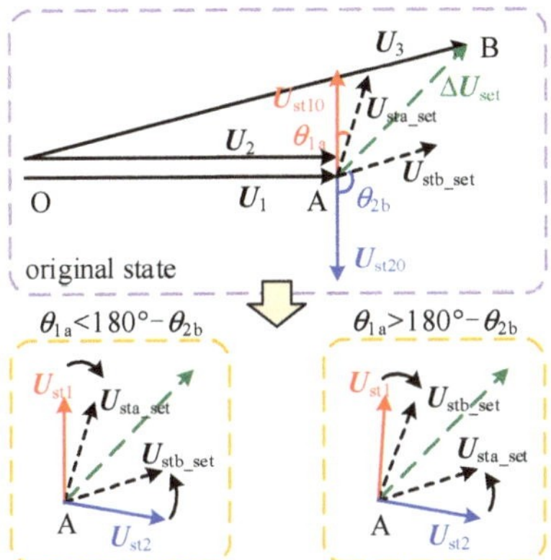

Figure 4. Schematic diagram of rotational angle coordinated control process phases.

If the initial state satisfies $\theta_{1a} < 180°\text{-}\theta_{2b}$, $\boldsymbol{U}_{sta\_set}$ and $\boldsymbol{U}_{stb\_set}$ are chosen as the stator voltage setpoints for RPST1 and RPST2, respectively. Accordingly, the rotation angle setpoints α_{1_set} and α_{2_set} are determined as follows:

$$\begin{cases} \alpha_{1\_set} = \alpha_{a\_set} = \varphi_{set} + \delta_{set} \\ \alpha_{2\_set} = \alpha_{b\_set} = \varphi_{set} - \delta_{set} \end{cases} \tag{6}$$

Similarly, if $\theta_{1a} > 180°\text{-}\theta_{2b}$, $\boldsymbol{U}_{stb\_set}$ is assigned as the stator voltage setpoint for RPST1, while $\boldsymbol{U}_{sta\_set}$ is assigned for RPST2. As a result, the rotation angle setpoints α_{1_set} and α_{2_set} are given by:

$$\begin{cases} \alpha_{1\_set} = \alpha_{b\_set} = \varphi_{set} - \delta_{set} \\ \alpha_{2\_set} = \alpha_{a\_set} = \varphi_{set} + \delta_{set} \end{cases} \tag{7}$$

Depending on the operating conditions, select either Equation (6) or Equation (7) as the setpoint for the RPFC rotation angles, and adjust α_1 and α_2 accordingly. This ensures continuous control of the output voltage's magnitude and phase until the voltages at the loop-closing point satisfy the requirements of the distribution network.

In summary, the RPFC flexible loop-closing control strategy involves the following four steps, as illustrated in Figure 5.

- First-stage path planning. The RPFC aims to align the voltage phases at the loop-closing point, ensuring equal voltage magnitudes throughout the process.
- Second-stage path planning: The RPFC focuses on equalizing voltage magnitudes at the loop-closing point, maintaining consistent phases during the operation.

- Phase angle difference calculation: Determine the set angle δ_{set} between the stator and output voltages using Equation (1).
- Rotation angle coordination control: Compare the initial positions of the two stator voltages with their setpoints, and select the closest stator voltage setpoint as the target value.

By meticulously planning the RPFC output voltage's operating path, this strategy successfully eliminates the voltage difference at the loop-closing point, ensuring the distribution network operates reliably.

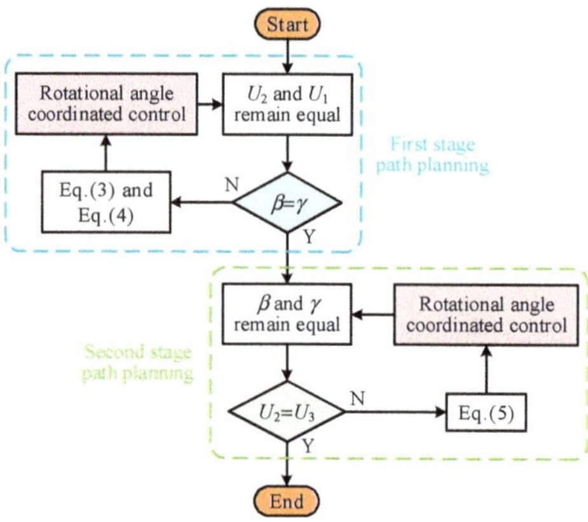

Figure 5. Block diagram of RPFC's combined loop strategy.

4. RPFC Flexible Loop-Closing Simulation

4.1. Loop-Closing Simulation Scenario

Taking a specific loop-closing network as an example, its network topology and load distribution are shown in Figure 6. The loop network consists of multiple feeders, each with an impedance of (0.11 + j0.15) Ω/km, with the lengths of the feeders marked in the figure. The load power at each node is expressed in MVA, representing the power demand at different nodes. The electromotive forces are denoted as E_1 and E_2, where E_1 is 10 kV at a 10° phase, and E_2 is 10 kV at a 40° phase. The specific parameters of the RPFC are listed in Table 1.

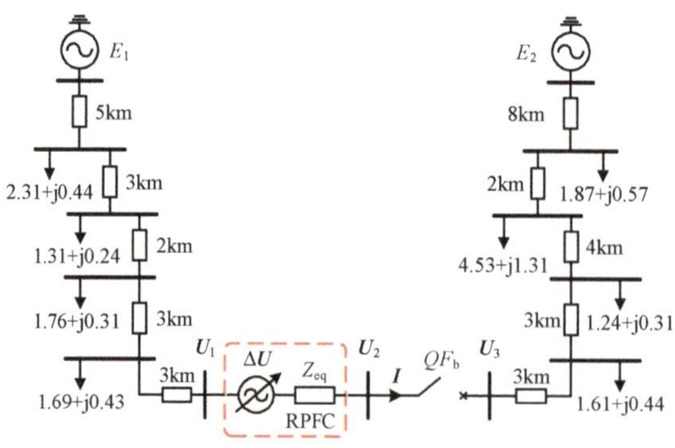

Figure 6. RPFC flexible combined loop operation scenario.

Table 1. Simulation parameters of RPFC.

	Capacity /MVA	Voltage Level/kV		Turn Ratio	Rated Speed °/s
		Primary	Secondary		
RPST1	2.0	10	2.9	3.44	6
RPST2	2.0	10	2.9	3.44	6
RPFC	4.0	10	5.8	1.72	6

4.2. Loop-Closing Performance Analysis

Figures 7 and 8 illustrate the voltage magnitude and phase angle waveforms at the loop-closing point under different control strategies, while Figure 9 depicts the loop-closing current waveforms. Figures 10 and 11 present the output voltage waveforms under various control strategies. These figures clearly demonstrate the changes in voltage, phase, and current under different strategies, further reflecting the effectiveness of each strategy. In the figures, D*q* represents the phase angle difference between the voltages at both ends of the loop-closing point. The subscripts "dr," "tr," and "op" represent direct loop-closing, the control method proposed in the literature [26], and the new control strategy proposed in this paper, respectively. The corresponding voltage and current data are summarized in Table 2, providing key performance indicators for different methods. A detailed comparative analysis of these three loop-closing methods is provided below, highlighting their respective strengths, weaknesses, and applicable scenarios.

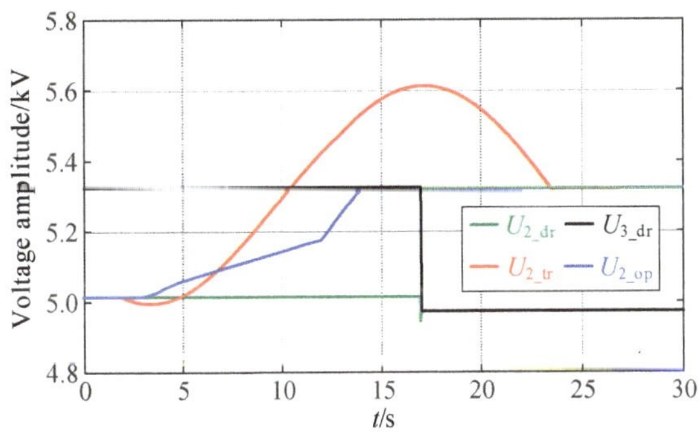

Figure 7. Voltage amplitude waveforms at the closing loop point under different control strategies.

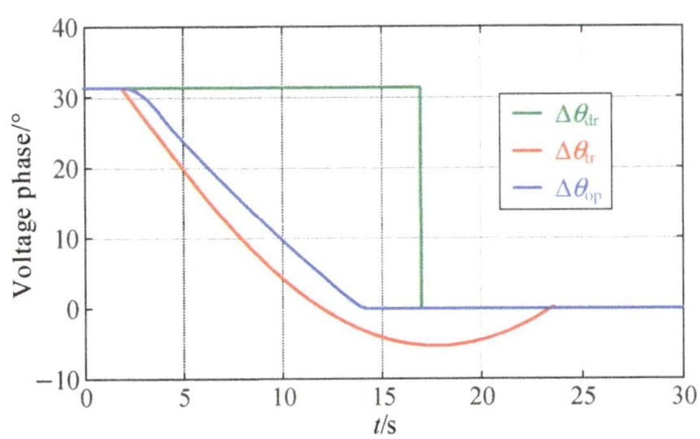

Figure 8. Voltage phase waveforms at the closing loop point under different control strategies.

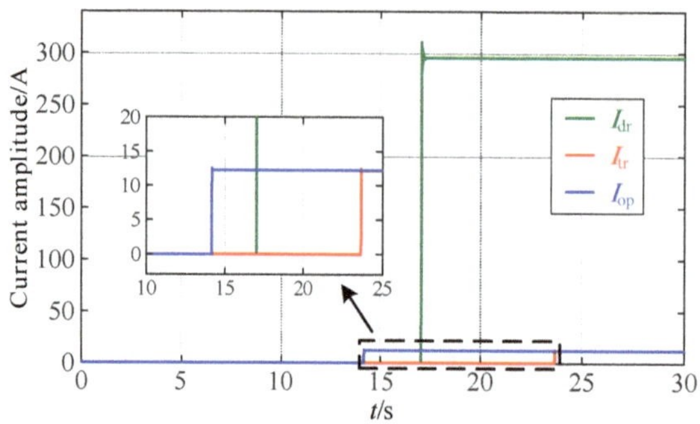

Figure 9. Closed-loop current variations across different control strategies.

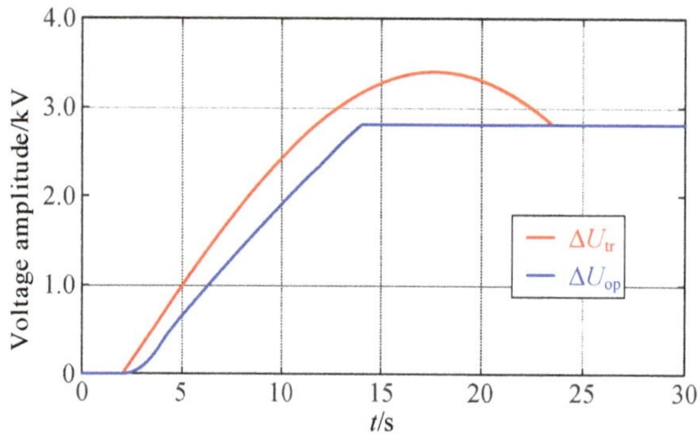

Figure 10. Output voltage amplitude variations across different control strategies.

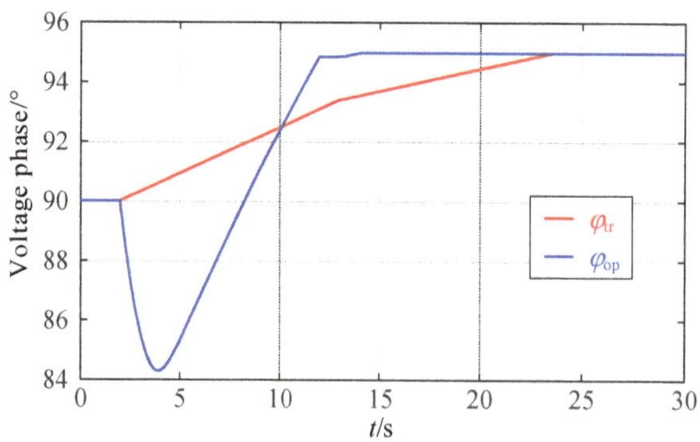

Figure 11. Output voltage phase waveforms under different control strategies.

Table 2. Voltage and current metrics across various control strategies.

Parameters	Direct Loop-Closing	Method in Reference [26]	Proposed Method
U_2/kV	5.01	5.31	5.32
U_3/kV	5.32	5.32	5.32
Dq/°	31.35	0.11	0.10
I/A	295.36	12.20	12.18

For the direct loop-closing method, the voltage magnitudes at the loop-closing point are U_{2_dr} = 5.01 kV and U_{3_dr} = 5.32 kV, with a phase difference of $\Delta\theta_{dr}$ = 31.35°. At t = 17.0 s, QF_b is directly closed, resulting in the voltages on both sides dropping to 4.97 kV and the loop-closing current I_{tr} reaching 295.36 A. This current value far exceeds the safety threshold, failing to meet the loop-closing conditions and causing system instability. The main issue with the direct loop-closing method is that it does not account for the interaction between voltage and current, leading to excessive voltage fluctuations and high loop-closing currents. This not only violates the stability requirements of the power system but also risks triggering relay protection malfunctions, further compromising system safety. Therefore, although the direct loop-closing method is simple to operate, its lack of coordinated voltage and current control severely limits its practical applicability, as it is prone to voltage violations.

When operating with the loop-closing control strategy from the literature [26], RPFC starts operating at t = 2 s. The voltage magnitude at the end U_{3_tr} reaches 5.32 kV, and the loop-closing condition is met at t = 23.64 s, at which point QF_b is closed, resulting in a loop-closing current I_{tr} of 12.20 A. The voltage U_{2_tr} increases from 5.01 kV to 5.31 kV, with a peak of 5.61 kV and a trough of 4.99 kV during operation. The phase difference $\Delta\theta_{tr}$ decreases from 31.35° to 0.11°, with a trough of −5.28° during the process. Although this method reduces the loop-closing current by approximately 95.87% compared to direct loop-closing, it experiences significant voltage fluctuations, which impact system stability.

With the loop-closing control strategy proposed in the literature [26], the RPFC begins operating at t = 2 s. The voltage magnitude U_{3_tr} gradually increases to 5.32 kV, and the loop-closing condition is met at t = 23.64 s, at which point QF_b is closed, resulting in a loop-closing current of I_{tr} = 12.20 A. During the process, the voltage U_{2_tr} increases from 5.01 kV to 5.31 kV, with a peak of 5.61 kV and a trough of 4.99 kV. The phase difference $\Delta\theta_{tr}$ decreases from 31.35° to 0.11°, with a minimum value of −5.28° during the process. Compared to the direct loop-closing method, this strategy significantly reduces the loop-closing current by approximately 95.87%. However, despite the significant reduction in current, the method in the literature [26] exhibits substantial voltage fluctuations during the loop-closing process, which negatively impact system stability. These fluctuations not only affect the normal operation of equipment but can also disrupt subsequent voltage regulation and power distribution. As such, although this method performs well in reducing current, its limitations in voltage control restrict its applicability.

Under the RPFC loop-closing control strategy proposed in this paper, the voltage magnitude U_{3_op} stabilizes at 5.32 kV, and the loop-closing condition is met at t = 14.15 s. After QF_b is closed, the loop-closing current I_{op} is 12.18 A. The voltage U_{2_op} increases from 5.01 kV to 5.32 kV, with minimal fluctuations, and the phase difference $\Delta\theta_{op}$ decreases from 31.35° to 0.10°. Compared to the method in the literature [26], the proposed strategy not only achieves a similar reduction in loop-closing current but also reduces voltage fluctuations by approximately 50.0% and shortens the operation time by 40.14%. This method not only improves operational efficiency but also ensures zero voltage difference at the loop-closing point, guaranteeing the stability and reliability of the distribution network.

Figures 10 and 11 further illustrate the differences in voltage and phase waveforms under the two methods. Notably, both methods share the same initial and final steady-state values for voltage and phase, but their adjustment processes differ significantly. In Figure 10, the traditional method exhibits significant voltage fluctuations, with the voltage rising from 0 V to a peak of 3400 V, then falling to a steady value of 2809.34 V. This indicates poor voltage control, as the method fails to achieve a smooth transition. In contrast, the proposed strategy uses an optimized control approach that allows the output voltage to directly stabilize at the target value without overshooting, ensuring fast and stable voltage regulation.

Similarly, in Figure 11, the phase adjustment process of the traditional method is slower due to a lack of optimized control over the phase angle. The method does not prioritize the nearest path for adjustment, resulting in inefficiencies. In contrast, the proposed strategy adjusts the phase more efficiently, even though it involves a temporary dip followed by a rise. Specifically, the phase decreases from 90° to 84.3° and then increases to 94.98°. During this process, the small output voltage amplitude ensures negligible impact on the terminal voltage, maintaining overall stability. Overall, the proposed strategy significantly accelerates the phase adjustment process while ensuring voltage stability, demonstrating its effectiveness in enhancing power system stability and reliability.

The key difference lies in the fact that the method in the literature [26] does not take voltage stability into consideration during the loop-closing process. As a result, it is highly susceptible to voltage violations, as evidenced by the significant voltage swings and the failure to maintain voltage within the acceptable limits. In contrast, the proposed RPFC method not only reduces the loop-closing current but also ensures that voltage remains within the required limits, thereby maintaining system stability. By effectively controlling both current and voltage, the RPFC strategy ensures a reliable and stable operation of the distribution network, making it a more robust solution compared to the method in the literature [26].

5. Conclusions

This paper presents a flexible loop-closing control strategy for the RPFC, utilizing a two-stage optimal path planning approach to mitigate potential voltage limit issues during active loop-closing operations. The main conclusions are as follows:

- The proposed strategy effectively minimizes voltage fluctuations and current surges during loop-closing by optimizing the transitions of voltage magnitude and phase. This approach helps to prevent voltage limit violations, ensuring stable operation during the loop-closing process and avoiding system instability caused by voltage fluctuations.
- As an independent device, the RPFC can perform loop-closing control without relying on other system equipment (such as auxiliary voltage sources or distributed energy sources) for coordination. This characteristic makes RPFC more flexible and convenient for practical applications, reducing dependency on other devices and enhancing system operational efficiency and stability.
- Compared to traditional methods, the proposed control strategy significantly shortens the loop-closing time. Through rotational angle coordination control, the adjustment path of RPFC output voltage is optimized, effectively improving the response speed of loop-closing operations and ensuring quick restoration of the power grid.
- As an electromagnetic device, the RPFC offers excellent tolerance and low maintenance costs, but its operational efficiency may be slightly slower compared to power electronic devices. This limitation makes RPFC more suitable for long-term stable operation in scenarios where extremely fast response times are not critical, though it remains a reliable solution for many applications requiring flexible loop-closing control.
- Future research could focus on strengthening the application of RPFC after loop-closing, especially in its role in power flow distribution and scheduling within the entire power system. Further research could also explore how to integrate RPFC with energy storage systems, distributed energy sources, and other power system technolo-

gies to optimize configuration and collaborative scheduling, thereby enhancing the flexibility and stability of the grid.

Author Contributions: Conceptualization, W.X.; Methodology, Y.Y., X.Z., H.C., J.L. and C.Z.; Validation, W.X., Y.Y. and J.L.; Investigation, X.Z.; Writing—Original Draft, W.X. All authors have read and agreed to the published version of the manuscript.

Funding: This work was conducted by the State Grid Jiangsu Electric Power Company Science and Technology Project (J2023122).

Data Availability Statement: The original contributions presented in this study are included in the article. Further inquiries can be directed to the corresponding author.

Conflicts of Interest: All authors were employed by the Power Science Research Institute of State Grid Jiangsu Electric Power Co., Ltd. The remaining authors declare that the research was conducted in the absence of any commercial or financial relationships that could be construed as a potential conflict of interest.

Nomenclature

Terms	optimal value
RPFC	Rotary power flow controller
B1-B5	Breaker1-breaker5
$\boldsymbol{U}_1$	Delivery voltage
$\boldsymbol{U}_2$	Left-side loop voltage
$\boldsymbol{U}_3$	Right-side loop voltage
$\Delta\boldsymbol{U}$	RPFC output voltage
$\boldsymbol{U}_{st1}, \boldsymbol{U}_{st2}$	Stator voltage of RPFC
$\boldsymbol{U}_{st10}, \boldsymbol{U}_{st20}$	Initial stator voltage
$\boldsymbol{U}_{sta\ set}, \boldsymbol{U}_{stb\ set}$	Setpoints of the stator voltage
α_1, α_2	Rotation angles of RPST
QF_b	Loop closing switch
Z_{eq}	Equivalent impedance of RPFC
φ	Output voltage phase of RPFC
δ	Angle between output voltage and stator voltage
β	Phase at the left end of the loop closing point
γ	Phase at the right end of the loop closing point
$\Delta\boldsymbol{U}_{set}$	Output voltage magnitude setpoint
φ_{set}	Output voltage phase setpoint
β_{set}	Phase setpoint at the left end of the point
δ_{set}	Angle setpoint of d
$\alpha_{a\_set}, \alpha_{b\_set}$	Rotation angle setpoint
t_1, t_2	Time of the first and second stages
k_r	Variation rate of b_{set}
θ_{1a}	Angle with $\boldsymbol{U}_{st10}$ and $\boldsymbol{U}_{sta\_set}$
θ_{2b}	Angle with $\boldsymbol{U}_{st20}$ and $\boldsymbol{U}_{stb\_set}$
α_{1_set}	Rotation angle a_1 setpoint
α_{2_set}	Rotation angle a_2 setpoint
E_1, E_2	Line electromotive force
Z_{L1}, Z_{L2}	Line impedance
I	Loop-closing current
Dq	Phase difference at both ends of the loop-closing point
dr	Direct loop-closing method
tr	Method from Reference [22]
op	Proposed method from this paper

References

1. Ruan, H.; Gao, H.; Liu, Y.; Wang, L.; Liu, J. Distributed voltage control in active distribution network considering renewable energy: A novel network partitioning method. *IEEE Trans. Power Syst.* **2020**, *35*, 4220–4231. [CrossRef]
2. Li, P.; Wu, Q.; Yang, M.; Li, Z.; Hatziargyriou, N.D. Distributed distributionally robust dispatch for integrated transmission-distribution systems. *IEEE Trans. Power Syst.* **2021**, *36*, 1193–1205. [CrossRef]
3. Hayes, B.P.; Gruber, J.K.; Prodanovic, M. A closed-loop state estimation tool for MV network monitoring and operation. *IEEE Trans. Smart Grid* **2015**, *6*, 2116–2125. [CrossRef]
4. Vardani, B.; Tummuru, N.R. A single-stage bidirectional inductive power transfer system with closed-loop current control strategy. *IEEE Trans. Transp. Electrific.* **2020**, *6*, 948–957. [CrossRef]
5. Wang, Z.; Xie, J.; Wang, Y. Analytical of loop closing operation based on multi-dimensional index system. In Proceedings of the 2020 International Symposium on Computer Engineering and Intelligent Communications (ISCEIC), Guangzhou, China, 7–9 August 2020; pp. 76–81.
6. Wang, H.; Yu, X.; Zhang, Z.; Yao, Y. Calculation, measurement and engineering application of loop closing current in power grid. In Proceedings of the 2021 IEEE 5th Conference on Energy Internet and Energy System Integration, Taiyuan, China, 22–25 October 2021; pp. 2237–2241.
7. Wang, H.; Zhang, Z.; Dong, J.; Zhang, Y.; Wu, H.; Zhang, W. Application of incentive-type variable weight in decision of 500/220kV received electromagnetic looped grid decomposing operation. *IEEE Access* **2020**, *8*, 185169–185176. [CrossRef]
8. Vargas, R.; Macedo, L.H.; Home-Ortiz, J.M.; Romero, R. Optimal restoration of distribution systems considering temporary closed-loop operation. *IEEE Syst. J.* **2021**, *15*, 5483–5494. [CrossRef]
9. de Groot, R.J.W.; Morren, J.; Slootweg, J.G. Reliable and efficient operation of closed-ring distribution grids supported by distribution automation. *Sustain. Energy Grids Netw.* **2018**, *15*, 53–62. [CrossRef]
10. Hafez, A.A.; Omran, W.A.; Hegazy, Y.G. A decentralized technique for autonomous service restoration in active radial distribution networks. *IEEE Trans. Smart Grid* **2018**, *9*, 1911–1919. [CrossRef]
11. Marques, L.T.; Delbem, A.C.B.; London, J.B.A. Service restoration with prioritization of customers and switches and determination of switching sequence. *IEEE Trans. Smart Grid* **2018**, *9*, 2359–2370. [CrossRef]
12. Singh, P.; Tiwari, R.; Sangwan, V.; Gupta, A.K. Optimal allocation of thyristor-controlled series capacitor (TCSC) and thyristor-controlled phase-shifting transformer (TCPST). In Proceedings of the 2020 International Conference on Power Electronics & IoT Applications in Renewable Energy and Its Control, Mathura, Uttar Pradesh, India, 28–29 February 2020; pp. 491–496.
13. Xu, Z.; Qin, R.; Ma, H.; Lu, J.; Tang, J.; Yang, Y. Research on the optimal regulation strategy of loop closing device based on phase shifting transformer. In Proceedings of the 2021 IEEE 5th Conference on Energy Internet and Energy System Integration, Taiyuan, China, 22–25 October 2021; pp. 1203–1208.
14. Xu, Z.; Qin, R.; Ma, H.; Lu, J.; Niu, C.; Yang, Y. Comparative study of AC/DC side loop-closing strategy based on back-to-back VSC-HVDC flexible loop-closing device in distribution network. In Proceedings of the 2021 5th International Conference on Power and Energy Engineering (ICPEE), Wuhan, China, 2–4 December 2021; pp. 53–60.
15. Zhenping, Z.; Chaoqun, N.; Yuanyuan, M.; Jianxiong, T.; Yongchun, Y. Research on SVG participating in voltage regulation and non-power outage closing loop transfer control strategy of distribution network. In Proceedings of the 2022 6th International Conference on Power and Energy Engineering (ICPEE), Shenzhen, China, 25–27 November 2022; pp. 72–78.
16. Haddadi, A.; Kazemi, A. Optimal power flow control by rotary power flow controller. *Adv. Electr. Comput. Eng.* **2011**, *11*, 79–86. [CrossRef]
17. Larsen, E.V. A classical approach to constructing a power flow controller. In Proceedings of the 1999 IEEE Power Engineering Society Summer Meeting, Edmonton, AB, Canada, 18–22 July 1999; Volume 2, pp. 1192–1195.
18. Tan, Z.; Zhang, C.; Jiang, Q. Research on characteristics and power flow control strategy of rotary power flow controller. In Proceedings of the International Youth Conference on Energy (IYCE), Pisa, Italy, 27–30 May 2015; pp. 1–8.
19. Ba, A.O.; Peng, T.; Lefebvre, S. Rotary power flow controller for dynamic performance evaluation—Part I: RPFC modeling. *IEEE Trans. Power Deliv.* **2009**, *24*, 1406–1416. [CrossRef]
20. Ba, A.O.; Peng, T.; Lefebvre, S. Rotary power-flow controller for dynamic performance evaluation—Part II: RPFC application in a transmission corridor. *IEEE Trans. Power Deliv.* **2009**, *24*, 1417–1425. [CrossRef]
21. Jia, J.; Zhou, J.; Gao, Y.; Shao, C.; Lu, J.; Jia, J. A rotating tidal current controller and energy router siting and capacitation method considering spatio-temporal distribution. *Energies* **2024**, *17*, 5919. [CrossRef]
22. Yan, X.; Shao, C.; Jia, J.; Aslam, W.; Peng, W.; Yang, R. An Accurate Power Control Strategy for Electromagnetic Rotary Power Controllers. *IET Gener. Transm. Distrib.* **2023**, *17*, 3524–3540. [CrossRef]
23. Yan, X.; Deng, W.; Wang, G.; Rasool, A.; Qi, S. Research on novel var compensator based on dual-rotary phase-shifting transformers and its control strategy. *Int. J. Electr. Power Energy Syst.* **2023**, *144*, 108632–108642. [CrossRef]

24. Yan, X.; Wu, W.; Shao, C.; Peng, W. Comprehensive compensation method for co-phase power supply in electrified railways based on V/v transformer and electromagnetic single-phase var compensator. *IET Gener. Transm. Distrib.* **2024**, *18*, 3170–3186. [CrossRef]
25. Shao, C.; Yan, X.; Yang, Y.; Aslam, W.; Jia, J.; Li, J. Multiple-zone synchronous voltage regulation and loss reduction optimization of distribution networks based on a dual rotary phase-shifting transformer. *Sustainability* **2024**, *16*, 1029. [CrossRef]
26. Yan, X.; Peng, W.; Wang, Y.; Aslam, W.; Shao, C.; Li, T. Flexible loop closing control method for an active distribution network based on dual rotary phase shifting transformers. *IET Gener. Transm. Distrib.* **2022**, *16*, 4204–4214. [CrossRef]
27. Shao, C.; Yan, X.; Siddique, A.; Aslam, W.; Jia, J.; Xie, W.; Alqahtani, S.A.; Alamri, A.M. Two-stage power control method of rotary power flow controller for flexible interconnection of distribution network. *IET Renew. Power Gener.* **2024**, *early view*, 1–14. [CrossRef]

electronics

MDPI

Article

Reducing Noise and Impact of High-Frequency Torque Ripple Caused by Injection Voltages by Using Self-Regulating Random Model Algorithm for SynRMs Sensorless Speed Control

Yibo Guo, Lingyun Pan, Yang Yang *, Yimin Gong and Xiaolei Che

College of Physics, Jilin University, Changchun 130000, China; guoyb22@mails.jlu.edu.cn (Y.G.); ply@jlu.edu.cn (L.P.); gongym@jlu.edu.cn (Y.G.); chexl@jlu.edu.cn (X.C.)
* Correspondence: jlu_yang@jlu.edu.cn

Abstract: For the sensorless control in a low-speed range of synchronous reluctance motors (SynRMs), injecting random high-frequency (HF) square-wave-type voltages has become a widely used and technologically mature method. It can solve the noise problem of traditional injection signal methods. However, all injection signal methods will cause problems such as torque ripple, which causes speed fluctuations. This article proposes a self-regulating random model algorithm for the random injection signal method, which includes a quantity adaptive module for adding additional random processes, an evaluation module for evaluating torque deviation degree, and an updated model module that is used to receive signals from the other two modules and complete model changes and output random model elements. The main function of this algorithm is to create a model that updates to suppress the evaluation value deviation based on the evaluation situation and outputs an optimal sequence of random numbers, thereby limiting speed bias always in a small range; this can reduce unnecessary changes in the output value of the speed regulator. The feasibility and effectiveness of the proposed algorithm and control method have been demonstrated in experiments based on a 5-kW synchronous reluctance motor.

Keywords: HF voltage injection; SynRM; sensorless control; random number

1. Introduction

Synchronous reluctance motors (SynRMs) exhibit advantages such as simple structure, ruggedness, high efficiency, wide speed range, and low cost. The price of rare earth metals for making permanent magnets is becoming increasingly high. In the future, SynRMs will receive a wider range of applications. In order to fully utilize these advantages, it is necessary to achieve sensorless speed control. The main methods to achieve sensorless control include using high-frequency signal injection, extended back electromotive force (EMF) or flux observers, Kalman Filter, etc. These methods can estimate the rotor position and speed, achieving sensorless control of the motor.

In [1–5], back EMF or flux observers are used; these methods have good performance in high-speed ranges. For SynRMs, because there is no permanent magnet on the rotor, the back EMF and flux linkages are very small at low-speed or zero-speed ranges, which means that these methods require additional operations [6] or need to be used in scenarios with extremely high sampling accuracy, which is impossible to adapt to all application conditions. And for the methods of using the Kalman filter [7,8], the calculations are enormous. So, these methods are difficult to achieve in most cases. Based on the above reasons, when SynRMs run in the low-speed range, the method of using high-frequency (HF) injection signals is usually chosen. This method has been widely used in speed sensorless control of various types of motors, with some injecting current signals [9–11] and others injecting voltage signals [12–20].

For general speed-control methods, the injection port can be set at the output end of the current regulator of the methods, which use voltage signals; this means that the injection voltage value will be more accurate. This is the reason that voltage signals are used more frequently than current signals. In [12], a square-wave-type voltage injection incorporated into the associated signal processing method is proposed which has been widely accepted. But, there is a very serious problem with the control method of injecting a fixed voltage signal, which is the noise problem. In order to solve this problem, it is no longer appropriate to inject fixed signals, and many methods of injecting random voltages have been proposed [13–16]. The method of pseudo-random voltage injection is more commonly used because it does not require changing the carrier frequency of pulse width modulation (PWM), making the control period more stable and the regulator coefficients easier to design. This article will improve the sensorless speed-control method based on the pseudo-random voltage injection method and prove that this method also has the function of reducing noise.

Another issue with the injection voltage method is that it introduces current harmonics, which cause additional torque ripple [21], further causing fluctuations in speed, thereby affecting sensorless speed control. This is because, unlike permanent magnet synchronous motors (PMSMs), the electromagnetic torque of SynRMs is the only reluctance torque. Therefore, the $i_d = 0$ control strategy commonly used in PMSMs is no longer applicable. During this operation, there is current on the *d-q* axis of the motor, and any direction of current fluctuation may cause torque fluctuation. The ripple can be suppressed by optimizing motor design [22] but cannot be minimized completely. There are also methods to reduce torque ripple by compensating for current [21,23] or injecting voltages in a special way [18]. But, from the analysis process of these methods, it can be seen that they heavily rely on the accuracy of position estimation and the stability of motor parameters. However, based on the saturation phenomenon of the SynRMs, those sensorless control methods of SynRMs as [16–18] often require the use of methods such as those proposed in [24] or [25] to calibrate the motor model in advance and the use of look-up-table methods during control to ensure the control successfully. In fact, the traditional HF voltage injection method is a saliency detection method for running in a low-speed range; the saturation phenomenon does not seriously affect the successful operation of sensorless control [19,20]. Therefore, it is not necessary to identify the motor parameters when the motor operates in the low-speed zone, which can ensure the universality and simple structure of the control method used.

In summary, in order to make the sensorless speed control of SynRMs effective, simple, and with low noise, it is necessary to use random voltage injection and minimize the impact of torque ripple on speed. At present, there is no suitable method to simultaneously consider the speed fluctuation caused by torque ripple under the premise of using random injection voltage. This article will analyze in detail the impact of torque ripple on speed when using random voltage injection for sensorless speed control of SynRMs. The impact of torque ripple on speed can be quantified and controlled within a very small range by using the proposed self-regulating random model algorithm. This algorithm can minimize speed fluctuations under the premise of using random voltage injection in a low-speed range.

The main contribution of this article is to propose the self-regulating random model algorithm. This algorithm has only three parts, which are one quantity adaptive module, one evaluation module, and one updated model module. This algorithm has a simple structure, low computational complexity, and can run independently without changing the structure of the original random voltages injection control framework. It is worth mentioning that this algorithm can be applied in a wide range of scenarios. In this article, the speed deviation degree caused by torque ripple is considered as the evaluation object. When applied to other scenarios, it is only necessary to change the evaluation object and evaluation method; the algorithm can update the active elements in the model and select the optimal random output sequence.

The rest of this article is organized as follows. The basic introduction of the sensorless control, which uses random square-wave-type voltage, is illustrated in Section 2. The analysis of how injecting voltage can cause torque ripples and further cause speed bias, and the calculated effect of voltage used in the experiment on speed are presented in Section 3. In Section 4, the self-regulating random model algorithm is introduced in detail, and the operation mode and principle of each module are explained; the overall control diagram is shown. The power spectral density (PSD) calculation used to explain how this method can reduce noise is presented in Section 5. The effectiveness of algorithms, sensorless control methods, and the results of comparison of experimental effects on reducing speed fluctuation are shown in Section 6. Finally, Section 7 concludes this article.

2. Random Square-Wave-Type Voltage Injection Sensorless Control Strategy for SynRMs

2.1. The Mathematical Model of a SynRM

The model of a SynRM in a *dq*-axis synchronous frame under the steady state can be expressed as

$$\begin{cases} u_d = R_s i_d + L_d \frac{di_d}{dt} - \omega_{re} L_q i_q \\ u_q = R_s i_q + L_q \frac{di_q}{dt} + \omega_{re} L_d i_d \end{cases} \tag{1}$$

where u_d, u_q, R_s, i_d, i_q, ω_{re}, L_d, L_q are *d*-axis stator voltage, *q*-axis stator voltage, stator resistance, *d*-axis stator current, *q*-axis stator current, motor electrical speed, *d*-axis inductance, and *q*-axis inductance, respectively.

The electromagnetic torque mathematic model and the motor motion equation can be expressed as

$$T_e = \frac{3P}{2}(L_d - L_q) i_d i_q \tag{2}$$

$$J\frac{d\omega_r}{dt} = T_e - T_L - B\omega_r \tag{3}$$

where P is the number of pole pairs; J is the rotor inertia; T_L is the load torque; ω_r is motor speed, which is calculated as $\omega_r = \omega_{re}/P$, and B is the damping coefficient.

2.2. Sensorless Control Based on Square-Wave-Type Voltage Injection

It has been proposed in [12] that when the injection voltage is described as

$$U_{d,inj} = \begin{cases} U_h & half\ duty \\ -U_h & otherwise \end{cases} \quad (U_h > 0), U_{q,inj} = 0 \tag{4}$$

as shown in Figure 1, the high frequency of the corresponding induced current in the $\alpha\beta$-axis can be described as

$$\begin{cases} \Delta i_{\alpha,h,k} = \frac{U_{d,inj,k-1}\Delta T}{L_d} \times \cos\theta_{re} \\ \Delta i_{\beta,h,k} = \frac{U_{d,inj,k-1}\Delta T}{L_d} \times \sin\theta_{re} \end{cases} \tag{5}$$

where the subscript k represents the k-th data; ΔT is the interval time between two adjacent data sampling; θ_{re} is the electrical angle of the rotor position; $\Delta i_{\alpha,h,k}$ and $\Delta i_{\beta,h,k}$ are high-frequency parts of the $\alpha\beta$-axis current variation, which can be obtained by

$$\begin{cases} \Delta i_{\alpha,h,k} = i_{\alpha,h,k} - i_{\alpha,h,k-1} \\ \Delta i_{\beta,h,k} = i_{\beta,h,k} - i_{\beta,h,k-1} \end{cases} \tag{6}$$

According to (5), if a matrix is established as

$$e_n = \begin{bmatrix} \Delta i_{\alpha,h} \\ \Delta i_{\beta,h} \end{bmatrix} \tag{7}$$

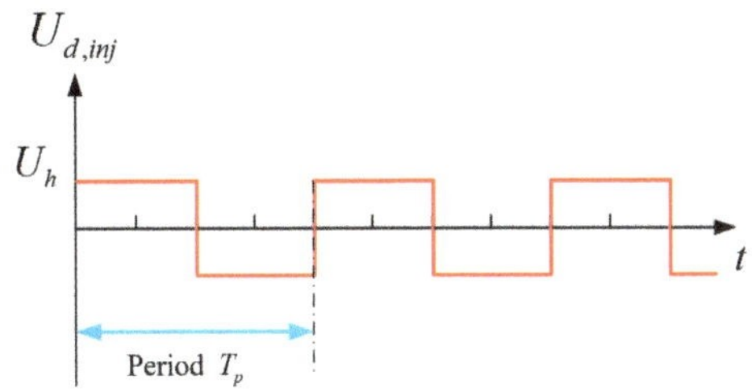

Figure 1. The schematic diagram of the injection voltage of the *d*-axis.

It is easy to obtain that

$$\frac{de_n}{dt} = \omega_{re} R e_n, \text{ where } R = \begin{bmatrix} 0 & -1 \\ 1 & 0 \end{bmatrix} \tag{8}$$

The motor's electrical speed ω_{re} can be estimated by an adaptive speed estimator, which is used in [3].

$$\frac{d\tilde{e}_n}{dt} = \omega_{re} R \tilde{e}_n + \begin{bmatrix} g & 0 \\ 0 & g \end{bmatrix} (\tilde{e}_n - e_n) \tag{9}$$

where the values with symbol ~ mean estimated values; g is the feedback gain of the adaptive estimator. For the convenience of designing the coefficients in the adaptive speed estimator, the matrix e_n can be normalized. Another advantage of normalization is that there is no need to consider changes in L_d, and there is no parameter sensitivity issue in the entire control process. The configuration of the adaptive speed estimator is shown in Figure 2; by obtaining the high-frequency current component of the α-β axis and combining it with formulae (6)–(9), the motor speed can be obtained. Further, the electrical angle of the rotor can be calculated by a phase-locked loop (PLL), which is shown in Figure 3. Unlike directly integrating the motor speed to obtain the angle, this method can quickly compensate for the error between the true angle and the estimated angle. Based on this advantage, this method can avoid estimating the initial angle of the motor, and even if there is an error in estimating the speed, it can ensure that all coordinate transformations can be successfully achieved. The sensorless control block diagram composed of an adaptive speed estimator and PLL is shown in Figure 4. This is the original voltage injection method (OVIM). The sensorless control method proposed in [16] is also applicable, but the method proposed in this article uses an adaptive speed estimator, which ensures that the estimated speed has smaller fluctuations. Due to further analysis of velocity fluctuations in the following text, it is not suitable to use angle error integration to obtain motor velocity.

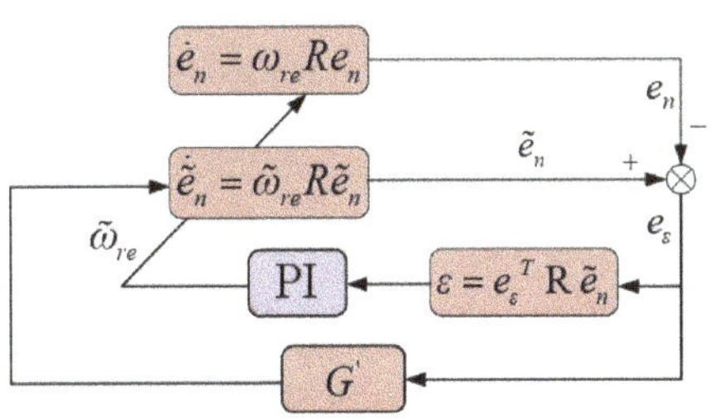

Figure 2. Configuration of adaptive speed estimator.

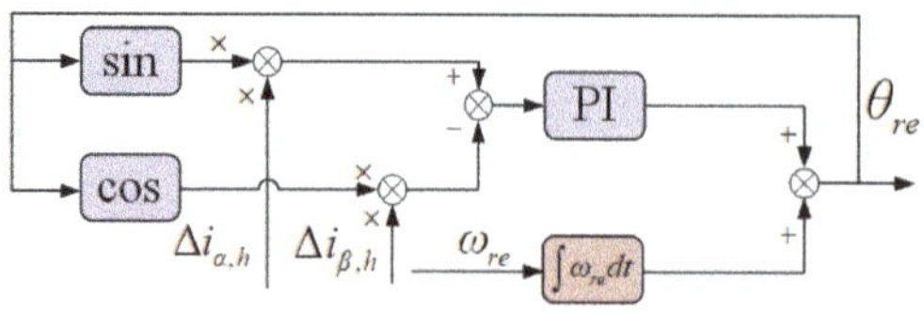

Figure 3. PLL for calculating electrical angle.

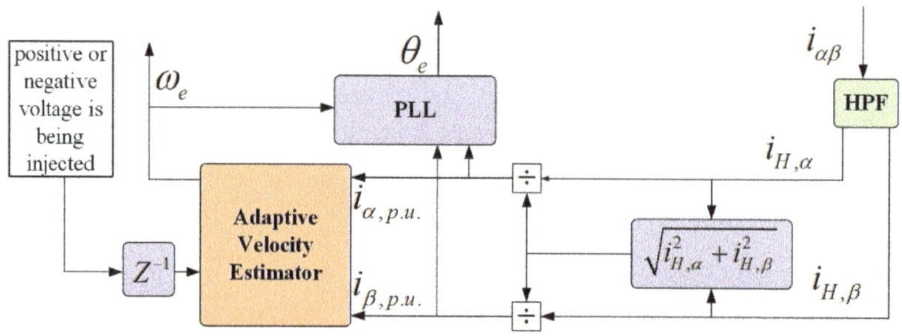

Figure 4. Using adaptive velocity estimator and PLL to obtain motor speed and electrical angle.

2.3. *Random High-Frequency Square-Wave-Type Voltage Injection Method Description*

The high-frequency SynRM model of the square-wave-type voltage injection method is expressed as

$$\begin{bmatrix} U_{d,inj} \\ U_{q,inj} \end{bmatrix} = \begin{bmatrix} L_d & 0 \\ 0 & L_q \end{bmatrix} \frac{d}{dt} \begin{bmatrix} i_{d,h} \\ i_{q,h} \end{bmatrix} \tag{10}$$

where $i_{d,h}$ and $i_{q,h}$ are high-frequency part of the current in dq-axis.

It is obvious that if the injection voltages in the d-axis are of the square-wave-type, the corresponding induced currents in the d-axis are of the triangular-wave-type. A square-wave form function $u_f(t, T_p, \varphi)$ and a triangular-wave form function $i_f(t, T_p, \varphi)$ can be defined as

$$\begin{gathered} u_f(t, T_p, \varphi) = \begin{cases} 1, 0 \leq t_{mod}\left(t - \frac{\varphi}{360^\circ}, T_p\right) \leq \frac{T_p}{2} \\ -1, \frac{T_p}{2} < t_{mod}\left(t - \frac{\varphi}{360^\circ}, T_p\right) < T_p \end{cases} \\ \begin{pmatrix} if\ t_{mod}(t, T_p) < 0, t_{mod}(t, T_p) = t_{mod}(t, T_p) + T_p \\ if\ t_{mod}(t, T_p) > T_p, t_{mod}(t, T_p) = t_{mod}(t, T_p) - T_p \end{pmatrix} \end{gathered} \tag{11}$$

$$i_f(t, T_P, \varphi) = \frac{1}{T_P} \int_0^t u_f(t, T_P, \varphi) dt \tag{12}$$

where T_p is the period of the injecting voltage; φ denotes the phase of the voltage, and in this article, it only has the right values, $\varphi = \{0^\circ, 45^\circ, 90^\circ, 135^\circ, 180^\circ, 225^\circ, 270^\circ, 315^\circ\}$ and $t_{mod}(t, T) = t \bmod T$. The voltage and induced current at these eight phases are shown correspondingly in Figure 5.

T_s is the period of the PWM wave and is also the period of the control, considering that for eliminating the effect of non-ideal characteristics of the inverter and the third harmonics in the position estimation error [26] and to ensure fast response of the adaptive speed estimator, the periods T_p of the injection voltages in this article are set as $8T_s$ and $4T_s$. Twelve voltages have been set up. In order to facilitate the calculation of torque deviation degree. The corresponding numbers for the amplitude, period, and phase of these twelve voltages are shown in Table 1. Randomly injecting these twelve voltages, combined with the aforementioned sensorless control strategy, can achieve sensorless control of SynRMs. This method is called the random voltage injection method (RVIM).

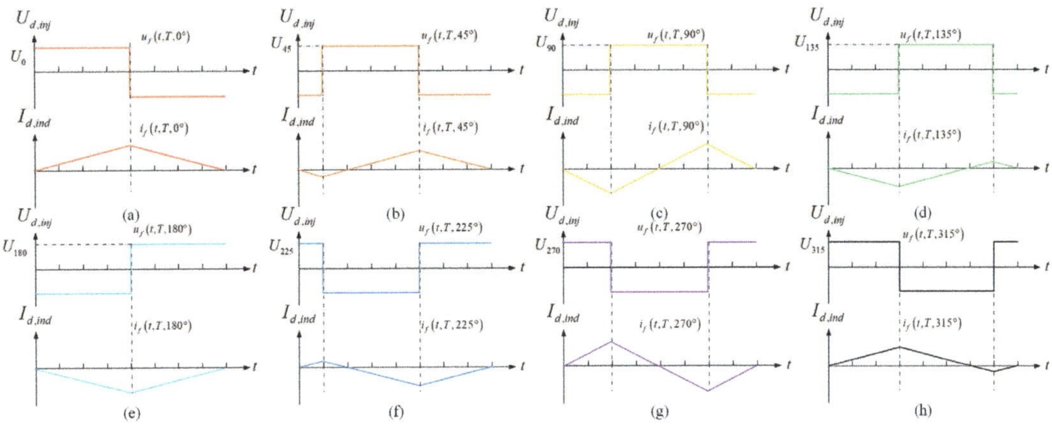

Figure 5. The schematic diagram of the eight basic injection voltages and corresponding induced currents in d-axis. (**a**) $\varphi = 0°$. (**b**) $\varphi = 45°$. (**c**) $\varphi = 90°$. (**d**) $\varphi = 135°$. (**e**) $\varphi = 180°$. (**f**) $\varphi = 225°$. (**g**) $\varphi = 270°$. (**h**) $\varphi = 315°$.

Table 1. Information of Twelve Voltages.

Number	1	2	3	4	5	6	7	8	9	10	11	12
Amplitude (V)	15		30		15		30	15	30	60	30	60
Period (s)					$8T_s$						$4T_s$	
Phase	0°	45°	90°	135°	180°	225°	270°	315°	0°	90°	180°	270°

3. Torque Ripple Analysis

3.1. The Torque Ripple Phenomenon Caused by Injected Voltage

The voltage injection sensorless control method is a method of determining the electrical angle and electrical speed of the rotor by analyzing the characteristics of the induced current and obtaining the saliency detection of the rotor. Induced current is inevitable, but it can cause high-frequency torque ripple, which affects the speed control of the motor. For the sensorless speed control of SynRMs, it is necessary to analyze torque ripple and minimize its impact on speed.

The torque ripple phenomenon caused by injected voltage has been analyzed in [19,22]. Due to the fact that the frequency of injected voltage is extremely high relative to the speed-control frequency, according to Equation (2), torque bias within an injection cycle can be expressed as

$$\Delta T_e = \frac{3P}{2}(L_d - L_q)i_q\Delta i_d \tag{13}$$

Neglecting changes in friction torque and load torque, the impact of this torque bias on motor speed can be calculated using Equation (3)

$$\Delta\omega_r = \frac{1}{J}\Delta T_e\Delta t = K(\Delta i_d\Delta t) \tag{14}$$

$$K = \frac{3P}{2J}(L_d - L_q)i_q \tag{15}$$

Within an injection cycle, the changes in current and inductance can be ignored, so K can be considered a constant. It can be seen that the speed change value is directly proportional to the product of the d-axis current change value and unit time, which means that the impact of torque ripple caused by the injected voltage on speed is proportional to the area enclosed by the induced current and the time axis. Taking injection voltages numbered 1 and 2 in Table 1 as examples to calculate their impact on speed, the reference figure is shown in Figure 6. S_1 and S_2 are the areas enclosed by the induced current and time axis for these two injection voltages.

$$S_1 = \frac{1}{2} \cdot 8T_s \cdot K\frac{U_1}{L_d} \cdot 4T_s = 16KT_s^2\frac{U_1}{L_d} \tag{16}$$

$$S_2 = -\frac{1}{2} \cdot 2T_s \cdot K\frac{U_2}{L_d} \cdot T_s + \frac{1}{2} \cdot 6T_s \cdot K\frac{U_2}{L_d} \cdot 3T_s = 8KT_s^2\frac{U_2}{L_d} \tag{17}$$

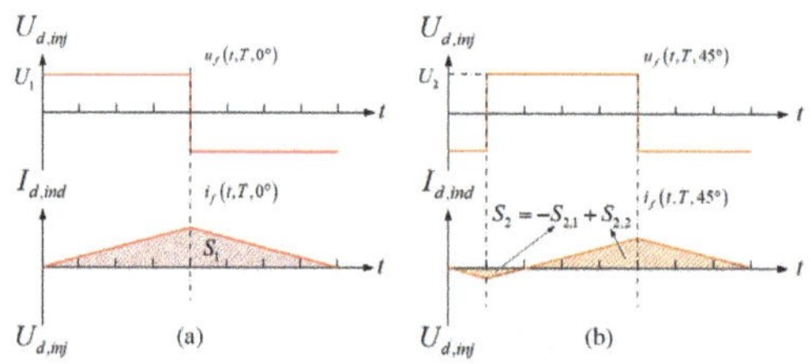

Figure 6. Reference diagram for calculating the impact of injected voltage on speed. (**a**) No.1. (**b**) No.2.

For coefficients shown in Table 1, $U_1 = U_2$; therefore, $S_1 = 2S_2$. If the speed bias value caused by the injection voltage numbered 2 is the unit of 1, the remaining bias values should be calculated, as shown in Table 2.

Table 2. Speed bias caused by twelve voltages.

Speed Bias	−2	−1	0	1	2
Number	5	4, 6, 11	3, 7, 10, 12	2, 8, 9	1

In order to ensure sensorless speed control and minimize the impact of injected voltage, it is necessary to ensure that the total speed bias is around 0. This can ensure a stable output of the speed loop regulator and prevent overall speed fluctuations and oscillations.

3.2. Analysis of the Torque Ripple Phenomenon in SynRMs

According to (15)–(17), the magnitude of the speed bias caused by injected voltage is related to the inductance values of L_d and L_q, and q-axis current i_q. And when the operating state of the motor changes, such as acceleration, deceleration, and changing the load, the current value changes. Due to the saturation effect of SynRMs, as shown in Figure 7, its dq-axial inductance value will change with the variation in axial current. This means that the relative ratio of the influence of each voltage on speed remains unchanged, but the absolute value always changes.

The above characteristics impose stricter restrictions on control. In order to ensure that the speed does not fluctuate even when the motor's operating state changes, it is necessary to maintain the speed bias around 0 at all times. This is difficult for RVIM to achieve. This control method only needs to randomly generate one random number and then select the corresponding voltage for injection. However, random numbers generated by most programming languages exhibit oscillations and periodicity over a long period of time. If there is also oscillation in the torque deviation caused by the corresponding injection voltage, it may cause speed fluctuations. Figure 8 shows the speed bias calculated based on Table 2 when operating the motor using RVIM. From Figure 8, it can be seen that the injection voltage has a significant impact on the speed.

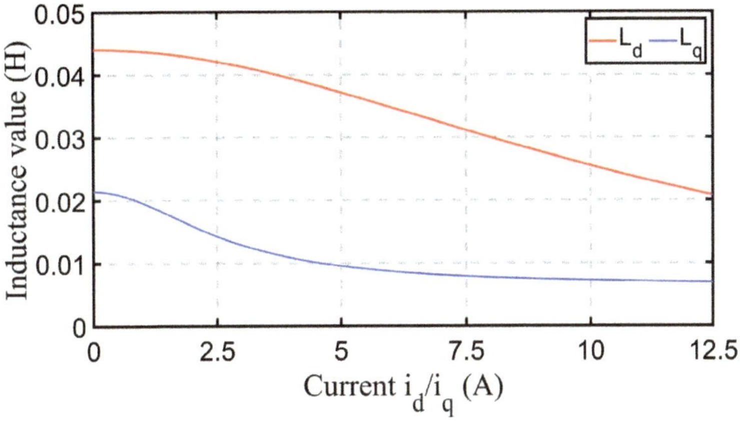

Figure 7. The variation in SynRM *dq*-axis inductance due to saturation effect.

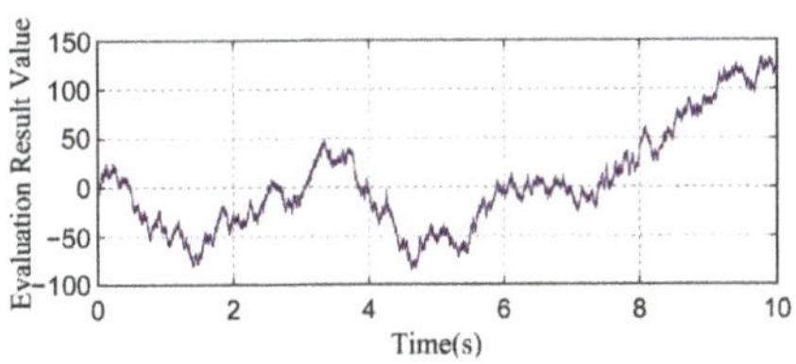

Figure 8. The summed-up value of the speed bias of using RVIM according to Table 2.

4. Introduction of the Proposed Self-Regulating Random Model Algorithm

In this section, the proposed self-regulating random model algorithm will be introduced in detail. The proposed self-regulated random model algorithm is created as a Simulink model and uploaded to Supplementary Materials. The use of a self-regulating random model algorithm with multiple random processes can eliminate the fluctuation impact of the generated random numbers and the injection voltage they represent on speed control. This algorithm can increase the randomness of injection voltage usage and ensure that the speed bias is around 0 in every short time duration.

4.1. Operation Mode of the Self-Regulating Random Model Algorithm

The overall diagram of the self-regulating random model algorithm is shown in Figure 9. The controller needs to generate three random numbers to help the algorithm process.

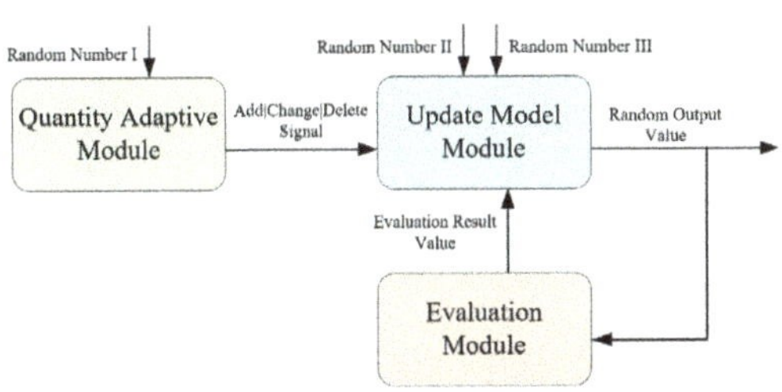

Figure 9. Overall diagram of self-regulating random model algorithm.

The main function of the algorithm is to create a model with 12 elements; some are in an activated state, while others are in an inactive state. This model continuously updates the activated elements in the model based on the combination of evaluation result value and model-activated element quantity adjustment signals and finally outputs a random number from these activated elements to ensure that the evaluation result value is always maintained near the expected value. This can meet our requirement that the speed bias is always around 0.

At the initial condition, it is necessary to set the model with 3 to 7 activated elements; elements are the numbers from 1 to 12. Each number corresponds to its assigned voltage.

4.2. Operation Mode of the Quantity Adaptive Module

The principle diagram of the quantity adaptive module is shown in Figure 10, which is the law on how to update the model. This module is used to randomly generate signals of the way how to change the model. As can be seen from Figure 10, the module is divided into three regions; when the Random Number I, ranging from 0 to 1, belongs to three different regions, the module will output Add signal, Change signal, and Delete signal separately. The probability of generating various signals varies with the number of activated elements in the model. For example, if the number of activated elements in the model is five, there is a 50% probability that the module will generate a Delete signal, and when a probability is 25%, the module will generate an Add signal or a Change signal.

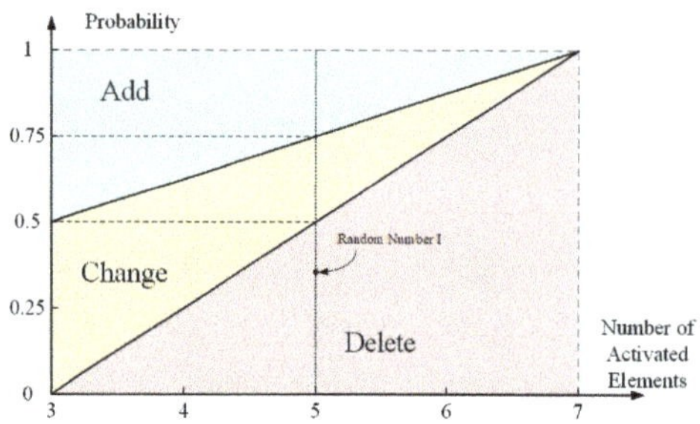

Figure 10. Principle diagram of the quantity adaptive module.

4.3. Operation Mode of the Evaluation Module

The theme structure of the evaluation module is a simple accumulator, which receives the random output value of the algorithms of the previous cycle, finds it in Table 2, records the speed bias generated by this output, accumulates the previous results, and outputs the new evaluation result value (V) to the Update Model Module. For other methods of suppressing torque fluctuations, complex mixed calculations of motor parameters are required, and control through the design of regulators is necessary. The computational complexity of the method proposed in this article only exists in this module, consisting of simple addition and subtraction of positive integers.

4.4. Operation Mode of the Update Model Module

The operation mode diagram of the Update Model Module is shown in Figure 11. Firstly, receive the signal from the quantity adaptive module and judge how to change the model. If an Add Signal is received, run Add Module; if a Delete Signal is received, run Delete Module; if a Change Signal is received, run both Add Module and Delete Module.

Every time the Add Module is run, it receives the state of the model, which is shown in Figure 11b; the red numbers mean elements that are activated, and the others are inactive. Add Module is used to select one of those inactive elements and change it to an activated state. Receive evaluation result value (V) from the evaluation module and make judgments. If $V > 0$, run Reduce Evaluation Value Module (REVM) with ①; if $V < 0$, run Increase Evaluation Value Module (IEVM) with ②; if $V = 0$, run Random Module (RM) with ③. The numbers ①, ②, ③ indicate the collections of inactive elements in the model. But ① removes 1, 2, 8, 9, which can increase the value of V; ② removes 4, 5, 6, 11.

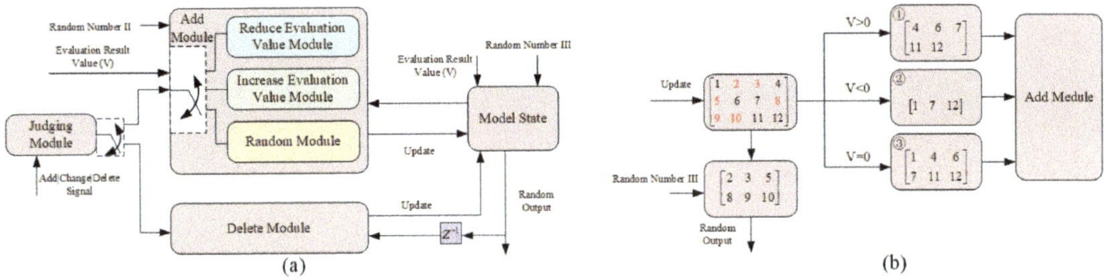

Figure 11. Update Model Module. (**a**) Overall logic diagram. (**b**) Schematic diagram of model state changes.

Every time the Delete Module is run, it receives random output from the algorithm of the previous cycle and changes this element to inactive to complete the update.

From a macro perspective, there are two trends in the changes in the elements within the model, adding elements that reduce the deviation of evaluation values and the repetition of the same element. The model is constantly changing toward the optimal model at the moment of using this algorithm, so combining random numbers to randomly output values for elements within the model can also be considered the optimal random output.

4.5. Sensorless Control Using Self-Regulating Random Model Algorithm

The overall flowchart of using the output of the self-regulating random model combined with Table 1 for pseudo-random voltage injection is shown in Figure 12. The output of this algorithm will be used to control which voltage is injected, and the whole sensorless speed-control system is shown in Figure 13. This method can be called a self-regulating random model method (SRRMM). Speed and current controls are implemented using Proportional–Integral (PI) regulators. A Low-Pass Filter (LPF) is used to filter out induced current to ensure signal stability. A High-Pass Filter (HPF) is used to extract induced current for sensorless control. The cutoff frequencies of LPF and HPF can both be set to $\frac{\pi}{32T_s}$.

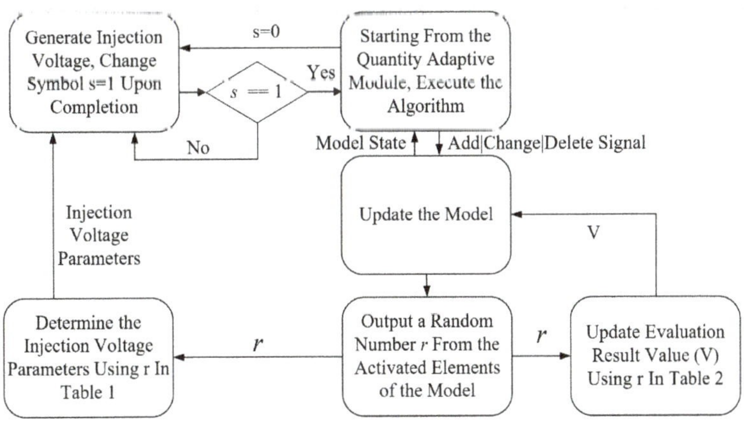

Figure 12. Flowchart of using the self-regulating random model algorithm to output injection voltage.

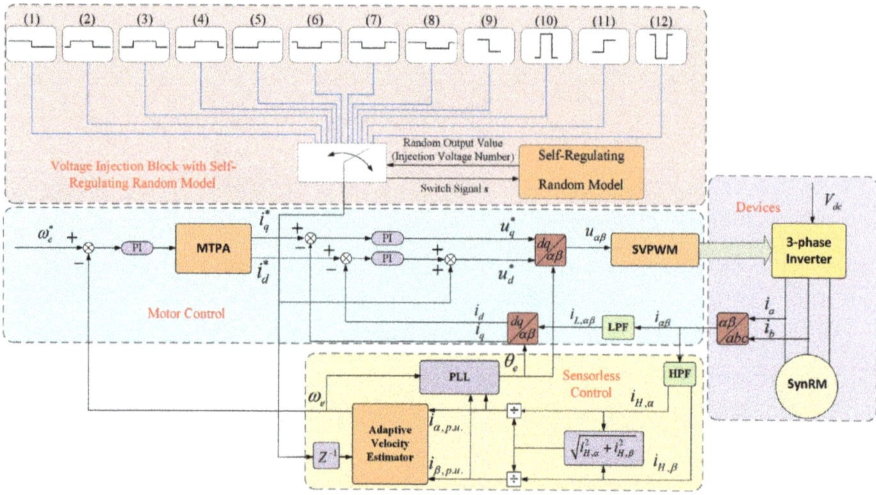

Figure 13. Block diagram of speed sensorless control system for SynRMs using self-regulating random model algorithm.

5. Analysis of PSD

Using a fixed frequency injection voltage can cause sharp noise, while the SRRMM proposed in this article uses pseudo-random injection voltage, which can achieve the function of reducing noise. Usually, the calculation of current PSD is used to demonstrate that PSD can effectively represent the distribution of energy in the frequency spectrum. The PSD values can be calculated using the following equation [27,28]:

$$S(f) = \frac{1}{E_T[T_p]}\left\{ \overbrace{E_{T,\varphi}\left[|I(f)|^2\right] - |E_{T,\varphi}[I(f)]|^2}^{continuous\ component} \quad \overbrace{\frac{1}{E_T[T_p]}|E_{T,\varphi}[I(f)]|^2 \sum_{k=-\infty}^{k=+\infty} \delta(f - kf_i)}^{discrete\ component} \right\} \tag{18}$$

where operator $E[]$ denotes mathematical expectation; $\delta()$ stands for the unit impulse function; $I(f)$ is the Fourier transform during one cycle.

For the twelve voltages, regardless of whether the period is $4T_s$ or $8T_s$, they all have a voltage with the phases of $\varphi = 0°, 90°, 180°, 270°$. The Fourier transform results of their induced current are as follows:

$$\begin{cases} I_{0°}(f,T_p) = \frac{1}{\pi^2 f^2 T_p}\left[2\left(1 - e^{-j2\pi f T_p}\right) + (2j\pi f T_p + 1)\left(e^{-j2\pi f T_p} - e^{-j\pi f T_p}\right)\right] \\ I_{90°}(f,T_p) = \frac{-1}{\pi^2 f^2 T_p}\left[2\left(e^{-j\frac{\pi}{2} f T_p} - e^{-j\frac{3\pi}{2} f T_p}\right) + \left(e^{-j2\pi f T_p} - 1\right)\right] \\ I_{180°}(f,T_p) = \frac{-1}{\pi^2 f^2 T_p}\left[2\left(1 - e^{-j2\pi f T_p}\right) + (2j\pi f T_p + 1)\left(e^{-j2\pi f T_p} - e^{-j\pi f T_p}\right)\right] \\ I_{270°}(f,T_p) = \frac{1}{\pi^2 f^2 T_p}\left[2\left(e^{-j\frac{\pi}{2} f T_p} - e^{-j\frac{3\pi}{2} f T_p}\right) + \left(e^{-j2\pi f T_p} - 1\right)\right] \end{cases} \tag{19}$$

where $T_p = \{4T_s, 8T_s\}$.

For the voltages with the period of $8T_s$, there are also four phases of φ = 45°, 135°, 225°, 315°. The amplitudes of the extreme values reached by the induced current are 0.5 and 0.25. The Fourier transform results of their induced current are as follows:

$$\begin{cases} I_{45^\circ}(f,T_p) = \frac{1}{2\pi^2 f^2 T_p}\left[1-e^{-j2\pi f T_p}+2\left(e^{-j\frac{5\pi}{4}fT_p}-e^{-j\frac{\pi}{4}fT_p}\right)\right] \\ I_{135^\circ}(f,T_p) = \frac{1}{2\pi^2 f^2 T_p}\left[1-e^{-j2\pi f T_p}+(2-j\frac{3\pi}{2}fT_p)\left(e^{-j\frac{7\pi}{4}fT_p}-e^{-j\frac{3\pi}{4}fT_p}\right)\right] \\ I_{225^\circ}(f,T_p) = \frac{-1}{2\pi^2 f^2 T_p}\left[1-e^{-j2\pi f T_p}+2\left(e^{-j\frac{5\pi}{4}fT_p}-e^{-j\frac{\pi}{4}fT_p}\right)\right] \\ I_{315^\circ}(f,T_p) = \frac{-1}{2\pi^2 f^2 T_p}\left[1-e^{-j2\pi f T_p}+(2-j\frac{3\pi}{2}fT_p)\left(e^{-j\frac{7\pi}{4}fT_p}-e^{-j\frac{3\pi}{4}fT_p}\right)\right] \end{cases} \quad (20)$$

where $T_p = 8T_s$.

It has been proved many times that if the injection signal frequencies only have odd times Least common multiple, the PSD will have discrete harmonics [14]. The period time of the voltages selected in this article are $4T_s$ and $8T_s$, respectively. So, for the system proposed in this article, there must be no discrete harmonics. Moreover, $I_{0^\circ}(f) = -I_{180^\circ}(f)$; $I_{90^\circ}(f) = -I_{270^\circ}(f)$; $I_{45^\circ}(f) = -I_{225^\circ}(f)$; $I_{135^\circ}(f) = -I_{315^\circ}(f)$. And from the operation mode of the algorithm, it can be seen that the probability of voltage injection with numbers 3, 7, 10, and 12 is the same, with a probability of about 10.530%, and the probability of the other eight voltages is about 7.235%. Equation (18) is finally reduced as

$$S(f) = \frac{1}{E_T[T_p]} E_{T,\varphi}\left[|I(f)|^2\right] \quad (21)$$

The specific result of the PSD of the induced current in the SRRMM can be expressed as

$$\begin{aligned} S(f) = {} & \tfrac{1}{4T_s}\left(0.1447|I_{0^\circ}(f,4T_s)|^2+0.2106|I_{90^\circ}(f,4T_s)|^2\right)+ \\ & \tfrac{1}{8T_s}(0.1447|I_{0^\circ}(f,8T_s)|^2+0.1447|I_{45^\circ}(f,8T_s)|^2+ \\ & 0.2106|I_{90^\circ}(f,8T_s)|^2+0.1447|I_{135^\circ}(f,8T_s)|^2) \end{aligned} \quad (22)$$

For OVIM which uses fixed frequency voltage, the result is

$$S(f) = \frac{1}{T_p}|I_{0^\circ}(f,T_p)|^2 \quad (23)$$

The control frequency used in this article is 10 kHz, which means T_s = 0.0001 s. The three PSD results of SRRMM and two OVIM with the period of $4T_s$ and $8T_s$, with voltage numbers 1 and 9 in Table 1, can be plotted. Usually, due to the small amplitude of PSD results, they take the logarithm based on ten and multiply it by ten. The processed results are shown in Figure 14. From the figure, it can be seen that the SRRMM smoothly disperses the induced current energy to various frequencies without any outliers appearing. This means that there will be no high-decibel noise of a certain frequency.

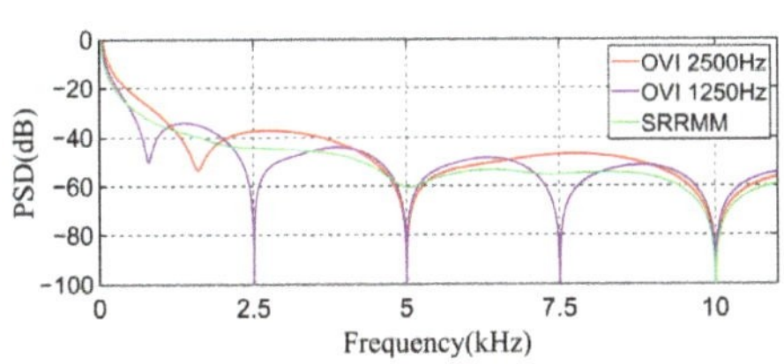

Figure 14. PSD results of the induced current of three different control methods.

6. Experimental Research

The experimental platform used is shown in Figure 15, where an eddy current dynamometer is used to apply load and verify the accuracy of the sensorless control of the motor speed estimation; the current probe is used to sample the phase current; a signal oscilloscope is used to record and display current signals, and the PC is used to communicate and control with the motor controller. The controller uses the chip RXR5F524T8ADFM. The parameters of the 5-kw motor are displayed in Table 3. The experimental verification is mainly divided into four steps. Firstly, it will be verified that the proposed self-regulating random model algorithm can be operated effectively. Secondly, various experiments are used to verify that this method can reduce the impact of torque ripple on sensorless speed control. Finally, it will be verified that the method used in this paper can effectively reduce noise.

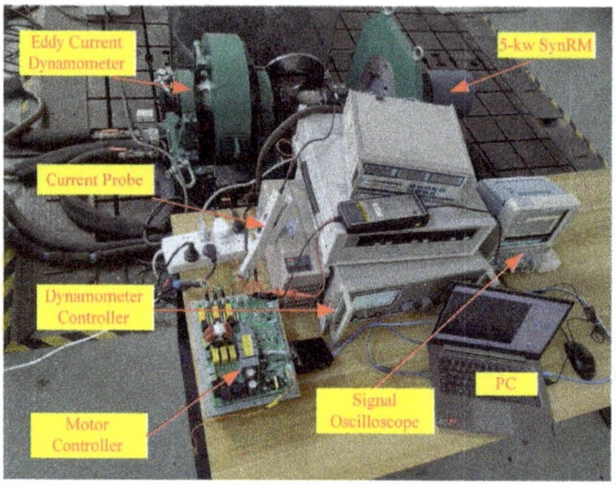

Figure 15. Experimental platform display diagram.

Table 3. Parameters of the 5-kW SynRM.

Parameter	Value
Pole Pairs	2
Resistance	0.25 Ω
Rated Current	12.5 A
Rated Voltage	380 V
D-Axis Inductance	43.9 mH
Q-Axis Inductance	22.1 mH

6.1. Validation of Self-Regulating Random Model Algorithm Performance

The specific operation mode of the self-regulating random model algorithm proposed in this article has been introduced in detail in Section 4. In order to prove the successful operation of this algorithm, running the motor using SRRMM for about 25 s, important parameters of each module of the algorithm are recorded based on the operating principles of the three modules included in the algorithm.

The main function of the quantity adaptive module is to output signals on how the model changes while also ensuring that the number of elements in the model has a certain degree of randomness. The results between 12 s and 12.1 s of the recorded output signal of the module and the number of elements in the model are shown in Figure 16. Regarding the Add | Change | Delete signal shown in Figure 16, if it is 1, it means the output is an Add signal; if it is 2, it means the output is a Delete signal, and if it is 3, it means the output is a Change signal. The corresponding data on the number of activated elements in the model below are also adjusted according to the output signal.

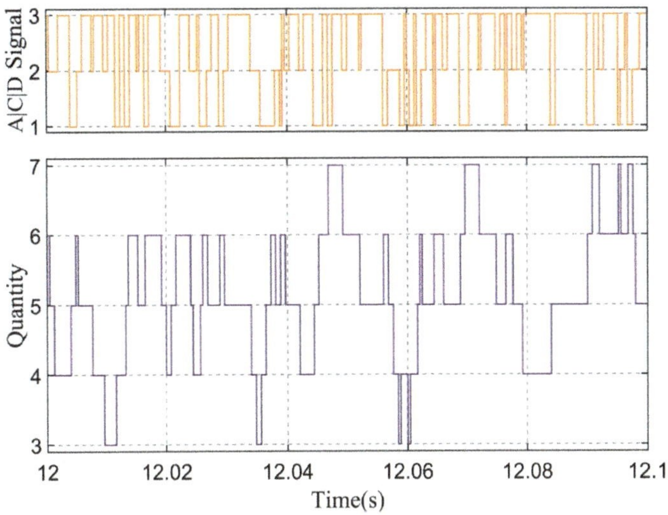

Figure 16. The output signal from quantity adaptive module and the number of active elements.

For the evaluation module and the update module, the parameters, which are worth paying attention to are the evaluation result value and the output random sequence. The model output and evaluation result values between 12 s and 12.1 s are also recorded, which are shown in Figure 17; the output of the model will be the number of injected voltage.

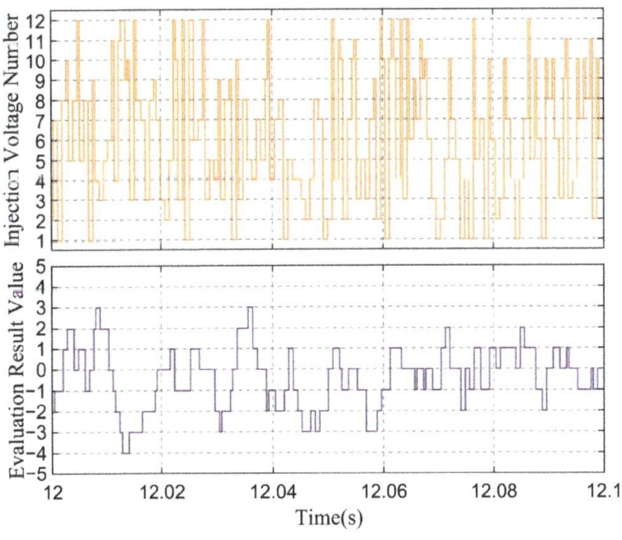

Figure 17. The evaluation result value and the injection voltage number.

From the figure, it can be seen that the evaluation value has remained stable around 0, even for an extremely long duration; there are no large or small values. The existence of such a phenomenon can ensure the rationality of the analysis of torque ripple in Section 3. The motor does not always operate under fixed operating conditions, and the parameters may also change. Maintaining an estimation result at 0 can ensure that the speed control is not subject to oscillations caused by random numbers generated by programming languages, which can cause oscillations in the output of the speed regulator. Therefore, using this method can ignore the inductance variation in the SynRM.

6.2. Reduction in Torque Ripple and Speed Fluctuation

To illustrate, the random voltage injection sensorless control method using a self-regulating random model algorithm has a smaller amplitude on torque ripple and a smaller impact on speed control. Many comparative experiments between RVIM and SRRMM are conducted, mainly including steady state under different load conditions, start-up in stationary and reverse rotation states, and acceleration and deceleration experiments.

We used these two methods to run the motor at 100 rpm with no-load, 50% rated-load, and rated-load conditions, respectively, and record the phase current, torque, and speed. The sampling results are shown in Figure 18. In order to better display the details of each quantity, the phase current is displayed using 100 ms/div, and there is a Zoom that displays the details of the induced current caused by the injection voltages. The corresponding amplitudes for the three load conditions are 2 A/div, 3.5 A/div 5 A/div. The speed displays the duration of 5 s, while the torque displays the duration of 20 s. It can be seen that when using SRRMM, the phase current is relatively stable, while when using RVIM, there is a noticeable jitter marked by the green circle in the figure, especially when the current amplitude is large. The speed fluctuation is also relatively small when using SRRMM, only within 1.5 rpm; while using RVIM, the speed fluctuation can reach 6 rpm. The torque fluctuation is also greater for RVIM, which is about three times that of using SRRMM. It can also be seen that there is a more macroscopic fluctuation in torque when using RVIM.

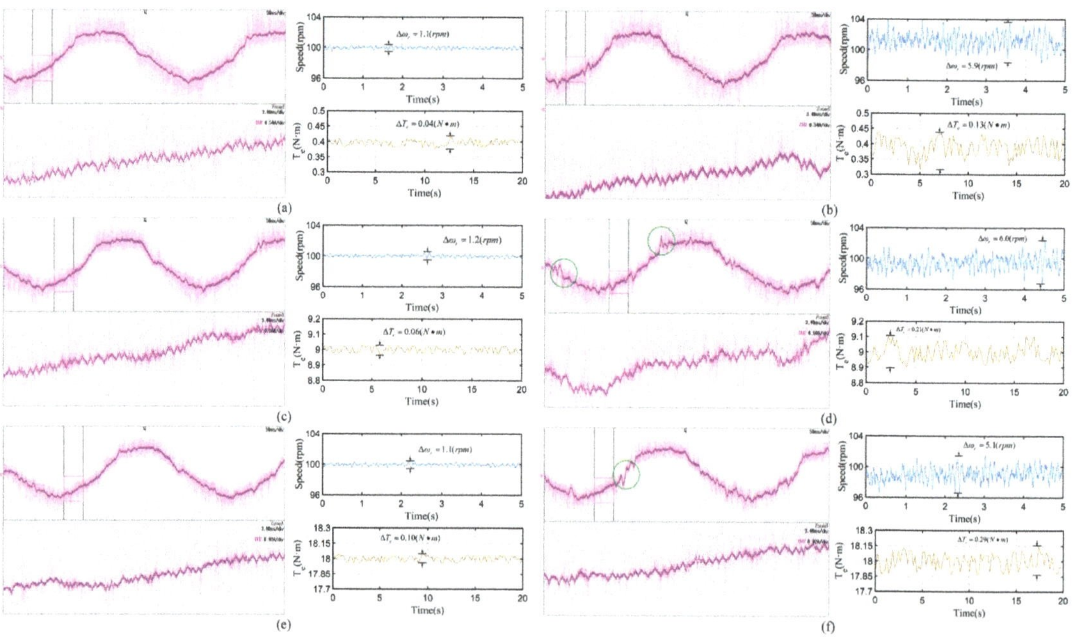

Figure 18. Sampling results of current, speed, and torque during motor running at 100 rpm. (**a**) SRRMM with no load. (**b**) RVIM with no load. (**c**) SRRMM with 50% rated load. (**d**) RVIM with 50% rated load. (**e**) SRRMM with rated load. (**f**) RVIM with rated load.

The sensorless control method proposed in this article does not require the use of methods such as I-f or V-f to start up from a stationary or negative speed state. Two control experiments are conducted, starting from a stationary state to 100 rpm and from −100 rpm to 100 rpm, respectively. The sampling images of phase current and motor speed are shown in Figures 19 and 20. The phase current is displayed as 10 s/div in time and 2 A/div in amplitude with no load, 5 s/div in time, and 5 A/div in amplitude with rated load. The current is limited at 3.5 A with no load and 12 A with rated load. For experiments starting from the stationary state, Zoom1 displays the current at the highest speed, while Zoom2 displays the current at the steady speed. The current using SRRMM is still relatively stable, while using RVIM not only causes current jitter in a small range, but also current oscillations in a large range. Large-range current oscillations can cause significant speed oscillations, all of which are marked by green circles. Small-range current oscillations are significant with a rated load, while large-range current oscillations are significant with no load. This is because the braking force brought by the eddy current dynamometer has some filtering effect, so the speed overshoot is not significant under rated-load conditions. For experiments starting from the reverse rotation state, Zoom1 displays the current at the zero speed position, while Zoom2 displays the current at the steady speed. As before, current jitter and oscillations, as well as speed oscillations, are all marked. The speed fluctuation, which we are most concerned about, is still below 1.5 rpm when using SRRMM, while using RVIM, it even reaches 7.3 rpm in Figure 19b, and the rest is also above 5 rpm.

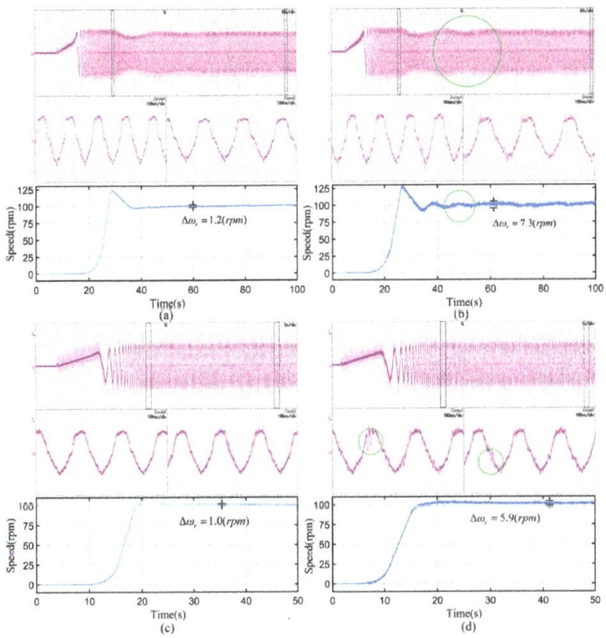

Figure 19. Sampling results of current and speed during motor start-up to 100 rpm. (**a**) SRRMM with no load. (**b**) RVIM with no load. (**c**) SRRMM with rated load. (**d**) RVIM with rated load.

Two different methods are used to conduct experiments on acceleration and deceleration between speeds of 100 rpm and 150 rpm. The experimental results are shown in Figure 21. The phase current is displayed as 10 s/div in time and 2 A/div in amplitude with no load, 5 s/div in time, and 5 A/div in amplitude with rated load. Zoom1 displays the current at 150 rpm, while Zoom2 displays the current at 100 rpm. The oscillations of speed and current are marked with green circles. For the same load situation using the same method, it can be seen that at higher speeds and currents, speed fluctuations are also greater. This confirms the analysis results of Equations (14) and (15). For using SRRMM, the speed fluctuation is about 2 rpm at the speed of 150 rpm, while for RVIM, it reaches around 7 rpm.

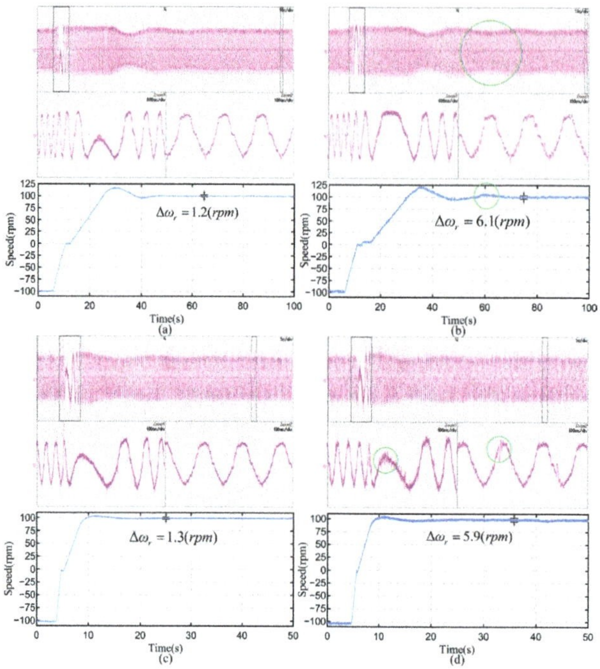

Figure 20. Sampling results of current and speed of the motor from −100 rpm to 100 rpm. (**a**) SRRMM with no load. (**b**) RVIM with no load. (**c**) SRRMM with rated load. (**d**) RVIM with rated load.

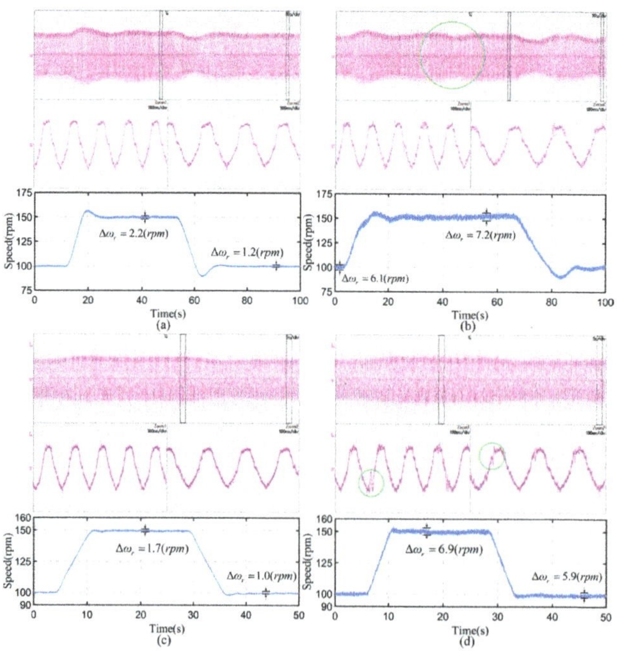

Figure 21. Sampling results of current and speed during acceleration and deceleration of the motor between 100 rpm and 150 rpm. (**a**) SRRMM with no-load. (**b**) RVIM with no-load. (**c**) SRRMM with rated load. (**d**) RVIM with rated load.

The above multiple experiments have shown that regardless of any operating state, the use of SRRMM can always maintain torque fluctuations caused by injected voltage near zero, which resists the influence of non-random or even periodic voltage selection. Therefore, the speed bias is always maintained near zero, which is beneficial for sensorless speed control.

6.3. Verification of Noise Reduction Effect Using SRRMM

In order to verify that the proposed SRRMM method can effectively suppress the noise caused by injected voltage, the following comparative experiment is designed, and all injection voltages used in the experiment are described in Table 1. The injection voltages have the frequencies of 2.5 kHz or 1.25 kHz, respectively, and to verify that the method used in this article can reduce noise, it is necessary to conduct OVIM experiments using these two different frequencies of voltages; the voltage numbers used in the two methods are 1 and 9, respectively. This method has high noise and is convenient for comparison. RVIM has been proven many times to have the function of reducing noise, and, as a commonly used method for comparison in this article, this method is conducted using all the injection voltages in Table 1. The above three methods, along with the SRRMM method proposed in this article, are all implemented under no-load and rated-load conditions. The use of two different load conditions can better demonstrate the effectiveness of the noise reduction function. When the motor runs stably at 100 rpm, acoustic noise is recorded by a sound meter Smart Sensor AS804B placed approximately 50 cm above the motor shaft; the experimental setup is shown in Figure 22, and the results are presented in Table 4. From the results, it can be seen that SRRMM, like RVIM, also has the function of reducing noise, while the noise generated by using OVIM is relatively large.

Figure 22. Collecting the noise level generated by the motor using Smart Sensor AS804B.

Table 4. The results of collecting acoustic noise levels using different methods.

Load State	No Load	Rated Load
OVIM (2.5 kHz)	76.3 dB	76.7 dB
OVIM(1.25 kHz)	76.5 dB	77.1 dB
RVIM	71.2 dB	71.4 dB
SRRMM	70.9 dB	70.9 dB

Similarly, when the motor runs stably at 100 rpm, the phase current is recorded, and the PSD and Fourier transform are calculated. As mentioned in Section 5, the calculated value of PSD is very small, and the results have been processed, they are taken the logarithm based on ten and multiplied by ten. The times of each numbered voltage are injected within 25 s and are shown in Figure 23.

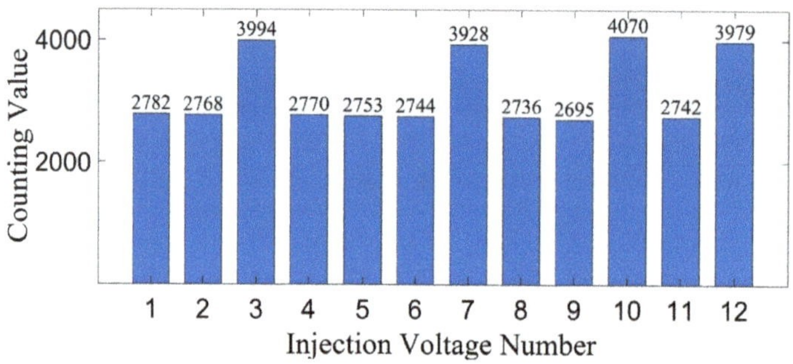

Figure 23. Record of voltage injection times for each number.

The sampled current image and the calculated PSD and Fourier transform results are shown in Figure 24. The time axis is 50 ms/div; the current axis for the no-load experiment is 2 A/div, and the rated-load experiment is 5 A/div. All current images have a Zoom to more clearly display the induced current caused by the injected voltage reflected in the phase current. The current amplitude is about 3.3 A under no-load conditions and about 9.3 A under rated-load conditions.

It can be seen from the figure that the PSD results of RVIM and SRRMM are below −60 dB in the frequency between 2 kHz and 8 kHz, which is the area where human hearing is more sensitive. From the results of the Fourier transform, it can also be seen that the current spectrum is spread out in the low-frequency range. As for the larger values at 10 kHz, they occur because the switching frequency of PWM is 10 kHz, and switching the inverter at this frequency will bring some current spikes.

The experimental results of two OVIMs show that there are significant peaks at the injection frequency and its harmonics, many of which are within the range of 2 kHz to 8 kHz. This will cause sharp noise, which is also what all random voltage methods want to eliminate.

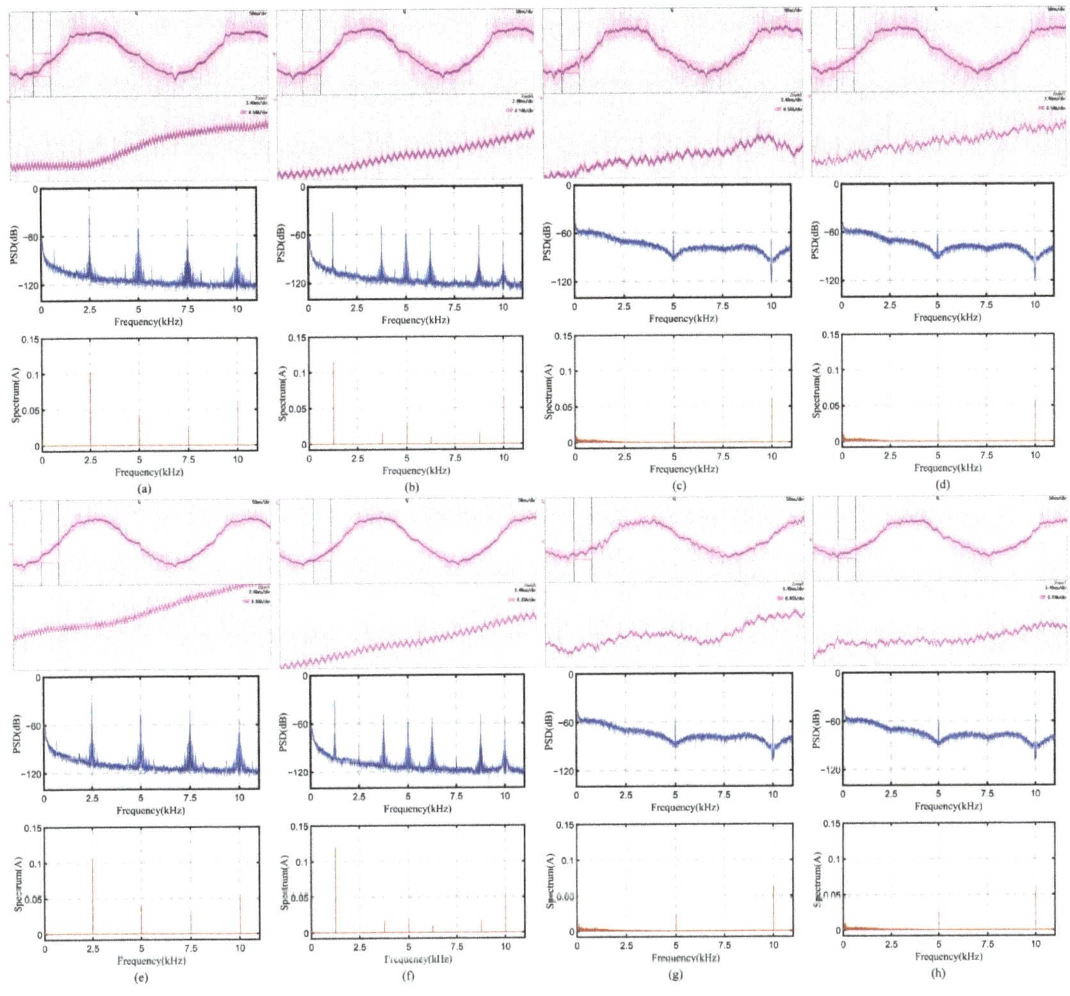

Figure 24. During steady state of the motor, phase current image, corresponding PSD calculation results, and Fourier transform results display diagram. (**a**) OVIM with 2.5 kHz injection voltage with no load. (**b**) OVIM with 1.25 kHz injection voltage with no load. (**c**) RVIM with no load. (**d**) SRRMM with no load. (**e**) OVIM with 2.5 kHz injection voltage with rated load. (**f**) OVIM with 1.25 kHz injection voltage with rated load. (**g**) RVIM with rated load. (**h**) SRRMM with rated load.

7. Conclusions

To solve the sensorless speed-control problem of synchronous reluctance motors (SynRMs) and weaken the speed bias and noise caused by injected voltage. The self-regulating random model algorithm is proposed in this paper, which can be used for sensorless control by injecting high-frequency square-wave-type voltages. This algorithm creates a model and continuously updates the activated elements within the model based on evaluation results, making the trend of the random number sequence output from the model decrease the bias of the evaluation value. Taking the torque deviation caused by induced current as the evaluation object and its impact on speed as the evaluation method, it is used to output twelve voltage numbers reasonably. This experiment has proven the feasibility of the algorithm, and the entire sensorless control method has good performance;

like other methods using random voltage, this method also has the function of reducing the noise caused by injection voltages.

Supplementary Materials: The following supporting information can be downloaded at: https://www.mdpi.com/article/10.3390/electronics13163327/s1, File S1: The Self-Regulating Random Model Algorithm proposed in this article is made into a Simulink model and compressed into zip format.

Author Contributions: Conceptualization, Y.G. (Yibo Guo); methodology, Y.G. (Yibo Guo) and L.P.; software, Y.G. (Yibo Guo) and X.C.; validation, Y.G. (Yibo Guo) and Y.Y.; formal analysis, Y.G. (Yibo Guo); investigation, Y.Y.; resources, Y.G. (Yimin Gong); data curation, Y.G. (Yibo Guo) and Y.G. (Yimin Gong); writing—original draft preparation, Y.G. (Yibo Guo); writing—review and editing, Y.G. (Yibo Guo); visualization, Y.G. (Yibo Guo) and L.P.; supervision, Y.G. (Yimin Gong); project administration, Y.Y.; funding acquisition, Y.Y. All authors have read and agreed to the published version of the manuscript.

Funding: This research received no external funding.

Data Availability Statement: The experimental data are encrypted. If there are any reasonable requests, please contact the author Yibo Guo (email: guoyb22@mails.jlu.edu.cn) for assistance. The self-regulating random model algorithm that can be used is created as a Simulink model and included in this manuscript, which can be opened through Matlab 2023a or higher versions.

Acknowledgments: Thanks to all the project team members.

Conflicts of Interest: The authors declare no conflicts of interest.

References

1. Wu, X.; Huang, S.; Liu, K.; Lu, K.; Hu, Y.; Pan, W.; Peng, X. Enhanced Position Sensorless Control Using Bilinear Recursive Least Squares Adaptive Filter for Interior Permanent Magnet Synchronous Motor. *IEEE Trans. Power Electron.* **2020**, *35*, 681–698. [CrossRef]
2. Yousefi-Talouki, A.; Boldea, I. Combined active flux and high-frequency injection methods for sensorless direct-flux vector control of synchronous reluctance machines. *IEEE Trans. Power Electron.* **2019**, *33*, 2447–2457. [CrossRef]
3. Ichikawa, S.; Okuma, S. Sensorless control of synchronous reluctance motors based on extended EMF models considering magnetic saturation with online parameter identification. *IEEE Trans. Ind. Appl.* **2006**, *42*, 1264–1274. [CrossRef]
4. Wang, D.; Lu, K.; Rasmussen, P.O. Improved Closed-Loop Flux Observer Based Sensorless Control Against System Oscillation for Synchronous Reluctance Machine Drives. *IEEE Trans. Power Electron.* **2019**, *34*, 4593–4602.
5. Varatharajan, A.; Pellegrino, G.; Armando, E. Sensorless synchronous reluctance motor drives: Auxiliary flux-based position observer. *IEEE J. Emerg. Sel. Top. Power Electron.* **2021**, *9*, 4330–4339. [CrossRef]
6. Zhang, Z.; Lamb, J. Active Q Flux Concept for Sensorless Control of Synchronous Reluctance Machines. *IEEE Trans. Ind. Electron.* **2023**, *70*, 4526–4536. [CrossRef]
7. Gao, F.; Yin, Z.; Bai, C.; Yuan, D.; Liu, J. A Lag Compensation-Enhanced Adaptive Quasi-Fading Kalman Filter for Sensorless Control of Synchronous Reluctance Motor. *IEEE Trans. Power Electron.* **2022**, *37*, 15322–15337. [CrossRef]
8. Pasqualotto, D.; Rigon, S.; Zigliotto, M. Sensorless Speed Control of Synchronous Reluctance Motor Drives Based on Extended Kalman Filter and Neural Magnetic Model. *IEEE Trans. Ind. Electron.* **2023**, *70*, 1321–1330. [CrossRef]
9. Foti, S.; De Caro, S.; Scimone, T.; Testa, A.; Tornello, L.D.; Scelba, G.; Cacciato, M. Rotor Position Error Compensation in Sensorless Synchronous Reluctance Motor Drives. *IEEE Trans. Power Electron.* **2022**, *37*, 4442–4452. [CrossRef]
10. Yousefi-Talouki, A.; Pescetto, P.; Pellegrino, G. Sensorless Direct Flux Vector Control of Synchronous Reluctance Motors Including Standstill, MTPA, and Flux Weakening. *IEEE Trans. Ind. Appl.* **2017**, *53*, 3598–3608. [CrossRef]
11. Chen, J.; Fan, Y.; Wang, W.; Lee, C.H.T.; Wang, Y. Sensorless Control for SynRM Drives Using a Pseudo-Random High-Frequency Triangular-Wave Current Signal Injection Scheme. *IEEE Trans. Power Electron.* **2022**, *37*, 7122–7131. [CrossRef]
12. Yoon, Y.-D.; Sul, S.-K.; Morimoto, S.; Ide, K. High-Bandwidth Sensorless Algorithm for AC Machines Based on Square-Wave-Type Voltage Injection. *IEEE Trans. Ind. Appl.* **2011**, *47*, 1361–1370. [CrossRef]
13. Li, C.; Wang, G.; Zhang, G.; Zhao, N.; Xu, D. Adaptive Pseudorandom High-Frequency Square-Wave Voltage Injection Based Sensorless Control for SynRM Drives. *IEEE Trans. Power Electron.* **2021**, *36*, 3200–3210. [CrossRef]
14. Zhang, G.; Xiang, R.; Wang, G.; Li, C.; Bi, G.; Zhao, N.; Xu, D. Hybrid Pseudorandom Signal Injection for Position Sensorless SynRM Drives With Acoustic Noise Reduction. IEEE Trans. Transport. *Electrification* **2022**, *8*, 1313–1325.
15. Zhang, Y.; Yin, Z.; Liu, J.; Zhang, R.; Sun, X. IPMSM Sensorless Control Using High-Frequency Voltage Injection Method With Random Switching Frequency for Audible Noise Improvement. *IEEE Trans. Ind. Electron.* **2020**, *67*, 6019–6030. [CrossRef]
16. Lv, L.; Hu, Z.; Li, S.; Guo, R.; Wang, J.; Wang, G.; Li, S. Variable-Angle Random High Frequency Voltage Injection Strategy with Cross-Saturation Effect Compensation for Sensorless Synchronous Reluctance Motor Drives. *Energies* **2024**, *17*, 725. [CrossRef]

17. Li, C.; Wang, G.; Zhang, G.; Xu, D.; Xiao, D. Saliency-Based Sensorless Control for SynRM Drives With Suppression of Position Estimation Error. *IEEE Trans. Ind. Electron.* **2019**, *66*, 5839–5849. [CrossRef]
18. Li, C.; Wang, G.; Zhang, G.; Zhao, N.; Gao, Y.; Xu, D. Torque Ripples Minimization of Sensorless SynRM Drives for Low-Speed Operation Using Bi-HFSI Scheme. *IEEE Trans. Ind. Electron.* **2021**, *68*, 5559–5570. [CrossRef]
19. Varatharajan, A.; Pellegrino, G.; Mariani, G.B.; Voyer, N.; Satake, A. LUT-Less Sensorless Control of Synchronous Reluctance Machines Using the Locus of Incremental Saliency Ratio Tracking (LIST). *IEEE Trans. Ind. Electron.* **2022**, *69*, 6530–6539. [CrossRef]
20. Bugsch, M.; Piepenbreier, B. High-Bandwidth Sensorless Control of Synchronous Reluctance Machines in the Low- and Zero-Speed Range. *IEEE Trans. Ind. Appl.* **2020**, *56*, 2663–2672. [CrossRef]
21. Lin, Z.; Li, X.; Wang, Z.; Shi, T.; Xia, C. Minimization of Additional High-Frequency Torque Ripple for Square-Wave Voltage Injection IPMSM Sensorless Drives. *IEEE Trans. Power Electron.* **2020**, *35*, 13345–13355. [CrossRef]
22. Chen, Q.; Yan, Y.; Xu, G.; Xu, M.; Liu, G. Principle of Torque Ripple Reduction in Synchronous Reluctance Motors With Shifted Asymmetrical Poles. *IEEE J. Emerg. Sel. Top. Power Electron.* **2020**, *8*, 2611–2622. [CrossRef]
23. Qu, J.; Zhang, C.; Jatskevich, J.; Zhang, S. Deadbeat Harmonic Current Control of Permanent Magnet Synchronous Machine Drives for Torque Ripple Reduction. *IEEE J. Emerg. Sel. Top. Power Electron.* **2022**, *10*, 3357–3370. [CrossRef]
24. Woo, T.-G.; Park, S.-W.; Choi, S.-C.; Lee, H.-J.; Yoon, Y.-D. Flux Saturation Model Including Cross Saturation for Synchronous Reluctance Machines and Its Identification Method at Standstill. *IEEE Trans. Ind. Electron.* **2023**, *70*, 2318–2328. [CrossRef]
25. Hinkkanen, M.; Pescetto, P.; Mölsä, E.; Saarakkala, S.E.; Pellegrino, G.; Bojoi, R. Sensorless Self-Commissioning of Synchronous Reluctance Motors at Standstill Without Rotor Locking. *IEEE Trans. Ind. Appl.* **2017**, *53*, 2120–2129. [CrossRef]
26. Hwang, C.-E.; Lee, Y.; Sul, S.-K. Analysis on Position Estimation Error in Position-Sensorless Operation of IPMSM Using Pulsating Square Wave Signal Injection. *IEEE Trans. Ind. Appl.* **2019**, *55*, 458–470. [CrossRef]
27. Tse, K.K.; Chung, H.S.-H.; Huo, S.Y.; So, H.C. Analysis and spectral characteristics of a spread-spectrum technique for conducted EMI suppression. *IEEE Trans. Power Electron.* **2000**, *15*, 399–410. [CrossRef]
28. Tse, K.K.; Chung, H.S.-H.; Hui, S.Y.R.; So, H.C. A comparative investigation on the use of random modulation schemes for DC/DC converters. *IEEE Trans. Ind. Electron.* **2000**, *47*, 253–263. [CrossRef]

electronics

MDPI

Article

Current-Prediction-Controlled Quasi-Z-Source Cascaded Multilevel Photovoltaic Inverter

Shanshan Lei, Ningzhi Jin * and Jiaxin Jiang

School of Electrical and Electronic Engineering, Harbin University of Science and Technology, Harbin 150080, China; leishanshanjy@outlook.com (S.L.); jianggazn@163.com (J.J.)
* Correspondence: sharon0716@hrbust.edu.cn

Abstract: To address problems that traditional two-stage inverters suffer such as high cost, low efficiency, and complex control, this study adopts a quasi-Z-source cascaded multilevel inverter. Firstly, the quasi-Z-source inverter utilizes a unique impedance network to achieve single-stage boost and inversion without requiring a dead zone setting. Additionally, its cascaded multilevel structure enables independent control of each power unit structure without capacitor voltage sharing problems. Secondly, this study proposes a current-predictive control strategy to reduce current harmonics on the grid side. Moreover, the feedback model of current and system state is established, and the fast control of grid-connected current is realized with the deadbeat control weighted by the predicted current deviation. And a grid-side inductance parameter identification is added to improve control accuracy. Also, an improved multi-carrier phase-shifted sinusoidal PWM method is adopted to address the issue of switching frequency doubling, which is caused by the shoot-through zero vector in quasi-Z-source inverters. Finally, the problems of switching frequency doubling and high harmonics on the grid side are solved by the improved deadbeat control strategy with an improved MPSPWM method. And a seven-level simulation model is built in MATLAB (2022b) to verify the correctness and superiority of the above theory.

Keywords: quasi-Z-source inverter; cascaded multilevel; deadbeat current control; multi-carrier phase-shifted sinusoidal PWM

1. Introduction

In recent years, there has been growing interest in research on photovoltaic power generation systems. The energy conversion efficiency and power generation stability of these systems are directly dependent on the selection of an inverter and its control method. Therefore, it is crucial to carefully choose an appropriate inverter and control strategy. The quasi-Z-source inverter replaces the boost circuit in the middle stage of the traditional two-stage inverter by adding an inductance and capacitance network between the inverter bridge and the DC side [1], thus using the shoot-through state of the upper and lower bridge arms that is not allowed by the traditional inverter to raise the DC side voltage and realize the single-stage boost inverter. Therefore, there is no need to place the dead zone in the control of the inverter, which avoids the harmonics caused by the existence of the dead zone, reduces the distortion rate of the output waveform, and improves the stability of the whole system. Moreover, the quasi-Z-source inverter, equipped with a series passive network before the inverter bridge, can function as both a voltage source inverter (VSI) and a current source inverter (CSI), eliminating the need for combining large capacitors or large inductors in series. Therefore, based on the above advantages, the quasi-Z-source inverter has been applied and developed in small- and medium-sized power applications such as new energy power generation, power batteries, and AC speed regulation systems.

A cascaded multilevel inverter (CMI), compared with a traditional two-level inverter, can improve the output current and voltage harmonics. Because the output waveform is a trapezoidal wave, when using PWM technology, its output waveform is closer to a sine wave [2]. Then, the cascaded multilevel inverter can realize high-voltage output by using low-voltage withstand devices and, compared with the clamped inverter with the series structure of power electronic devices, each H-bridge inverter is cascaded as the basic unit, which not only reduces the power burden of cascaded units but also avoids the problem of midpoint voltage balance [3,4]. In addition, the cascaded multilevel inverter, utilizing the H-bridge structure with an independent DC power supply as its fundamental power unit, can individually control each unit to operate at the maximum power point, thereby significantly enhancing operational efficiency [5,6].

In this study, a combination of the quasi-Z-source and cascaded multilevel techniques is adopted, leveraging the respective advantages of both. It can achieve a single-stage boost inverter. Also, there is no need to set the dead voltage, which solves the problem of complex control and high cost and eliminates the output voltage distortion caused by the addition of dead zone voltage. Moreover, integrating the application background of photovoltaic power generation, each output unit can operate independently at its maximum power point, enhancing efficiency [5,6].

For the control and modulation method of the system, due to the structural particularity of the quasi-Z-source cascaded multilevel inverter, there is a certain coupling relationship between the DC chain voltage rise and the inverter; that is, the sum of the modulation ratio and the shoot-through duty ratio must be less than or equal to one. Firstly, to address the issue of high harmonic content on the output side of the inverter bridge caused by the charging and discharging of inductors and capacitors in the qZS-CMI's quasi-Z-source network, this study employs an improved current-predictive control strategy that integrates mathematical modeling with state prediction [7–10]. This approach effectively mitigates output harmonics because of its benefits, such as fast response time, low computational requirements, and stable switching frequency [11–14]. Secondly, this study proposes an enhanced multi-carrier phase-shifted modulation method to address the issue of increased switching frequency and subsequent switching loss due to the insertion of the shoot-through zero vector [15–17]. Specifically, it is accomplished by altering the placement of the zero vector. Additionally, in order to prevent the control failure caused by the change in the grid-side parameters, the least squares method based on the forgetting factor is introduced to identify the parameters of the grid-side filter inductor [18,19].

In summary, this study first expounds the principle of a quasi-Z-source cascaded multilevel inverter and establishes a corresponding mathematical model. Then, aiming at the problem of high harmonic distortion in the qZS-CMI output and the multiplication of switching loss, an improved current-predictive control strategy of grid connected current is adopted. And the least squares method based on the forgetting factor is introduced to identify the parameters of the grid-side filter inductance to reduce grid-connected static error. Moreover, an improved multi-carrier phase-shifted modulation method is proposed to reduce the switching loss by changing the implantation time of the zero vector. Finally, a simulation model is used to verify its effectiveness.

2. Principle of Quasi-Z-Source Cascaded Multilevel Inverter

The topology of a quasi-Z-source cascaded multilevel inverter, shown in Figure 1, is a cascade of N such identical units which consists of a qZS_network in the front stage and an HBI in the back stage. A qZS_network is a quasi-Z-source inductance and capacitance network which completes the voltage boost on the DC side. An HBI is an H-bridge composed of electronic devices which completes the DC/AC inverter, L represents the filter inductance on the power grid side, and v_g represents the power grid voltage.

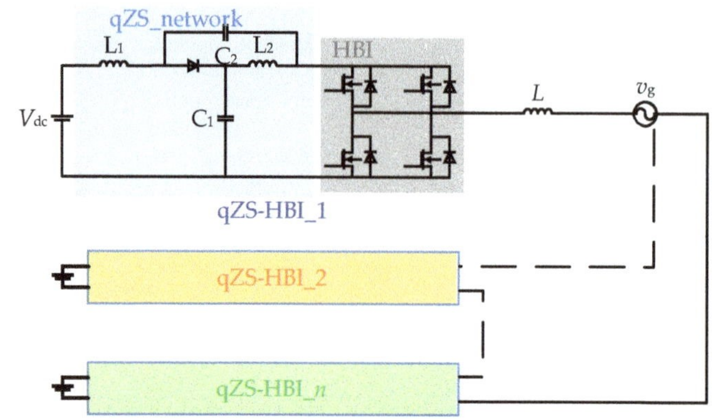

Figure 1. Diagram of the quasi-Z-source cascaded multilevel inverter.

2.1. Voltage Boost Principle of the Quasi-Z-Source

The quasi-Z-source inverter utilizes its two operational states, namely, the shoot-through state and non-through state, to achieve its intended functionality. In steady-state operation, the inverter bridge can be considered equivalent to a current source. Figure 2 below depicts the steady-state equivalent circuit diagram of the quasi-Z-source inverter.

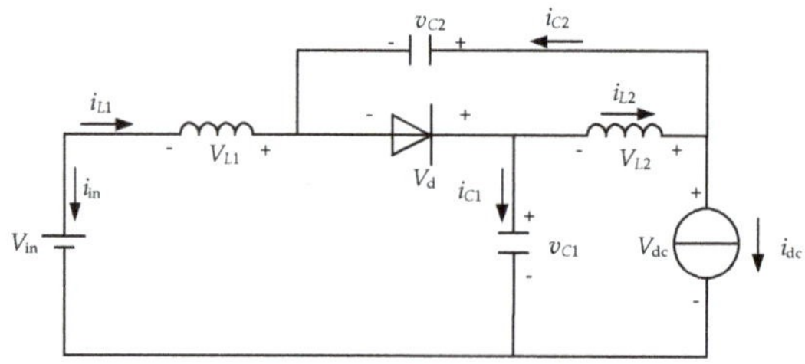

Figure 2. Equivalent circuit diagram of quasi-Z-source inverter.

When operating in the shoot-through state, the diode is in reverse cutoff mode, causing the capacitor to discharge. Simultaneously, both the DC power supply and the capacitor charge the inductor. In contrast, when functioning in the non-through state, the diode enters an on state while discharging the inductor. At this time, both the DC power supply and the inductor charge the capacitor while providing power to load. By analyzing these two states comprehensively, we can establish a relationship as shown by Equation (1):

$$\begin{cases} v_{C1} = \frac{1-D_0}{1-2D_0} \cdot V_{in} \\ v_{C2} = \frac{D_0}{1-2D_0} \cdot V_{in} \\ V_{\mathrm{dc}} = v_{C1} + v_{C2} = \frac{1}{1-2D_0} \cdot V_{in} \end{cases} \quad (1)$$

where the voltages at both ends of capacitors C_1 and C_2 are represented by v_{C1} and v_{C2}, respectively. D_0 denotes the shoot-through duty ratio, V_{in} represents the input voltage of the DC power supply, and V_{dc} signifies the DC chain voltage.

2.2. Principle of Cascaded Multilevel Inverter

In this system, the cascaded units of the quasi-Z-source cascaded multilevel inverter are independent of each other and cascaded by the circuit shown in Figure 3. This ensures that the cascaded superposition circuit can remain unimpeded when the H-bridge circuit outputs any current level, and the H-bridge circuit of each unit has four basic working states.

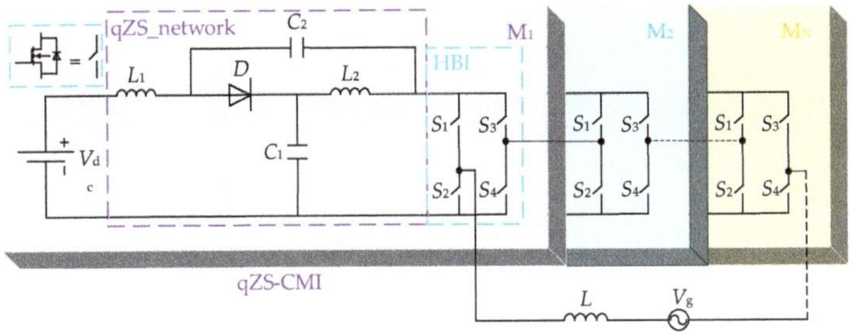

Figure 3. Quasi-Z-source cascaded multilevel inverter topology.

When switch tubes S_1 and S_4 are turned on, the H-bridge is in the forward conduction state, resulting in a positive output voltage of E; when switch tubes S_2 and S_3 are turned on, the H-bridge operates in reverse mode, generating a negative output voltage of $-E$. When switches S_1 and S_3 are turned on, the H-bridge functions as a forward bypass with zero output voltage, allowing current to flow forward; when switch tubes S_2 and S_4 are turned on, the H-bridge acts as a reverse bypass with zero output voltage while enabling current to flow in reverse. The CMI working states are shown in Figure 4. A CMI can work normally only when the above four working states exist simultaneously, which is also an important feature that distinguishes cascaded multilevel inverters from clamped inverters.

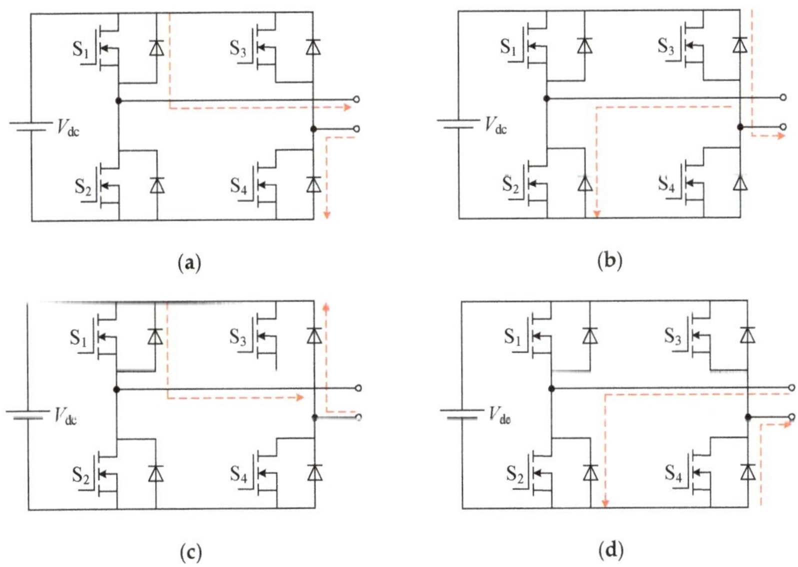

Figure 4. CMI working states: (**a**) forward conduction, (**b**) reverse conduction, (**c**) forward bypass, (**d**) reverse bypass.

2.3. QZS-CMI Mathematical Model

The mathematical model of the quasi-Z-source cascaded multilevel inverter, based on its principle, is established as follows. Firstly, the essence of the cascaded H-bridge inverter is to connect multiple H-bridges in series and realize multilevel output through the four-quadrant operation of the H-bridge inverter. Consequently, the output voltage v_o can be expressed as the summation of each module's output voltage:

$$v_o = \sum_{i=1}^{N} v_{dci} \cdot S_i \tag{2}$$

where v_{dci} is the DC chain voltage of module i, defined as S_i (i = 1, 2, 3... N), which is a switching function that meets the following condition:

$$S_i \in \{-1,0,1\} \tag{3}$$

Moreover, if the modulation ratio of each quasi-Z-source module is denoted as M_i, the expressions for both the output voltage v_{oi} and the total output voltage v_o of module i can be derived from Equation (4):

$$\begin{cases} v_{oi} = v_{dci} \cdot M_i \sin(\omega t) \\ v_o = \sum\limits_{i=1}^{N} v_{oi} \end{cases} \tag{4}$$

where ω is the angular frequency of the voltage on the grid side. Since qZS-CMI is generally used in small- and medium-power situations, it can be approximately considered that the output reactive power of the AC side is zero; that is, the power factor angle is equal to zero.

Set the grid voltage and current amplitude to V_g and I_g. Then, the instantaneous value of the grid voltage and current can be expressed as

$$\begin{cases} v_g = V_g \cdot \sin(\omega t) \\ i_g = I_g \cdot \sin(\omega t) \end{cases} \tag{5}$$

In addition, a single inductor filter is used at the grid side. The value of the filter inductance is set as L; then, the following can be obtained from KVL:

$$v_o = L\frac{d_{ig}}{d_t} + v_g \tag{6}$$

where the instantaneous values of power grid voltage and current are represented by v_g and i_g, respectively, while the output voltage of the inverter is denoted as v_o.

3. Current-Predictive-Control-Improved Deadbeat Current Control

3.1. Principle of Deadbeat Current Control

Deadbeat current control (DBC) is a control technique that utilizes a mathematical model of the circuit to accurately predict the value of the next sampling period based on the current system state. In comparison to other control methods, it offers advantages such as rapid response, minimal computation requirements, and stable switching frequency. By employing the forward Euler approximation formula and assuming a sufficiently small sampling period, we can derive the following equation by rewriting Equation (6):

$$v_o(k) = \frac{L}{T_s}(i_g(k+1) - i_g(k)) + v_g(k) \tag{7}$$

where T_s is the sampling period, $v_o(k)$ is the sampling value of the inverter output voltage at the current moment, $v_g(k)$ and $i_g(k)$ are the sampling values of the grid voltage and current at the current moment, and $i_g(k+1)$ is the sampling value of the grid current at the next moment.

When the fast-tracking of the grid current is realized in a sampling period,

$$i_g(k+1) = i_{ref}(k) \tag{8}$$

where $i_{ref}(k)$ is the grid current reference value.

The inverter output voltage reference value $v_{oref}(k)$ can be obtained by substituting Equation (8) into Equation (7):

$$v_{oref}(k) = \frac{L}{T_s}(i_{ref}(k) - i_g(k)) + v_g(k) \tag{9}$$

For single-phase inverters with SPWM, the modulation ratio M_i can be expressed as

$$M_i(k) = \frac{v_{\text{oref}}(k)}{v_{dc}(k)} \tag{10}$$

For the cascaded inverter, since the cascade is a topology that connects H-bridges by N in series to achieve multilevel output, if the input power of each H-bridge power module is balanced, the modulation ratio of each H-bridge module is

$$M_i(k) = \frac{v_{\text{oref}}(k)}{N \cdot v_{dci}(k)} \tag{11}$$

where $v_{dci}(k)$ is the output voltage of the H-bridge module i at the current time, and N is the total number of H-bridge modules.

The above process is the DBC principle of grid-connected current using cascaded inverters. The general idea is to predict the reference value of the inverter output voltage at the k'th moment by referring to the reference value of the grid-connected current, and then obtain the modulation signal M_i.

3.2. Improved Deadbeat Current-Predictive Control

The traditional DBC requires that the grid-connected current can track the reference without static error in a sampling period; that is, the current deviation i_e is zero. However, due to the inherent delay of the digital system, the current deviation cannot become zero in a sampling period. To realize the grid-connected current fast-tracking reference current, this study adopts an improved grid-connected current deadbeat predictive control.

Equation (7) can be deduced one step further:

$$v_{\text{oref}}(k+1) = \frac{T_s}{L}(i_g(k+2) - i_g(k+1)) + v_g(k+1) \tag{12}$$

The current deviation at time k and time $k + 1$ is defined as

$$\Delta i_g(k) = i_g(k+1) - i_g(k) \tag{13}$$

$$\Delta i_g(k+1) = i_g(k+2) - i_g(k+1) \tag{14}$$

The requirement of DBC $\Delta i_g(k + 1) = 0$ is relaxed, and $\Delta i_g(k + 1)$ is equal to the arithmetic mean of the root of the current deviation at two adjacent moments at $k + 1$; that is,

$$\Delta i_g(k+1) - \frac{1}{2}[\Delta i_g(k+1) + \Delta i_g(k)] - \frac{1}{2}(i_g(k+2) - i_g(k)) \tag{15}$$

The predicted value of the output voltage of the inverter can be obtained by substituting Equation (15) into Equation (12):

$$v_{\text{oref}}(k+1) = \frac{T_s}{2L}(i_g(k+2) - i_g(k+1)) + v_g(k+1) \tag{16}$$

The grid voltage at time $k + 1$ in Equation (16) can be obtained by linear extrapolation:

$$\hat{v}_g(k+1) = 2v_g(k) - v_g(k-1) \tag{17}$$

The modulation ratio M_i of each H-bridge module can be expressed as

$$M_i(k) = \frac{v_{\text{oref}}(k+1)}{N \cdot v_{dci}(k)} \tag{18}$$

The current closed-loop control block diagram, as depicted in Figure 5 below, is established for the aforementioned enhanced deadbeat current control strategy.

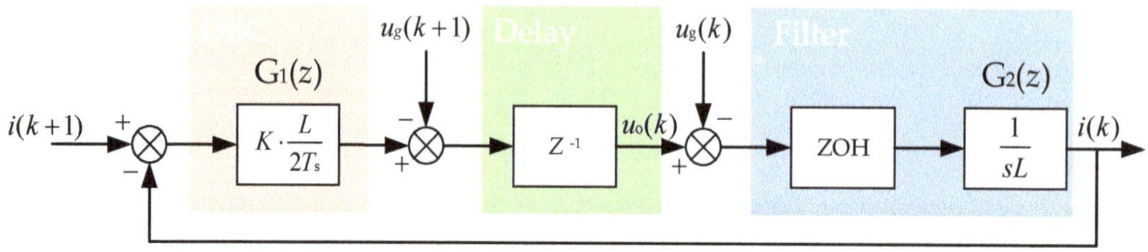

Figure 5. System control block diagram.

The system's responsiveness is analyzed using the above-mentioned system control block diagram. In the figure, T_s represents the sampling period of the system; $G_1(z)$ denotes the deadbeat controller, with K being the ratio between actual and given inductance values; Z^{-1} signifies inherent delay in digital systems; ZOH stands for zero-order holder; and $G_2(z)$ represents the filter inductance on the output side. The role of the zero-order holder within this system is to substitute discrete digital signals with continuous analog signals; its transfer function can be expressed as

$$G_Z(s) = \frac{1 - e^{-sT_s}}{s} \tag{19}$$

where T_s indicates the system sampling period.

The frequency domain transfer function of the zero-order holder and the actuator are discretized together in discrete systems, resulting in $G_2(z)$ in the discrete domain. The transformation process is illustrated by Equation (20).

$$G_2(z) = \frac{T_s}{L}(1 - z^{-1}) \cdot Z[\frac{1}{s^2}] = \frac{T_s}{L} \cdot \frac{1}{(z-1)} \tag{20}$$

According to the aforementioned conclusions, the forward transmission function of the system can be derived as follows:

$$G(z) = G_1(z) \cdot z^{-1} \cdot G_2(z) = \frac{K}{2z(z-1)} \tag{21}$$

The characteristic equation of the system is obtained:

$$F(z) = 2z^2 - 2z + K = 0 \tag{22}$$

The Jury stability criterion for second-order discrete systems is as follows:

$$\begin{cases} a_2 + a_1 + a_0 > 0 \\ a_2 - a_1 + a_0 > 0 \\ |a_0| < a_2 \end{cases} \tag{23}$$

The above equation features the leading coefficients of the characteristic equation arranged in descending order of power, namely, a_2, a_1, and a_0.

Based on the substitution of $a_2 = 2$, $a_1 = -2$, and $a_0 = K$ into the equation of the Jury stability criterion, the conclusion can be drawn that the system is stable at $0 < K < 2$. Under the same conditions, the stability region of the traditional deadbeat current control is $0 < K < 1$.

The rapidity and stability of the improved deadbeat algorithm were verified in the MATLAB/Simulink (2022b) simulation environment, disregarding the disturbance of the load voltage at the rear stage. The simulation results are presented in Figure 6.

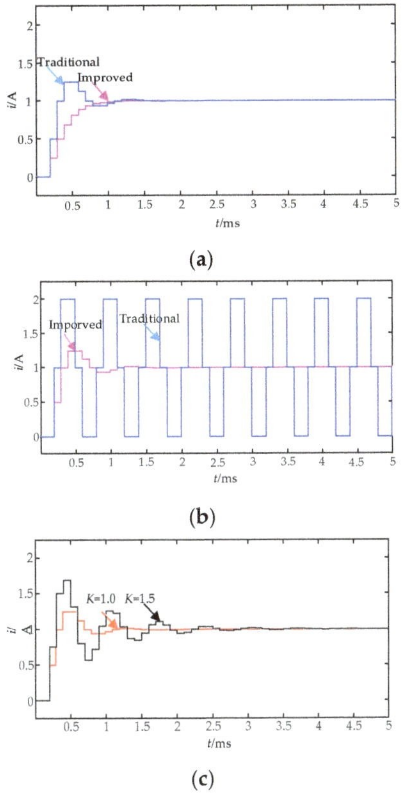

Figure 6. Simulation results of the enhanced deadbeat algorithm: (**a**) $K = 0.5$; (**b**) $K = 1.0$; (**c**) $K = 1.0$ and 1.5.

According to the simulation results in Figure 6a, it is evident that the improved deadbeat control algorithm demonstrates faster reference tracking without any overshoot, thereby significantly improving its speed compared with the traditional deadbeat algorithm. Figure 6c validates the stability of the improved deadbeat algorithm. However, as K increases, both systems overshoot and transition time also increases. Hence, this indirectly demonstrates that the deadbeat algorithm imposes higher accuracy requirements on the system model. As depicted in Figure 6b, when $K = 1$, the traditional deadbeat algorithm achieves critical stability while the improved deadbeat algorithm attains a state of basic stability, indicating an enhancement in stability relative to the traditional one.

3.3. Parameter Identification of Grid-Side Inductance Based on the Least Squares Method

From the analysis of the previous section, it can be seen that as a special model of predictive control, the effect of deadbeat current-predictive control is closely related to the accuracy of the established mathematical model. If the actual inductance is too large, there will be a steady-state error i_e between the grid-connected current and the reference current, and the value of i_e will increase with the increase in the inductance value. If the actual inductance value is small because the inductance mainly starts to filter, a too-small inductance value will cause the harmonic distortion rate of the grid-connected current to be too large, which cannot meet the requirements of grid connection. The above two situations will lead to the control failure of the grid-connected inverter, so when the above two situations occur, the inverter needs to stop working to prevent its impact on the large power grid. In this study, by adding the parameter identification of the filter inductance value, the change in the inductance value is monitored in real time to avoid the above situation.

The least squares method is currently widely used in system parameter estimation because of its simple and easy-to-understand advantages. However, in practical applications, the recursive least squares method will have the phenomenon of data saturation, leading to algorithm failure. To overcome these limitations, and to dilute the influence of old data on new data while improving the influence of new data, this study adopts recursive least squares with forgetting factor (FRLS) and its value function is

$$J = \sum_{k=1}^{n} \lambda^{(n-k)} \cdot [y(k) - \phi^{\mathrm{T}}(k)\hat{\theta}(k)]^2 \tag{24}$$

where $y(k)$ is the output matrix, $\phi(k)$ is the input matrix, and $\theta(k)$ is the parameter matrix to be identified, among them $y(k) = \phi(k)\theta(k)$. In this study, $y(k) = [i_g(k+1) - i_g(k)]$, $\phi(k) = [v_o(k)\ v_g(k)]$, $\theta(k) = [G\ \text{-}G]T$, where $G = L/T_s$.

Combined with the derivation of the recursive least squares method, the FRLS iteration equation is

$$\begin{cases} \hat{\theta}(k) = \hat{\theta}(k-1) + K(k)[y(k) - \phi^{\mathrm{T}}(k)\hat{\theta}(k-1)] \\ K(k) = \frac{P(k-1)\phi(k)}{\lambda + \phi^{\mathrm{T}}(k)P(k-1)\phi(k)} \\ P(k) = [I - K(k)\phi^{\mathrm{T}}(k)]P(k-1)/\lambda \end{cases} \tag{25}$$

where $K(k)$ represents the number of iterations, and λ is the forgetting factor, generally 0.95~1.0. Because the FRLS algorithm needs to use the prior values of the covariance matrix P and the parameter estimation matrix $\hat{\theta}$ in the iterative process, $P(0)$ and $\hat{\theta}(0)$ are usually given in advance. Generally, we let $P(0) = \delta^{-1} \cdot I$, where δ is a small real number, I is the unit matrix, and $\hat{\theta}(0) = 0$.

It can be seen from Equation (25) that when the value of λ is less than 1, as the data continue to iterate, the old data will decay exponentially, which is equivalent to adding a weight factor to the old data that decays exponentially.

4. Multi-Carrier Phase-Shifted Sinusoidal Pulse Width Modulation

Since the Z-source/quasi-Z-source inverter was proposed, modulation methods have emerged continuously. At present, for the single-phase qZS-CMI, the combination of simple boost and PSPWM is generally used as the modulation method of single-phase qZS-CMI, as shown in Figure 7a. The idea is to insert the shoot-through zero vector into the traditional zero vector in the inverter. However, the insertion of the shoot-through zero vector will double the switching frequency, resulting in an increase in switching losses, thereby reducing the efficiency of the system. To address these issues, this paper adopts an improved multi-carrier phase-shifted sinusoidal pulse width modulation (MPSPWM) approach. By adjusting the insertion time of the shoot-through vector, the switching frequency is reduced to mitigate switching losses. The improved multi-carrier phase-shifting SPWM scheme is illustrated in Figure 7b below.

The basic principle, as illustrated in the above figure, involves translating the original carrier to ensure that the two switching tubes of each bridge arm utilize distinct carriers with an amplitude difference of D_0. This extension of on-time for a specific switching tube within the same bridge arm fulfills the voltage boost requirement. The driver signals G_1 and G_2 share a modulated wave but employ different carriers. The amplitudes of these two carriers are $1 - D_0/2$ and $1 + D_0/2$, respectively, where D_0 represents the shoot-through duty ratio. The same applies to G_3 and G_4. While G_1 and G_4 share a common carrier, they adopt different modulated waves with a phase difference of 180° between them; the same is true for G_2 and G_3. The specific modulation method is shown in Figure 8.

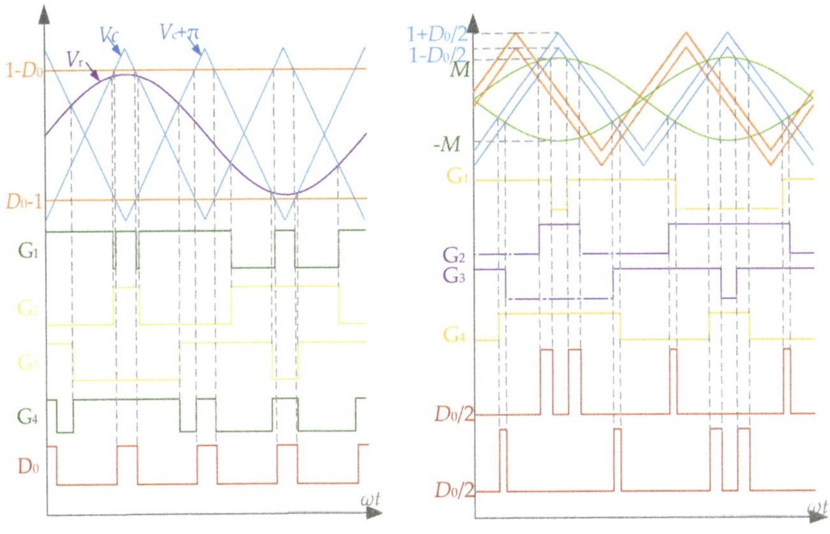

Figure 7. (a) Simple boost modulation and PS-PWM, (b) multi-carrier phase-shifted SPWM.

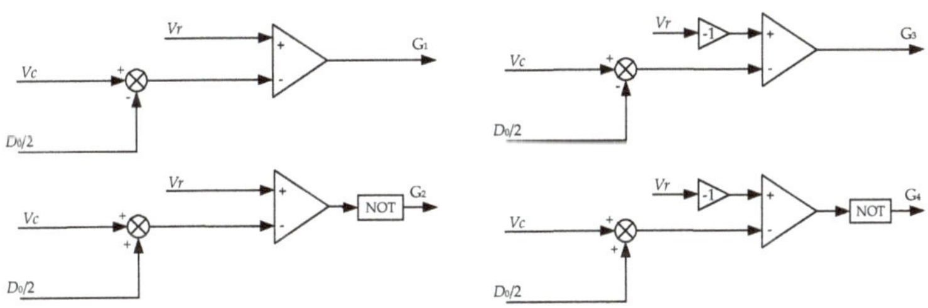

Figure 8. Multi-carrier phase-shifted SPWM modulation schematic.

In the diagram, V_c and V_r are triangular carriers and sinusoidal modulation waves with an amplitude of 1, respectively; D_0 is a shoot-through duty cycle; and G_1, G_2, G_3, and G_4 are four driving signals. When $V_r > (V_c - D_0/2)$, the S_1 output is positive, and vice versa; when $-V_r > (V_c - D_0/2)$, the S_3 output is positive, and vice versa; when $V_r > (V_c + D_0/2)$, S_2 output is positive, and vice versa; and when $-V_r > (V_c + D_0/2)$, the S_4 output is positive, and vice versa.

5. Results of the System

In order to validate the accuracy of the proposed system in this study, the MATLAB/Simulink (2022b) simulation platform was utilized to construct a seven-level qZS-CMI system simulation model. The control block diagram of the model is depicted in Figure 9 and the seven-level qZS-CMI system simulation model is depicted in Figure 10.

The power level, depicted in the figure, is a cascaded multilevel quasi-Z-source inverter module; DBC refers to an enhanced grid-connected deadbeat current controller. Its operational procedure can be outlined as follows: Firstly, the load voltage, current, and DC bus voltage are sampled. To ensure that the output current remains in phase with the grid voltage, phase information of the specified current is obtained with PLL. Subsequently, four signals are transmitted to the deadbeat controller for acquiring modulation signal M_i. Finally, multi-carrier phase-shifted SPWM is employed to modulate both shoot-through signal D_0 and modulated signal M_i, resulting in the generation of 12 driver signals. The specific simulation parameters are shown in Table 1.

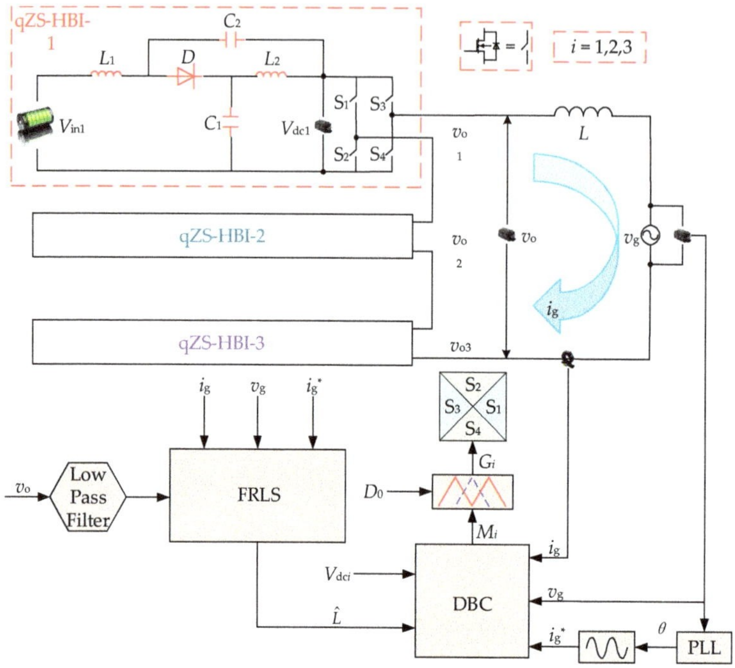

Figure 9. Block diagram of the qZS-CMI control strategy.

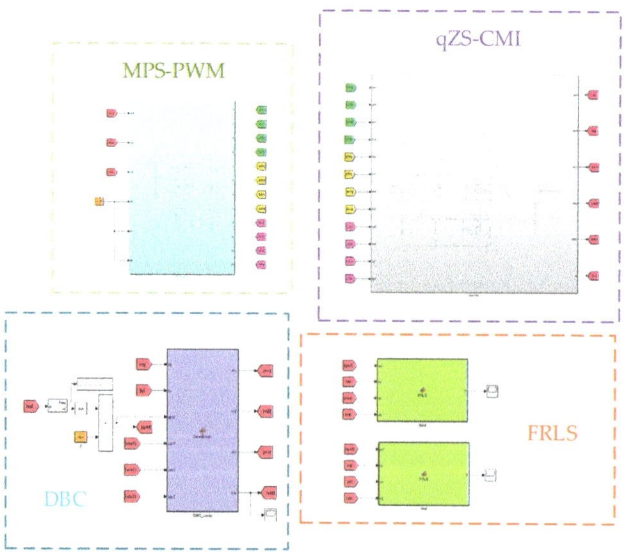

Figure 10. Seven-level qZS-CMI system simulation model.

Table 1. Main simulation parameters of qZS-CMI.

Main Parameters	Values
Quasi-Z-source Inductance (H)	0.003
Quasi-Z-source Capacitance (F)	0.004
Input Voltage (V)	35
Grid Voltage (V)	150
Grid Current (A)	2
Filter Inductance (H)	0.01
Maximum Shoot-through Duty Ratio	0.25
Switching Frequency (kHz)	10

5.1. Analysis of System Steady-State Simulation Results

In this study, given an input DC voltage of 35 V, the DC chain voltage can be calculated as V_{dc} = 70 V using Equation (1). Consequently, it is evident that the peak value of the inverter's output voltage is $V_o = N \times V_{dc}$ = 210 V. For the simulation, Figure 11 illustrates the output seven-level AC voltage of the inverter, while Figure 12 depicts the waveform diagram of the DC chain voltage for any Z-source module. The obtained values align with theoretical expectations, thereby validating the accuracy and feasibility of the qZS-CMI mathematical model.

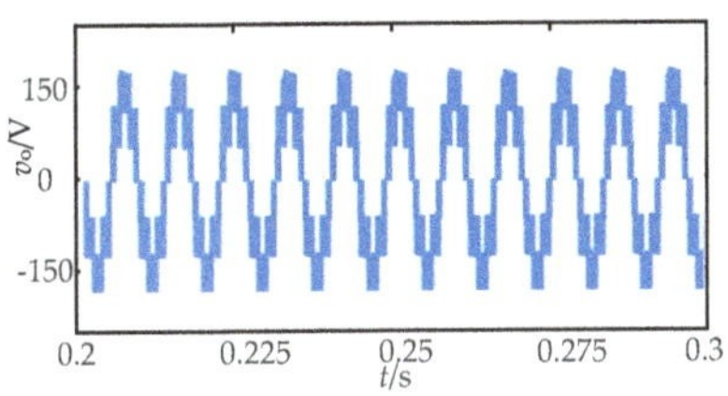

Figure 11. Inverter output voltage.

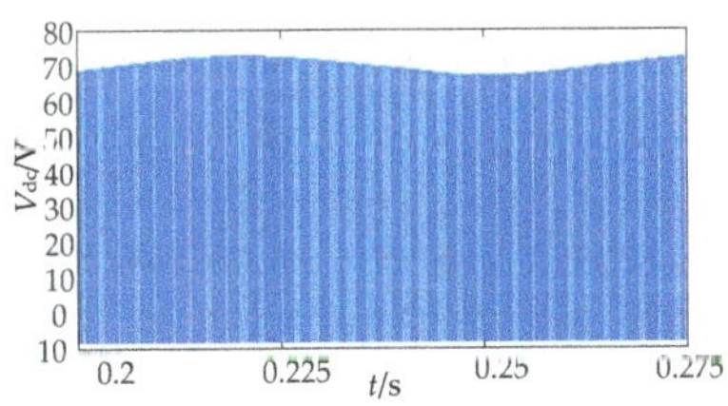

Figure 12. DC link voltage.

The driving signal of a qZS-CMI module is illustrated in Figure 13 below, with a fixed duty ratio set at 0.25. Following multi-carrier phase-shifted modulation, the shoot-through zero vector is divided into four segments and incorporated into the switching state. As a result, there will be four shoot-through zero states within each switching cycle, with each direct-zero state lasting for $D_0/4$ duration.

The current waveform at the grid side is illustrated in Figure 14, demonstrating that the current successfully tracks the set value after approximately 0.35 s, thereby validating the efficacy of the proposed enhanced deadbeat current control method in this study. Additionally, the current at the grid side exhibits phase alignment with the grid voltage, as depicted in Figure 15.

Figure 16 illustrates the total harmonic distortion (THD) of the grid side using both traditional and improved deadbeat control algorithms, which resulted in rates of 2.55% and 0.86%, respectively. It is evident that the deadbeat control algorithm effectively reduces the THD of the grid-connected current.

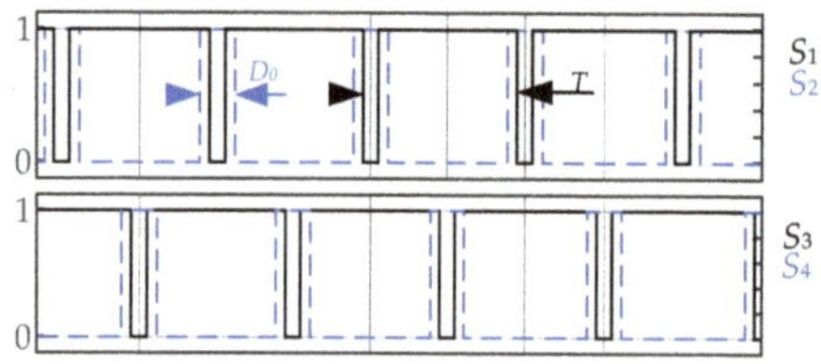

Figure 13. Drive signals.

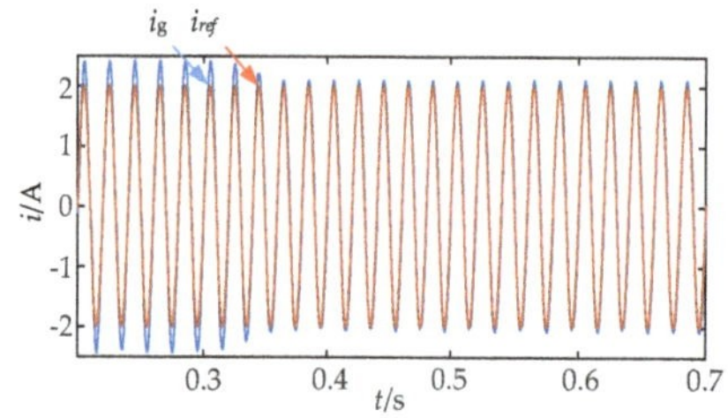

Figure 14. Grid current waveform.

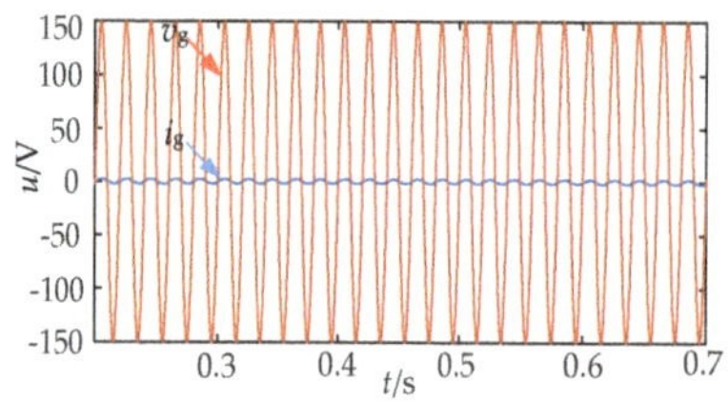

Figure 15. Grid voltage and grid current waveforms.

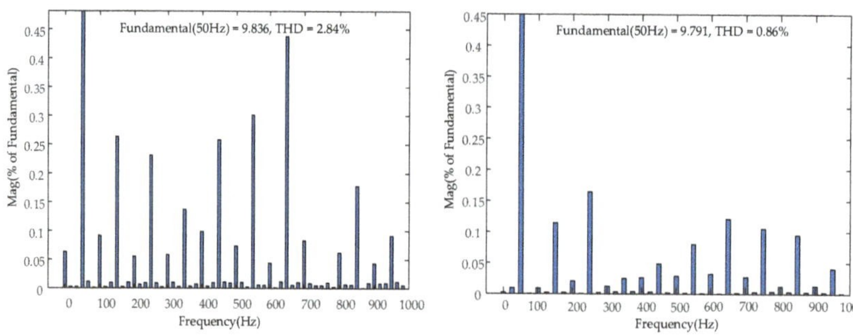

(a) (b)

Figure 16. Harmonic distortion rate of grid-connected current: (**a**) traditional control; (**b**) improved control.

Figure 17 shows the estimated value of the filter inductance. Combined with Figure 14 and Table 1, it can be seen that the FRLS algorithm can undertake the parameter identification of the qZS-CMI filter inductance.

According to Equation (7), the output voltage $v_{oref}(k + 1)$ predicted by the deadbeat algorithm, after the system reaches a steady state, should be equal to the grid voltage $v_g(k + 1)$. Figure 18 illustrates the waveform diagram of the output voltage predicted by the deadbeat algorithm and the grid-side voltage.

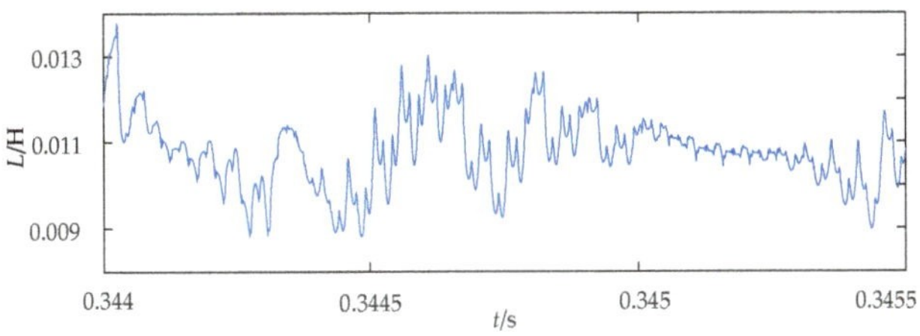

Figure 17. Filtering inductance estimation.

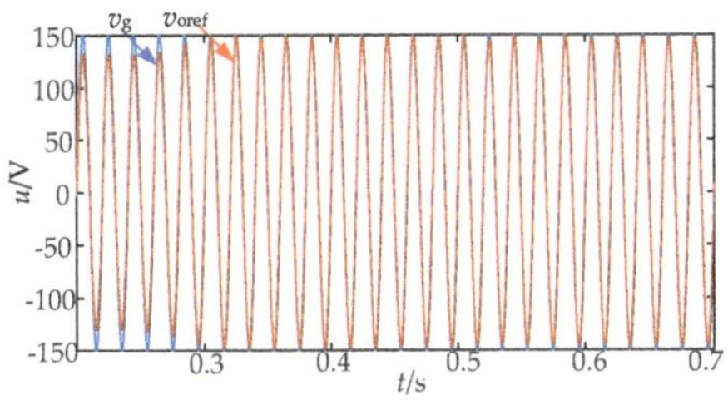

Figure 18. Predicted voltage waveform.

The voltage v_{oref}, as depicted in Figure 18, exhibits excellent tracking capability with the grid-side voltage v_g once the system reaches a steady state. This ensures precise modulation ratio M and satisfies the coupling relationship between the modulation ratio and shoot-through duty ratio.

5.2. Analysis of System Dynamic Simulation Results

To verify the stability of the improved deadbeat current-predictive control, the grid-side inductance is reduced at 0.5 s, and the grid-side inductance is increased at 0.6 s; that is, at 0.5 s, $K = 0.5$, and at 0.6 s, $K = 1.5$. Figures 19 and 20 provide the grid-connected current waveform and the grid-side inductance value identified using the FRLS algorithm.

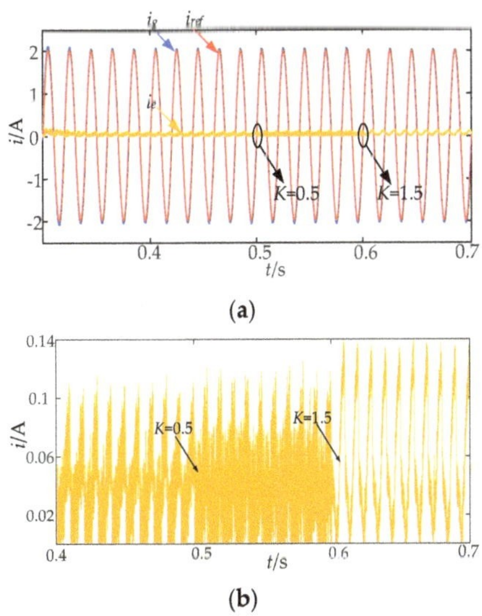

Figure 19. Current waveform when the grid-side inductance changes: (**a**) grid-connected current; (**b**) local amplification of current dynamic error.

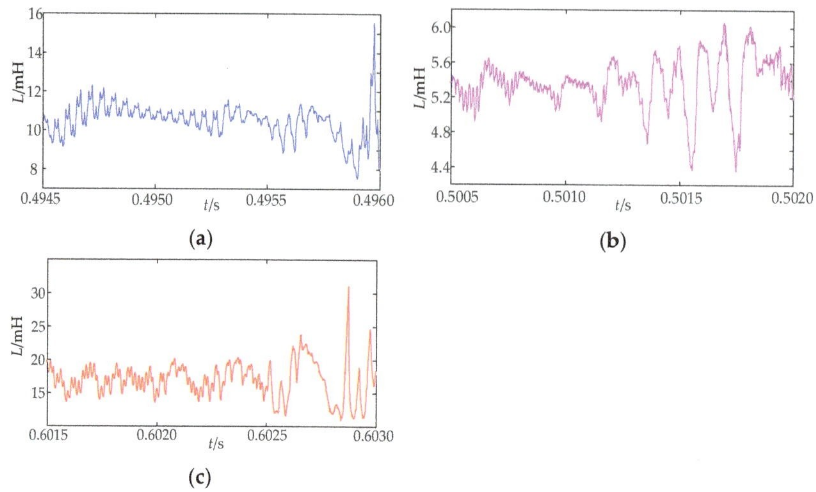

Figure 20. Network-side inductance parameter identification waveform: (**a**) 10 mH; (**b**) 5 mH; (**c**) 15 mH.

Figure 19a,b shows the process of grid-connected current tracking the given current under the fluctuation of grid-side inductance and the absolute value of the difference from the grid-connected current, respectively. It can be seen that the grid-connected current can effectively track the given value, and the grid-connected current has no obvious fluctuation in the case of grid-side inductance fluctuation, and its dynamic error fluctuates between 0 and 0.14 A. In addition, between 0.5 s and 0.6 s in the local amplification diagram, the harmonic component of the grid-side current increases because the filter inductance value is reduced to half of the original value. After 0.6 s, because the filter inductance value increases to 1.5 times the original, i_e increases; that is, the system overshoot increases and tends to be critically stable. The above simulation results verify well the content discussed in Section 3.2.

Figure 20 shows the inductance values identified using the FRLS algorithm in the process of grid-side inductance fluctuation. From the change curves in Figure 20a–c, it can be found that the FRLS algorithm has a large overshoot; in particular, when the inductance value increases, the overshoot is more obvious, about 10 mH. This also shows that the FRLS algorithm is sensitive to noise.

In addition, since the improved DBC strategy in this study does not add a voltage loop, it is necessary to verify the change in grid-connected current when the input voltage changes.

After the system enters the steady state, a 2.5 V step signal is applied to the input voltage V_{dc} at 0.5 s. Figures 21 and 22 present the waveforms of the grid-connected current and the DC link voltage when the input voltage changes.

From the simulation results in Figure 21a,b, it can be seen that when the input voltage fluctuation system re-transitions to a steady state, the grid-connected current produces a small overshoot with a value of about 0.06 A, but after one quarter of the cycle, that is, 0.005 s, it can re-enter the steady state. Figure 22 shows that the DC link voltage jumps to 75 V at this time. In addition, since the capacitor voltage cannot mutate and the inductor current cannot mutate, the DC link voltage does not jump immediately when the input voltage fluctuates. Therefore, for a grid-connected system with a relatively stable input voltage, the improved DBC does not need to adopt double closed-loop control, which also simplifies the design of the control link to a certain extent.

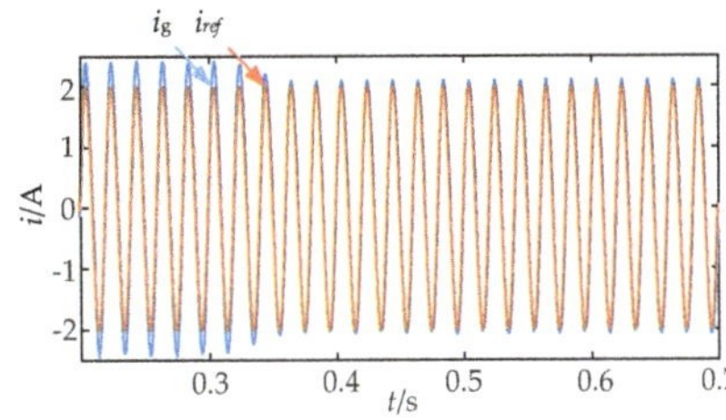

(a)

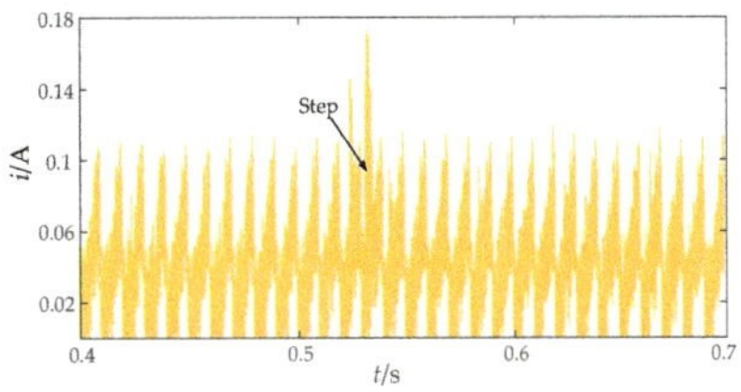

(b)

Figure 21. Current waveform when input voltage fluctuates: (**a**) grid-connected current; (**b**) local amplification of current dynamic error.

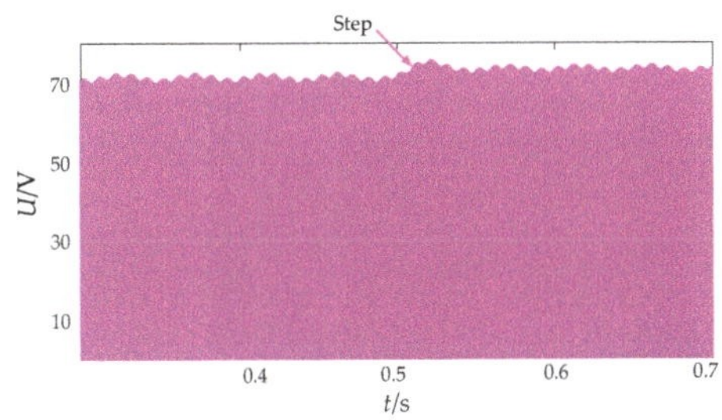

Figure 22. DC link voltage when input voltage fluctuates.

5.3. Analysis of Experimental Results

According to the design parameters listed in Table 1, this study constructed a seven-level qZS-CMI experimental platform with a power rating of 0.75 kW. The obtained experimental data were essentially consistent with the simulation results, which further validated the accuracy and feasibility of the system. The main parameters of the experimental platform are shown in Table 2; for other parameters, refer to the simulation section.

Table 2. Main parameters of qZS-CMI.

Main Parameters	Values
Input Voltage (V)	35
DC Link Voltage Peak (V)	70
Peak Output Voltage After Filtered (V)	150
Peak Output Current (A)	10
Switching Frequency (kHz)	10

Figure 23a presents the DC link voltage waveforms of three independent qZS-HBI modules. The three DC link voltages can remain stable and the peak values are equal to about 70 V. When the shoot-through duty cycle is 0.25, as mentioned in Section 5.1, the peak value of the DC link voltage should be 70 V. The waveform in Figure 23a is just consistent with the theoretical calculation value. Figure 23b shows the voltage waveform of the capacitor C_1 in the quasi-Z-source network of three independent qZS-HBI modules. According to the setting value of the shoot-through duty cycle and the input voltage parameters in the program and Equation (1), the peak value of the voltage of the capacitor C_1 is 60 V, while the peak value of the capacitor voltage waveform shown in Figure 23b is basically stable at 60 V; therefore, the theoretical analysis is consistent with the experimental results, and the capacitor voltage waveform is continuous.

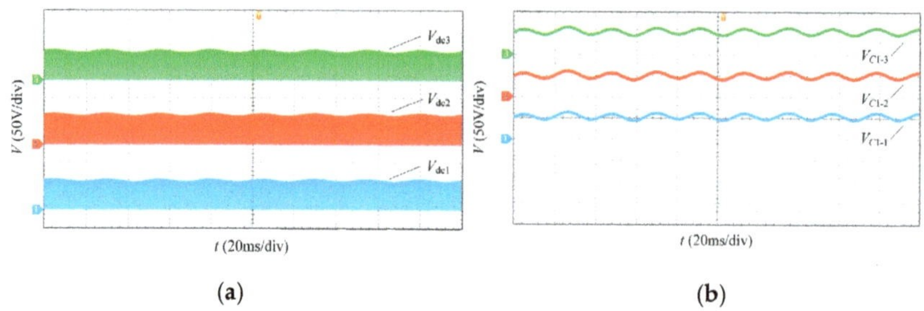

Figure 23. DC link voltage waveform: (**a**) DC link voltage waveform; (**b**) voltage waveform of the capacitor C_1.

Figure 24 shows the measured waveforms of the driving signals of the upper and lower switches S_1 and S_2 of the same bridge arm in a qZS-HBI module. MPSPWM control inserts the shoot-through zero vector into the switching moment, that is, the shadow part in Figure 24, which is roughly one-eighth of the switching period. The amplitude of its high level is 10 V, which can effectively drive MOS.

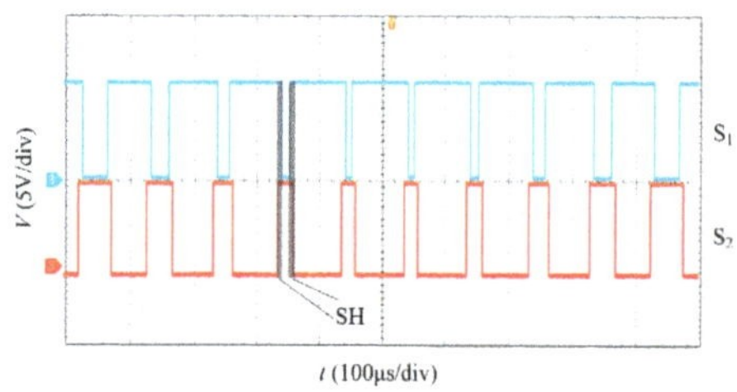

Figure 24. Driving signal waveform.

The output voltage waveform of the qZS-CMI prototype system is depicted in Figure 25a prior to filtering. Upon reaching steady state, the peak voltage waveform stabilizes at approximately 210 V, exhibiting a seven-level sine wave pattern. By referring to Table 2 for input voltage parameters and observing the waveform in Figure 25a, it can be inferred that the qZS-CMI prototype system effectively achieves a six-fold boost output. Furthermore, to ensure continuous current flow within the cascade structure, it is essential to ascertain the output level of an individual qZS-HBI module. Figure 25b illustrates the output voltage waveform of qZS-HBI module 1. As in the Section 2.2 analysis, maintaining current path necessitates four distinct operating states for each qZS-HBI module; thus, its output voltage should manifest as a three-level waveform. The voltage waveform presented in Figure 25b indeed exhibits this three-level characteristic with a peak voltage around 70 V, aligning with both theoretical analysis and the DC chain voltage of an individual qZS-HBI module.

Figure 26a,b below show the grid-connected current waveforms under the control of the traditional deadbeat algorithm and the improved deadbeat algorithm, respectively. According to the requirement of dynamic stability in a power system, the power system should be able to restore the original operation state or transition to a new stable operation state after being disturbed. As shown in Figure 26, when the load fluctuates, the system transitions to a new stable operating state. The output current under the traditional deadbeat control enters a stable state after about 40 ms, while the output current under the improved DBC enters a stable state after about 4 ms. This proves well the rapidity of the improved deadbeat control. Also, the current fluctuation in Figure 26a is large, whereas the current fluctuation in Figure 26b demonstrates a minor amplitude and closely resembles a sine wave. Therefore, it can be concluded that the power quality obtained under the improved DBC is higher than that under the traditional control, which can meet the requirements of grid connection. Therefore, the superiority of the current-predictive control method used in this paper is further proved.

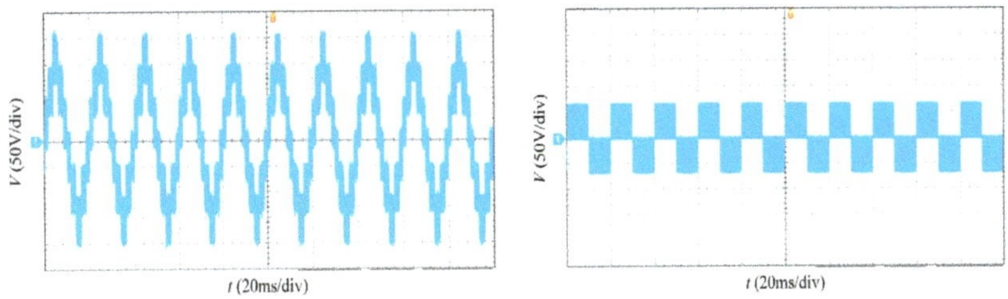

(**a**) (**b**)

Figure 25. Inverter output waveform: (**a**) CMI output voltage waveform; (**b**) single HBI module output voltage waveform.

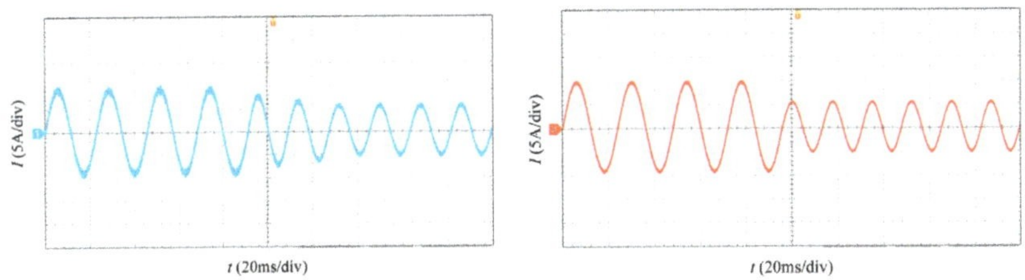

(**a**) (**b**)

Figure 26. The grid-connected current waveform: (**a**) traditional DBC; (**b**) improved DBC.

Figure 27 below provides the voltage and current waveforms of the load after filtration. In this figure, both voltage and current have amplitudes approximately equal to 150 V and 10 A, respectively, while their effective values are measured as 106.1 V and 7.1 A. The output power is determined as 0.753 kW, with voltage and current being in phase exhibiting sinusoidal variations.

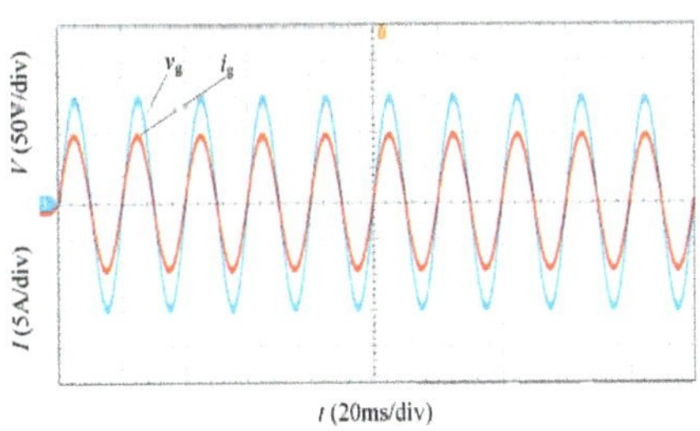

Figure 27. The voltage and current waveform after filtering.

6. Conclusions

In this article, the principle of a quasi-Z-source cascaded multilevel photovoltaic inverter is expounded firstly, and the mathematical model of a qZS-CMI is established. Then, an improved deadbeat control strategy is proposed to solve the problems of switching frequency doubling and high harmonic rate on the grid side. The main features are as follows:

(1) It replaces the traditional 'Boost + inverter' two-stage structure with a quasi-Z-source inverter to achieve a single-stage boost inversion.

(2) Considering the characteristics of photovoltaic distributed power generation systems and the susceptibility of photovoltaic cells to environmental factors, a cascaded multilevel inverter (CMI) is adopted to significantly reduce the power burden on all joint units, while the overall system's quality and efficiency are enhanced with the independent maximum power point tracking of each power unit.

(3) The current-predictive control strategy is introduced and improved to mitigate current harmonics caused on the grid side by the constant charging and discharging of the quasi-Z-source network.

(4) An improved multi-carrier phase-shifted sinusoidal PWM method is proposed as a solution to the problem of frequency doubling due to shoot-through zero vector insertion for the purpose of single-stage boost. This method inserts the shoot-through zero vector into switching time intervals, aiming to reduce switching frequency and minimize switching losses.

(5) In order to ensure the reliability and stability of the control, the FRLS is introduced to identify the parameters of the grid-side filter inductor, which can prevent the control failure caused by the change of the grid-side parameters, so as to realize the real-time monitoring of the grid-side current.

In forthcoming research, system efficiency and quality will be further enhanced in the method's application to photovoltaic power generation systems. The method can also be applied to motor drive and control to improve the operational properties.

Author Contributions: Conceptualization, N.J. and J.J.; methodology, N.J. and J.J.; software, S.L. and J.J.; validation, S.L. and J.J.; supervision, N.J.; writing—original draft, S.L.; writing—review and editing, S.L. and N.J. All authors have read and agreed to the published version of the manuscript.

Funding: This research was funded by China's Training Program of Innovation and Entrepreneurship for Undergraduates, grant number 202310214160.

Data Availability Statement: Data are contained within the article.

Conflicts of Interest: The authors declare no conflicts of interest. The funders had no role in the design of the study; in the collection, analysis, or interpretation of data; in the writing of the manuscript; or in the decision to publish the results.

References

1. Peng, F.Z. Z-source inverter. *IEEE Trans. Ind. Appl.* **2003**, *39*, 504–510. [CrossRef]
2. Lu, H.C. *Research on Permanent Magnet Synchronous Motor System Based on Three-Level Inverter and LC Filter*; Huazhong University of Science and Technology: Wuhan, China, 2022.
3. Zhang, C.H.; Zhou, J.W.; Du, C.S.; Chen, A.L. Review of Control Strategies of Single-phase Cascaded H-Bridge Multilevel Inverter for Grid-connected Photovoltaic Systems. *J. Power Supply* **2017**, *15*, 1–8.
4. Xu, J.; Xu, T.L. Research on NPC Single-Phase Inverter Based on Five-Level Control. *Sol. Energy* **2021**, *15*, 1–8.
5. Dong, S.; Zhang, Q.F.; Wang, R.; Wang, H.L.; Cheng, S.K. Development of the Z-Source Inverters and Its Key Technologies: A Review. *J. Electr. Eng.* **2016**, *11*, 1–12.
6. Su, H.S.; Lin, H.J.; Che, Y.L. Comparative Analysis of All Kinds of Z-source Inverters in Properties and Their Application Background. *Control Eng. China* **2017**, *24*, 188–196.
7. Li, S.; Si, W.X.; Chen, Y.; Cui, S. Research on Isolated Quasi-Z-Source Single Phase Photovoltaic Grid Connected Inverter Based on PR Control. *Acta Energiae Solaris Sin.* **2018**, *39*, 3081–3089.
8. Zhang, R.; Li, Y. System Parameter Design and Grid-connected Control Methods for Quasi-Z-Source Cascaded Multilevel Inverter Considering Power Difference of Modules. *Proc. CSEE* **2018**, *48*, 4836–4845.
9. Wang, J.; Zhang, R.; Liu, Y.; Li, Y. A Multi-Modulation Waveform Phase-Shift Sinusoidal Pulse Width Modulation Method of QZS-CHI and Its Output Voltage Analysis. *Trans. China Electrotech. Soc.* **2020**, *35*, 3470–3477.
10. You, Y.F.; Li, Y.; Xiao, X.Y.; Fang, F. Finite control set-model predictive control of single-phase quasi-Z-source cascaded multilevel inverter. *Power Syst. Prot. Control* **2020**, *48*, 1–8.
11. Niu, L.; Yang, M.; Liu, K.S.; Xu, D.G. A Predictive Current Control Scheme for Permanent Magnet Synchronous Motors. *Proc. CSEE.* **2012**, *32*, 131–137.
12. Yang, H.; Yang, L.H.; Chen, Y.M.; Shi, J.Z.; Wang, S.K. Deadbeat Predictive Current Control Strategy Based on Disturbance Compensation for Grid Connected Inverter with LCL Filter. *Proc. CSEE* **2023**, *43*, 1–16.
13. Chen, Y.D.; Luo, A.; Zhou, L.M.; Xie, N.; Jin, G.B.; Lu, Z.P. A Robust Predictive Deadbeat Grid-connected Control Method Based on Power Feed-forward Control. *Proc. CSEE* **2013**, *33*, 62–70.

14. Jiang, Y.; Xu, W.; Mu, C.; Liu, Y. Improved deadbeat predictive current control combined sliding mode strategy for PMSM drive system. *IEEE Trans. Veh. Technol.* **2017**, *67*, 251–263. [CrossRef]
15. Qu, A.W.; Chen, D.L.; Su, Q. Simple Boost Modified Space Vector Modulation Strategy for Three-Phase Quasi-Z-Source Grid-Connected Inverter. *Trans. China Electrotech. Soc.* **2018**, *33*, 826–836.
16. Guo, H.S. Improvement of SVPWM Direct Modulation Strategy for Quasi-Z Source Inverter. *Electr. Eng.* **2022**, *05*, 67–70.
17. You, Y.F.; Li, Y.; Fang, F.; Yang, X.Y. Deadbeat grid-connected control strategy of quasi-Z-source cascaded multilevel inverter and System Parameter Design. *Power Syst. Prot. Control* **2019**, *2022*, 47. [CrossRef]
18. Zhang, Y.C.; Li, B.Y.; Liu, J. Online Inductance Identification of a PWM Rectifier Under Unbalanced and Distorted Grid Voltages. *IEEE Trans. Ind. Appl.* **2020**, *56*, 3879–3888. [CrossRef]
19. Chen, Z.G.; Yang, L.Y.; Zhang, Y.L.; Li, Z.X. *A Control Method of PMSM Current-Loop Based on Online Parameter Identification*; IEEE Computer Society: Singapore, 2012; pp. 417–422.

electronics

MDPI

Review

A Review of Model Predictive Control for Grid-Connected PV Applications

Yousef Alharbi [1,2,*], Ahmed Darwish [1] and Xiandong Ma [1]

1 School of Engineering, Lancaster University, Lancaster LA1 4YW, UK; a.badawy@lancaster.ac.uk (A.D.); xiandong.ma@lancaster.ac.uk (X.M.)
2 Department of Electrical Engineering, University of Hafr Al Batin, Hafr Al Batin 31991, Saudi Arabia
* Correspondence: y.alharbi@lancaster.ac.uk

Abstract: This paper presents the latest advancements in model predictive control (MPC) for grid-connected power inverters in renewable energy applications. It focuses on grid-connected PV systems employing MPC techniques. Two main categories of MPC are introduced: continuous control MPC (CC MPC) and predetermined control MPC (PC MPC). In CC MPC, a modulator is required to generate the control signal, whereas in PC MPC, the MPC controller directly performs the control process and generates the control pulses. Consequently, PC MPC is preferred for most power converter applications due to its lower computational complexity and simpler implementation. However, ensuring a fixed switching frequency remains a significant challenge when using this control strategy for power electronic converters. Moreover, the computation requirements of MPC strategies are challenging, due to the large number of online calculations. Even with the significant improvements in DSPs, computation complexity is still a continuing issue, especially for applications requiring a high frequency. This paper also examines the design considerations for both types of MPC in PV applications. Lastly, it reviews recent developments in grid-connected inverters utilizing MPC strategies, with a focus on system stability, converter topology, and control objectives.

Keywords: model predictive control MPC; DC_AC inverter; photovoltaic (PV); grid connected system

1. Introduction

Over the last two decades, integrating various renewable energy sources with existing electrical systems has become a topic of interest for power system researchers [1]. Grid-connected renewable energy systems integrate different energy sources and loads into a single electrical system, which is typically linked to the main electrical network [2]. Power flow in such a system is managed according to the generated power and electricity demand. As renewable energy sources have different power forms, the power conversion process becomes a critical part of the electrical system. Solar and wind are currently the most popular and mature natural sources. Among these, photovoltaic (PV) energy is commonly used for distributed applications due to its simple implementation and low maintenance requirements [1]. Solar energy resources are connected to the main electrical network through electronic inverters, which are responsible for power conversion. These inverters also help maintain voltage levels at the common bus. Various control approaches are employed to manage the system. Typically, two control loops are used in grid-connected PV inverters: internal controllers maintain the normal operation of the electronic converters, and external controllers ensure that the PV energy source does not cause issues with the grid [3].

In the last few years, there has been increasing attention to improving the performance of unconventional energy sources, including solar, wind, and marine energy [4–12]. Power electronic converters are used in most electrical power applications (such as energy systems, rail transportation, and motor drives) to convert power from one form to another [13–19]. Existing electrical systems integrate multiple energy resources with varying power forms into a single system. As a result, power electronic converters are essential for unifying the power forms at a common electrical point [20]. Power electronic circuits typically convert electrical power from DC to AC or from AC to DC. However, power converters can also be used for other purposes, such as stepping up or stepping down voltage in DC systems or adjusting the magnitude or frequency of pure AC systems [21–24]. Several power electronic circuits have been proposed to effectively utilize renewable energy sources [25–31]. When designing these power electronic converters, key considerations include small size, reduced design complexity, low installation cost, and simple control strategies [32–34]. The development of fast-switching transistors, such as Gallium Nitride (GaN)-based transistors, plays a crucial role in meeting these requirements [35].

Power converter control technology has continuously improved over the last few decades [36–39]. The conventional control strategy for power converters primarily relies on linear controllers in combination with modulation techniques like Pulse Width Modulation (PWM) [38]. While this control method has been effective for many years, the operating range of the controller can be limited by a narrow bandwidth, leading to reduced performance outside of this range [39]. The hysteresis comparator control strategy, widely used in power electronic circuits, was developed to take advantage of advances in Digital Signal Processors (DSPs) technology [40–45]. As DSPs have become more capable, newer and more advanced control strategies have been introduced, including fuzzy control, sliding mode control, and Model Predictive Control (MPC) [46]. Each control strategy has its strengths and weaknesses, and the choice of controller is typically based on system requirements. However, these methods are still under investigation, and further improvements are expected. In particular, MPC has emerged as a simple and effective choice for controlling power electronic circuits in renewable energy systems [47–53].

In the past decade, MPC has gained significant interest in control theory research [54]. Originally introduced in the 1960s for large petrochemical plants [55,56], MPC later attracted attention from researchers in other industries due to its ability to control multivariable systems effectively [57]. One challenge in implementing MPC is the large number of calculations required in short time intervals; however, this issue has been reduced by improvements in digital microprocessors [58,59]. Recently, MPC has gained more attention in the power electronics industry [60]. This control strategy offers several advantages over traditional methods, including simpler control implementation for multivariable converters, the ability to handle converter nonlinearity, and the capacity to include constraints [61]. MPC is an optimization method that uses the model of power electronic converters to predict the converter's output. The optimal action that minimizes error is determined according to the cost function [59].

In the power electronics field, MPC can be a better alternative to traditional proportional-integral-derivative (PID) controllers, offering superior dynamic response [62]. The advancements in digital signal processing over the past decade have directed research toward more complex control strategies like MPC [63]. MPC provides a fast response to power converter dynamics, precise tracking, and simplified implementation, making it an ideal choice for nonlinear control of power electronic converters. MPC is a predictive control method that uses an identified minimization function to influence power converter variables in line with desired quantities. The MPC strategy typically employs a discrete-time model of the converter for control actions [64]. The optimal switching state of the converter is determined based on the output of the optimization process. In the prediction stage, the MPC uses the discrete-time model to generate future predicted values, which are then compared to the current control variables to determine the control action. The optimization block solves the minimization problem based on the predicted values and provides the optimal switching state to the converter [64,65].

Power control has an important role in grid-connected PV applications. The main objective of controlling the AC power at the grid side is to enforce the controlled variables to follow the desired values. Different controlling methods can be used to control grid PQ power in either national or d-q frames [66]. In [67] a stationary-frame perdition controlling strategy is used to control the active and reactive power of a grid-connected system with a steady-state error cancelation feature. A direct MPC controller is presented in [68] to regulate the active and reactive power of the phase grid-connected system. The controlling action directly controls grid power by obtaining the optimal switching state that achieves minimum errors.

The Four-legged grid-connected inverters (FLGCI) are commonly used as three-phase grid-connected inverters to increase the grid inverter's capability to deal with unbalanced load scenarios by carrying the current of the neutral point. A novel MPC controlling method is employed in [69] to effectively operate the FLGCI and track the grid current during balanced and unbalanced load scenarios. A spilled source inverter (SSI) has been introduced to overcome the issues of the Z source inverter (ZIS) including stress on the active switches and a high number of components. Unlike ZSI, the SSI is operating without an extra operating state which can minimize the loss and complexity of the system. The hierarchical control strategy used in [70] combines the advantages of conventional PI and MPC control to successfully operate the SSI.

MPC for power converters can be divided into two main categories: continuous control MPC (CC MPC) and predetermined control MPC (PC MPC). In CC MPC, a separate modulator is required to determine the switching states of the power converter, while in PC MPC, the switching states are predetermined and incorporated into the control stage. PC MPC is commonly used for power electronic circuits because it requires a finite number of switching states to solve the optimization function. Since the modulation stage is not needed in PC MPC, the control action is directly applied to the power converters, making the control process more straightforward [71]. Unlike CC MPC, adding more terms to the optimization cost function in PC MPC does not require redesigning the control system. However, a key disadvantage of PC MPC is that maintaining a fixed switching frequency is challenging due to the absence of a modulator. Some strategies have been proposed to address the issue of variable switching frequency by adding more control actions [72]. Figure 1a shows the CC MPC block diagram, where modulation occurs after the control action. Figure 1b illustrates the PC MPC method, where both the control action and modulation process are evaluated and solved in a single stage. The x* is the system input, u is the modulator input, y is the system output, and x is the feedback signal.

Multilevel grid-connected inverters have been introduced as a practical and robust alternative for high- and medium-power grid-tied applications over the last few decades. The operation of this technology is based on employing several DC sources in parallel with semiconductor devices for stepping the waveform of output voltage. The DC sources used in multilevel inverters (MIs) can be one of or a combination of the following capacitors: fuel cell, batteries, or PV module [73]. The main objective in presenting the MIs is tackling the common drawbacks of conventional single-level inverters including high- voltage stress on the active switches, total harmonics distortion, and high-switching frequencies.

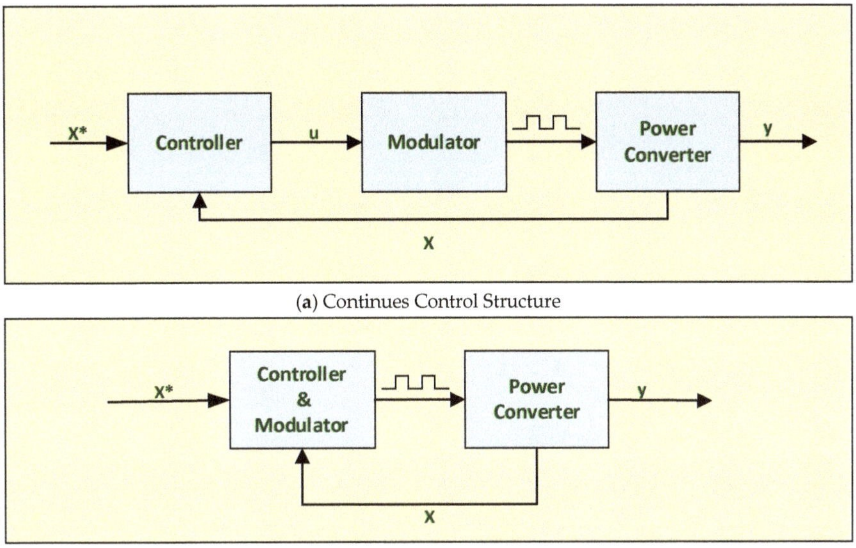

Figure 1. MPC controller schemes with the controlling pulse signals.

One of the main MI topologies is neutral point clamp (NPC), which is commonly used to generate output voltage with three voltage levels. In [74] the predictive control method is employed to control the 9-level inverter; however, the THD of the grid current exceeds the practical limit.

Packet E-Cell is an MI inverter which can operate with different voltage levels up to thirteen voltage levels. A CC MPC strategy is presented in [75]. The simulation results illustrate a practical operation of the proposed CC-MPC controlling method with 1.5% THD. Nonetheless, experimental investigations lack sufficient analysis, which makes it open for future research to perform further investigations.

Another important type of MI are Packet U Cells (PUC), which are commonly used as inverters or rectifiers for grid-connected applications. A 7-level grid-connected inverter is introduced in [76]. The simulation results show an acceptable THD percentage of about 3% (see Table 1).

Table 1. Illustrates a comparison of the common types of MI which employ MPC control technologies in terms of MPC category, voltage levels, active switches, and THD.

Multilevel Topology	MPC Type	Voltage Levels	No. of Switches	THD (%)	Ref.
Neutral Point Clamp (NPC)	PC MPC	3	12	≈1.36	[74]
Packet E-Cell (PEC)	CC MPC	5–9	8	≈1.45	[75]
Packet U-Cells (PUC)	PC MPC	7	6	≈3.12	[76]
Modular Multilevel Inverter (MMC)	PC MPC	$n + 1$	16	≈3.12	[77]
Isolated Modular Multilevel Inverter (IMMC)	PC MPC	7	16X	≈3.16	[78]

This paper reviews the latest advances in MPC technology for grid-connected PV inverters. It compares CC MPC and PC MPC applications for power electronic converters, with a particular focus on grid-tied inverters, controller models, inverter designs, and control objectives. Additionally, it provides a brief assessment of grid-connected inverter topologies using MPC approaches.

Section 2 presents the basic elements that are required to build the MPC controllers for grid-connected renewable energy applications. Section 3 discusses the recent applications which employ the MPC approaches to control the grid-connected applications. In Section 4, a detailed argument is made about the concerns of using MPC control for grid-tied power converters and prospective future trends of this controlling technology. A conclusion about the findings of this paper is presented in Section 5.

2. MPC Components for Power Converters

The MPC (Model Predictive Control) strategy usually consists of different components modified to the targeted controlled system. This section describes the most common MPC components used for controlling power electronic converters, which are arranged as follows:

2.1. Power Converter Model

The MPC method relies on obtaining accurate mathematical models of power converters. Deriving a precise model plays a crucial role in the MPC strategy. Modeling power converters to precisely capture their dynamics ensures straightforward implementation of the system controller. The predicted switching state of power converters at future moments is determined based on their mathematical models.

Power converter prediction models for MPC are generally represented as continuous-time modules. In these systems, storage elements, such as inductors and capacitors, describe system dynamics. Hence, the differential equations of inductor current and capacitor voltage are derived to model most power converter circuits. These continuous-time models must be converted into discreet-time models to be compatible with discrete-time control methods. Several strategies can be used to convert continuous-time systems into discreet-time models [79]. The first-order forward Euler technique is among the most common approaches for discretizing power converter models. Two key considerations for this technique are that the sampling frequency should be high enough to closely approximate the continuous model, and that no filtering circuits should be present in the system [80].

The general equation for the first-order Euler method is shown below:

$$\frac{dx}{dt} = \lim_{t=0} \frac{x_{k+1} - x_k}{T} \tag{1}$$

Using this method, the approximate value for the next time interval can be calculated as:

$$x_{k+1} = x_k + T\left(\frac{dx}{dt}\right) \tag{2}$$

2.2. Cost Function

The cost function in MPC can take various forms with different levels of complexity [81]. Although the control designer can freely select the cost function, the performance of the control system is significantly influenced by the complexity of the objective function [82]. Typically, the MPC cost function is a minimization problem that seeks to minimize the error within the function. Commonly, the absolute or quadratic error value is used. Quadratic error functions penalize errors more heavily, producing better performance but potentially increasing the switching frequency. In multi-objective cost functions, the error terms for different objective variables are added [83]. The equations below present the MPC cost function and the methods for calculating errors:

$$g = |x^* - x_n| \tag{3}$$

$$g = [x^* - x_n]^2 \tag{4}$$

$$\min(f_g[x^*, x_{pn}, S_n]),\ n = 1,\ \ldots, i \tag{5}$$

where g is the minimization cost function, x* is the targeted value, xn is the actual value, and, Sn is the switching state.

2.3. Objective Variables

Objective variables in MPC are typically represented in the cost function. The purpose of the cost function is to ensure that controlled variables follow their reference values. In power converter applications, currents are the most controlled variables. Quadratic error cost functions are widely used to maintain the quality of current waveforms with minimal ripples.

Voltage is another frequently controlled variable in MPC applications [84]. The inductor current and capacitor voltage, which represent power converter dynamics, are often chosen as control variables for many circuits [85]. For grid-connected inverters, the grid current and output voltage are commonly used as control objectives. The MPC approach is employed to control these variables, reducing control implementation complexity and improving control performance [86]. Figure 2 illustrates the process of applying the MPC control strategies for power converters

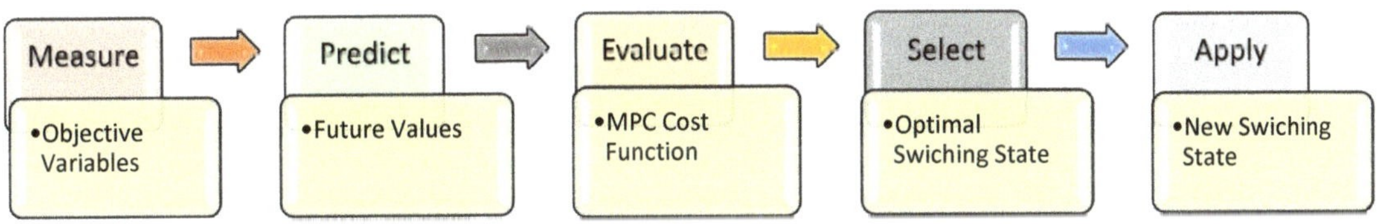

Figure 2. MPC flow chart of MPC process for power converters.

2.4. MPC Parametrization

Two major design considerations in MPC strategies for power electronics are sampling frequency and prediction horizon intervals. In PC MPC approaches, the absence of a modulator results in the switching process being performed at discrete time intervals. The stability of the discrete-time model and the quality of the modulated signal are directly related to the sampling period.

The controller's performance is also directly influenced by the sampling interval. The theoretical minimum sampling frequency is twice the switching frequency [63]. However, in practice, a higher sampling frequency is usually chosen to ensure signal quality. A proper time-domain discretization can be achieved by maintaining a sampling-to-switching frequency ratio of around 100 [87].

The computational complexity of MPC increases with longer prediction horizons and a larger number of control variables. Therefore, it is crucial to use short prediction horizons when designing MPC for power converters. Nonetheless, some applications require longer horizons to effectively predict system behavior and capture converter dynamics [88]. Table 2 illustrates common techniques used to minimize prediction horizon in MPC.

Table 2. Reducing horizon strategies.

Reducing Horizon Method	Computation Complexity	MPC Evaluation	Prediction Steps	Optimization Strategy	Ref.
Move Blocking	Medium	Online	Short & Long	Horizon division	[89]
Extrapolation	Medium	Online	Long	hysteresis bounds	[90]
Even Based	High	Offline	Long	Optimal Pulse Pattern	[91]

2.5. *Weighting Facror Tuning*

The cost function in MPC often involves multiple variables, which can be represented as a single function. In such cases, the weighting factors are introduced to prioritize significant factors over others. When the variables in the MPC cost function have the same characteristics, they are usually assigned equal weight, making weighting factors unnecessary. However, if the terms have different natures, weighting factors become essential to prioritize the most significant variables [92].

Although weighting factors are not always required in MPC for power converters, they can greatly influence the control system's performance, especially when cost function variables differ in importance [93]. Techniques such as ranking search methods are commonly used to appoint weights to variables [94].

More advanced strategies exist for assigning weights [95], but they are rarely used in power converters due to their complexity and time-consuming nature. Analytical methods for determining weighting factors provide a simpler and effective alternative for control designers [93].

3. MPC Applications for Grid-Connected PV Systems

The PC MPC approach in [96] is suggested for operating the grid-tied single-phase switching inverter. The investigated grid-connected PV system involves a PV array and storage battery system. The proposed microgrid sends the active and reactive power to the utility grid. The active power and reactive power that are sent to the main electrical network are controlled by a new PC MPC method. The recommended PC MPC specifies the optimal switching state of the grid-connected single-phase inverter that achieves the minimum error and enables the controller to follow the reference value of the targeted reactive power value. The controlling system has been studied in normal operation and under sag voltage issues. Under normal operation, active power is injected into the main electricity network.

In contrast, when the voltage sag occurs on the main AC bus the controller starts to send reactive power to mitigate the voltage sag concern. Thus, the low voltage ride-through ability is included in the controller to maintain normal operation at the common AC bus of the microgrid. The main grid policies and codes can be employed to determine the percentage of the voltage sag and consequently decide the reactive power needed to inject into the electricity network [96]. Figure 3 shows the circuit diagram and PC MPC controlling block of a single-stage single-phase grid-connected inverter.

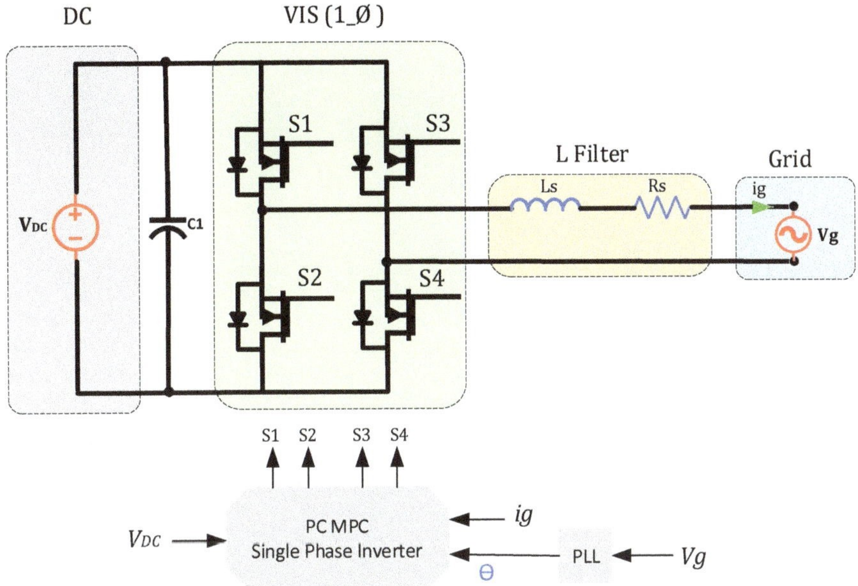

Figure 3. PC MPC control for single-phase single-stage grid-connected inverter.

In [97] a new CC MPC controller for a grid-tied PV system is introduced. The proposed MPC strategy is used to control the input voltage at the PV side of the grid-connected PV system to maintain a stable operation and obtain the maximum power. The controlling system involves two controlling loops. The outer loop of the cascaded inverter needs to be slower than the inner controlling loop of this cascaded system to maintain normal operation in the entire controlling process. The internal controlling system is employed to control the DC_DC boost converter input voltage of a three-phase grid-connected PV system. A disturbance-observing system was included in this study to eliminate the steady-state error and improve the robustness of the controlling system. Moreover, the disturbance observation leads to an accurate model during transient time and helps capture the dynamics of the power converter. Another controlling loop is used to examine the effectiveness of the proposed controlling strategy. A conventional PI controller is employed at the grid side of the grid-tied three-phase PV system. The PI controller is used to regulate the grid current of the three-phase systems. The proposed controller relies on a CC MPC; thus, a modulator is crucial [97]. Figure 4 illustrates the schematic diagram of a two-stage three-phase inverter with a CC MPC system. The input side of the inverter is VDC, at the grid side, the L filter is used to maintain the grid current quality ig and does not pose any harm to the grid voltage Vg. The Phase Locked Loop PLL is used in the controller to obtain the phase angle to perform the controlling process.

In [98] a typical PC MPC, a method is employed for regulating grid-tied voltage source inverter (VSI) configurations. Firstly, it has been applied to two-level VSIs to enhance the system performance and offer better system efficiency. Then, the same controlling strategy is used for three-level VSIs to provide better analysis and further investigate the proposed controller and compare the results of the two inverters. The Euler method is used to obtain the discrete-time model of the two VSI types. The proposed controlling strategy starts by studying the performance of the system with a single-step prediction horizon. After that, the two-step prediction horizon is used to enhance the prediction accuracy. The environmental conditions of the PV array are changed to examine the controller's performance under the disturbance effect. The THD percentage of the line current of both inverter configurations is calculated to study the system efficiency. The study concludes that the THD of the three-level inverter is relatively low and the MPPT process is more accurate. The MPC of this study aims to regulate the active and reactive power injected into the main electricity network [98]. Figure 5 shows the circuit configuration of a single-stage three-phase inverter with PC MPC.

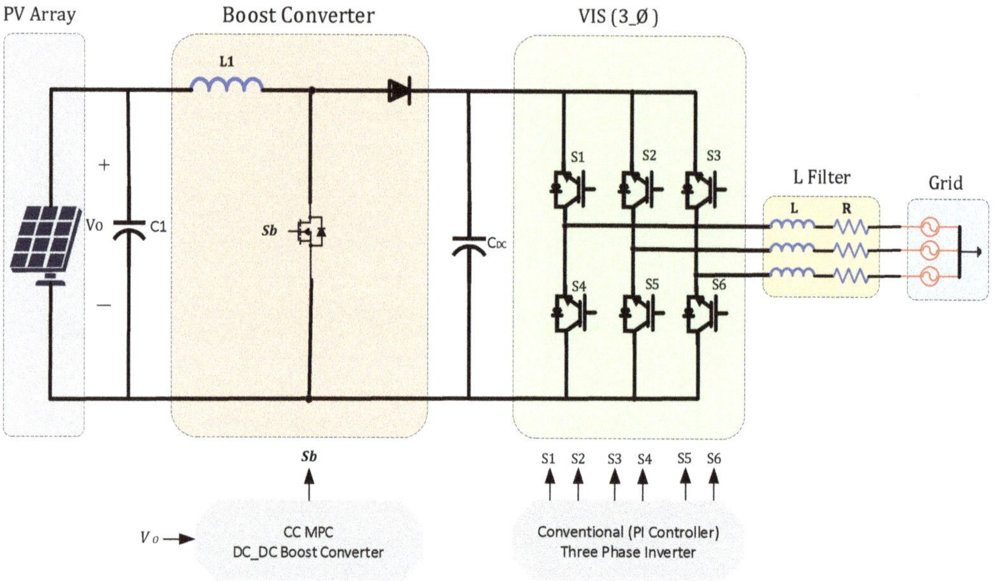

Figure 4. CC MPC for two-stage three-phase grid-tied PV inverter.

A proposed CC MPC strategy for regulating a twisted buck-boost inverter is introduced in [99]. It aims to study the feasibility of CC MPC for the grid-tied inverter. The recommended MPC controlling strategy in this study aims to mitigate the zero-crossing distortion problem of grid current which commonly occurs when using conventional PR controllers. The study provides a hardware implementation of the suggested controlling method for the twisted buck-boost inverter to examine the proposed controller's validity. The differential equations are used for obtaining the predicted quantities of the buck-boost inverter. Then the objective function is designed accordingly. The study employs the CC MPC approach to regulate the grid-connected inverter, thus a PWM modulator is responsible for generating the switching signals. A high prediction horizon plays an important role in obtaining the correct signal waveform. However, it can result in reducing microprocessor performance and increasing computation complexity. The proposed controlling method studies the performance of the system at different prediction horizons to choose the optimal horizon. The second horizon provides better performance and maintains the normal operation of the controller during transient time. A multi-core microprocessor has been employed in experimental implementation to deal with the large number of calculations. The proposed controlling method is suitable for LV applications where input voltage ranges from 100 to 400 [99]. Figure 6 shows the twisted buck-boost grid-tied inverter with a CC MPC block. The output voltage output corresponds to the maximum power from the PV side and is used to eliminate the steady state errors of IPV current and obtain the power reference values. The Clark transformation is used to obtain the d_q components. This can simplify the controlling process and allow the system to follow the desired values. The output of the controlling system is designed to operate switches.

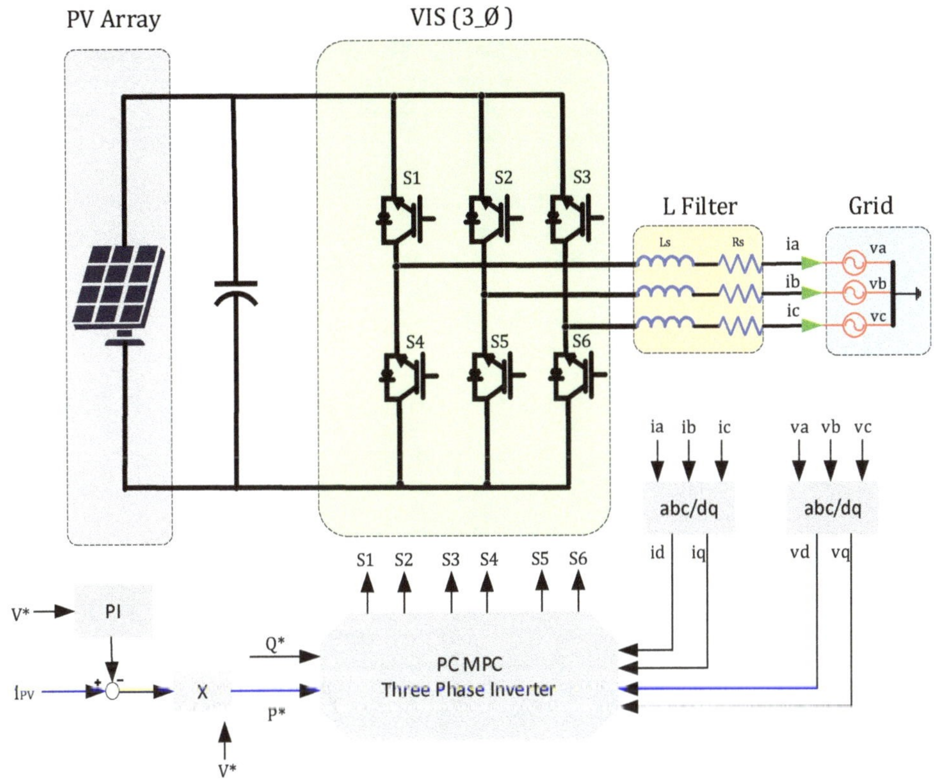

Figure 5. PC MPC control for single_stage three_phase grid_connected PV inverter.

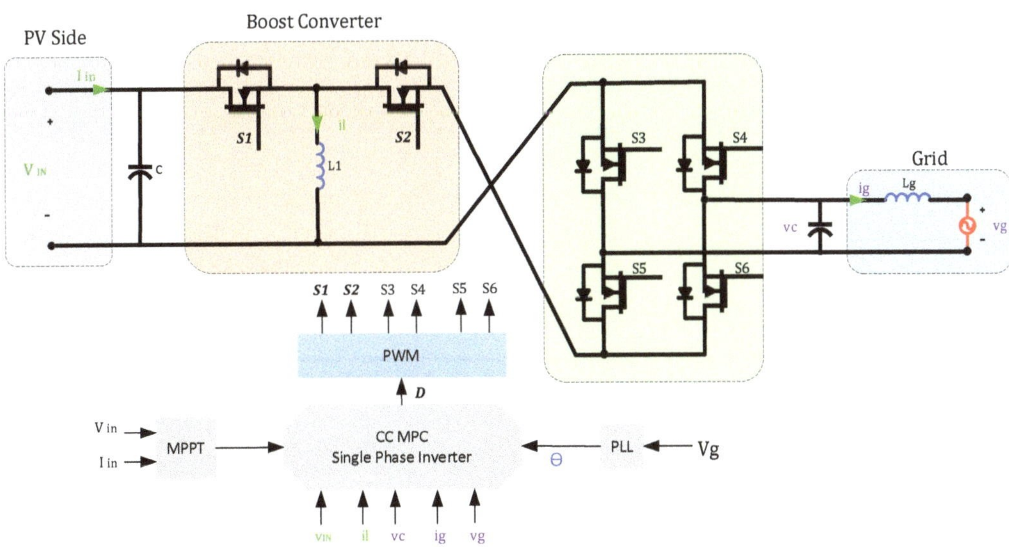

Figure 6. CC MPC for twisted buck-boost grid-connected inverter.

A new MPC approach is being investigated in [100] to regulate a multi-level grid-tied inverter. The proposed controlling technique is used to regulate a seven-level inverter that is connected to a single-phase power system. An inductor is used for filtering the grid current and enhancing the power converter efficiency. The proposed method is based on PC MPC; thus, the possible switching states are defined before obtaining the mathematical model of the inverter by calculating the differential equations of the inverter circuit. The proposed cost function consists of two objective variables for reducing the errors between actual and reference values. A weighting factor is introduced in the cost function to minimize the THD in the grid current and reduce the error in capacitor voltage. A new tuning strategy is presented to obtain the best value that achieves optimal regulating results. The tuning technique considers both grid current and capacitor voltage error and optimal value has been chosen accordingly. The performance of the controlling strategy has been examined under the effect of the disturbance. The robustness of the controller has been proved in both simulation and experimental results [100]. Figure 7 illustrates the PC MPC method for a MMC grid-connected PV inverter.

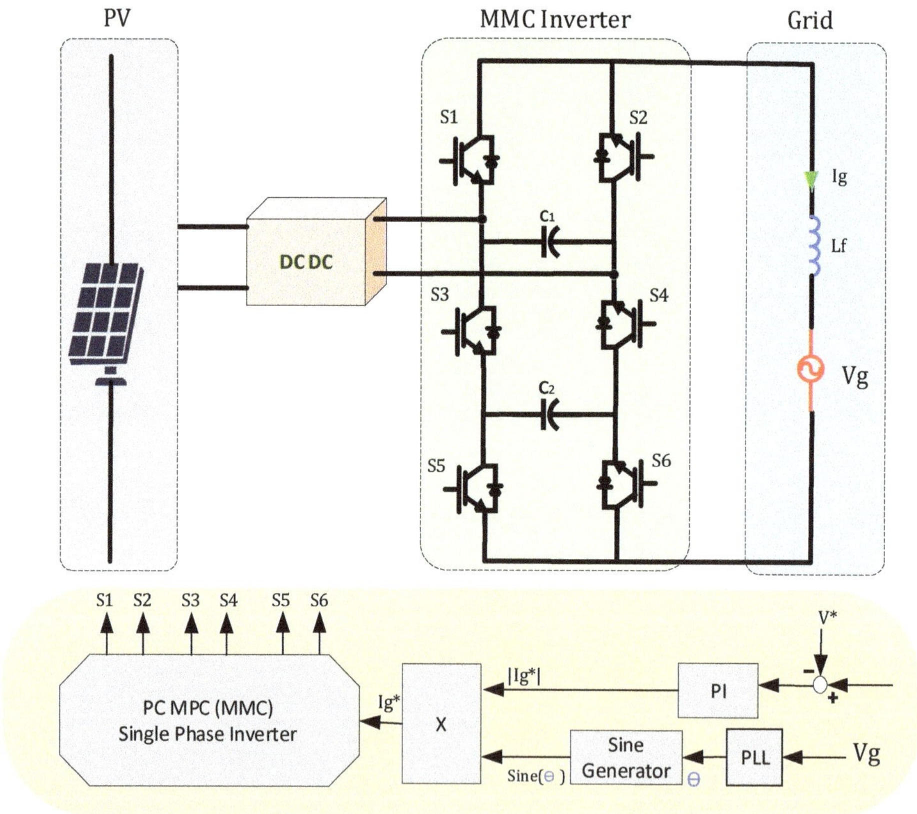

Figure 7. PC MPC of MMC grid-connected inverter.

The proposed PC MPC approach in [101] is employed to regulate a PV grid-connected single-phase inverter. Highly Efficient and Reliable Inverter Concept (HERIC) is an inverter topology that was recently introduced to enhance the efficiency of the grid-connected inverters. The suggested MPC algorithm is applied to the HERIC inverter to increase the controlling system robustness and performance. An LCL filter is employed at the grid side to perform the filtering process and maintain the signal quality of the grid current. The first-order Euler discretization is used for obtaining a discrete-time model of the HERIC inverter. The introduced PC MPC method increases the switching states of the inverter by inserting more virtual switching states to mitigate THD at the grid current and consequently improve the inverter efficiency. Two different scenarios are presented to validate the suggested controlling approach. One case study for investigating the contribution of the proposed controlling approach in THD mitigation is noted. The second scenario is to examine the controller robustness when sudden a change takes place in the system's parameters. A comparison study is presented to compare the proposed PC MPC technique and classical controlling strategy. An experimental validation for studying the effectiveness of the proposed controlling approach is presented in this study [101]. Figure 8 shows a single-stage grid-connected inverter with LCL filter employing the PC MPC approach.

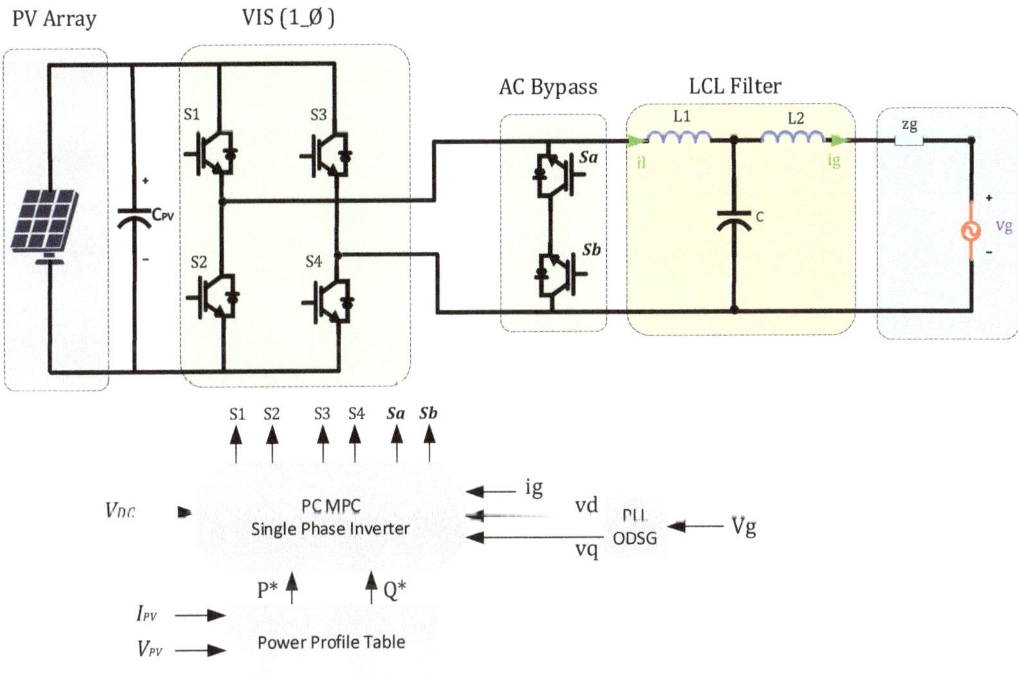

Figure 8. PC MPC of single-phase single-stage grid-connected PV inverter with LCL filter.

A two-stage grid-connected boost-buck inverter is controlled by a PC MPC strategy introduced in [102]. The proposed MPC strategy aims to mitigate the continuous voltage change at the input side and maintain the DC link voltage within the operational range. The controlling system is designed to regulate different variables in the investigated inverter. The switching states of the proposed two-stage grid-tied inverter are expressed before obtaining the mathematical module. Then the controlling variables are defined as an output grid current and DC link voltage. The suggested strategy relies on Taylor expansion to obtain the discrete-time module. Taylor expansion technique can lead to obtaining an accurate model of the modulated signal. However, it is less popular because the computational complexity is relatively high. The second-order Taylor expansion is employed in this study for defining the control variables of the MPC system with less computational burden. A weighting factor is presented in the cost function to achieve better performance of the proposed controlling system. The weighting factor in the cost function is tuned according to the grid current THD and the dynamic response of the inverter module. An experimental test and simulation results are presented to examine the validity of the proposed controller of a grid-connected two-stage inverter [102]. Figure 9 shows the PC MPC of a two-stage optimized grid-connected PV inverter.

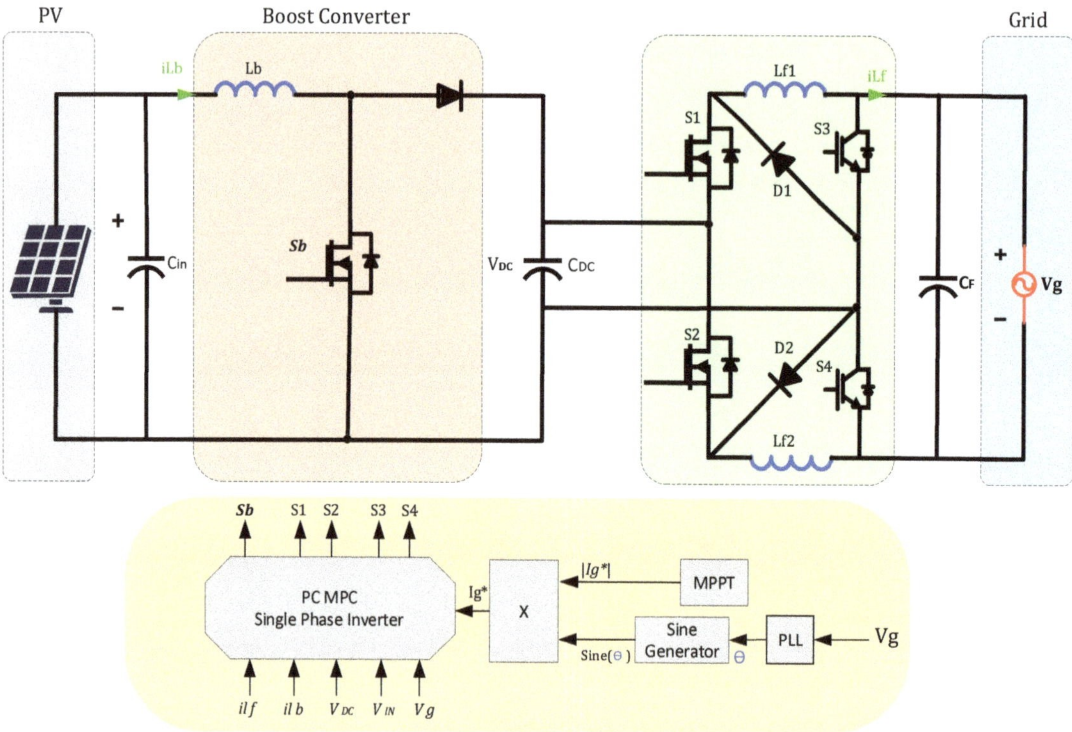

Figure 9. PC MPC of two-stage optimized grid-connected inverter.

In [103] a novel MPC technique is used for obtaining the maximum available power from the PV side of a grid-connected PV system. The proposed MPC method is employed for a Z-source grid-tied inverter. The features of the MPPT based on MPC include simple implementation, higher flexibility, and the capability of dealing with the nonlinear nature of power electronic converters. The MPC approaches provide better dynamic performance and enhance the controlling system robustness. Thus, it can be an appropriate controlling choice to deal with variable environmental conditions in PV-related applications. One crucial advantage of the presented MPC-based maximum power tracking system is operating with a fixed switching frequency which can improve tracking system accuracy. The proposed MPPT algorithm based on MPC is employed to reduce the oscillation around the maximum power point. This study uses two controlling systems to link the PV system with the main electricity network. On the PV side, an MPP tracking system based on MPC is employed to operate the Z source grid-connected inverter to improve the dynamic response and increase the stability margin of the grid-tied system. A grid-side controller is used for effectively regulating grid-injected power. Experimental results prove the validation of the proposed controlling system and its ability to minimize the associated THD of grid current [103]. Figure 10 shows the impedance source grid-connected inverter with a predictive MPPT approach.

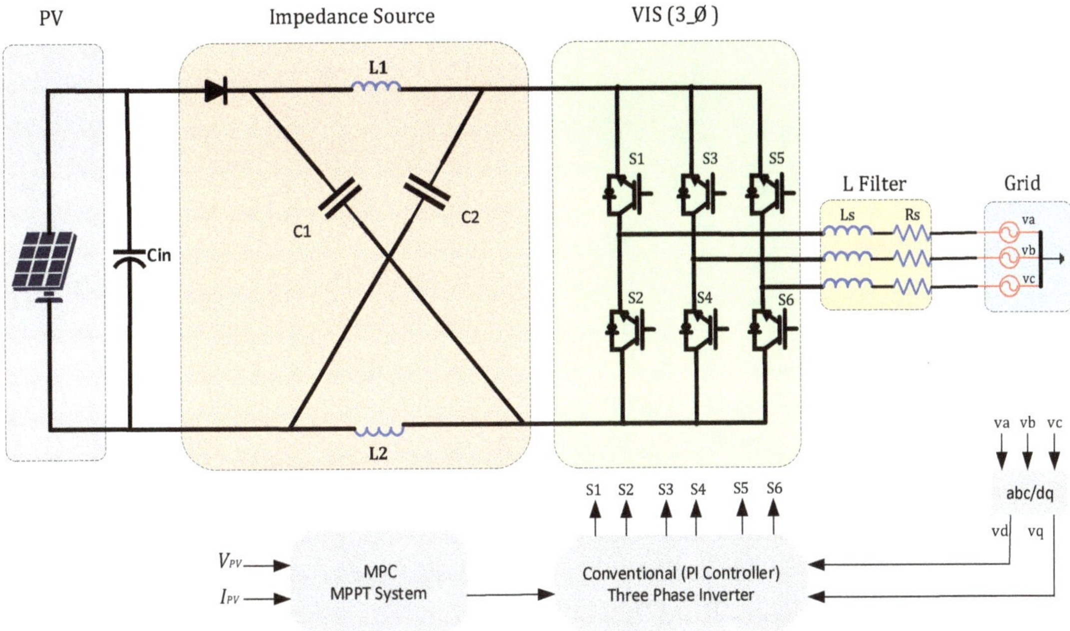

Figure 10. PC MPC of MPPT for single-stage three-phase impedance source grid-connected inverter.

In [104] the paper suggests a new PC MPC method for regulating a grid-connected multi-level inverter. The introduced MPC controlling strategy is employed for operating a nine-level single-phase inverter. The presented method is designed to control grid-connected inverters at different operations of the voltage levels. Thus, the proposed PC MPV controlling algorithm can operate as a five-, seven-, or nine-level inverter, which increases the reliability and flexibility of the system. The investigated MPC controlling approach for regulating the grid-connected inverter meets power quality requirements and achieves relatively low THD. The Euler forward-approximation method is employed for obtaining the discrete-time model of the power converter. A prediction horizon with a single step is used to capture the system dynamics and predict the future values of the objective variables. The robustness of the MPC controlling strategy is examined when the inverter switches from a five-level operation mood to a nine-level operation mood. The results prove that the proposed controlling strategy can inject the current into the utility network with low THD when the operating level of multilevel inverter changes suddenly. The validity of the suggested PC MPC controlling strategy was examined by both simulation results and experimental tests [104]. Figure 11 shows the MMC grid-connected single-phase inverter using PC MPC strategy.

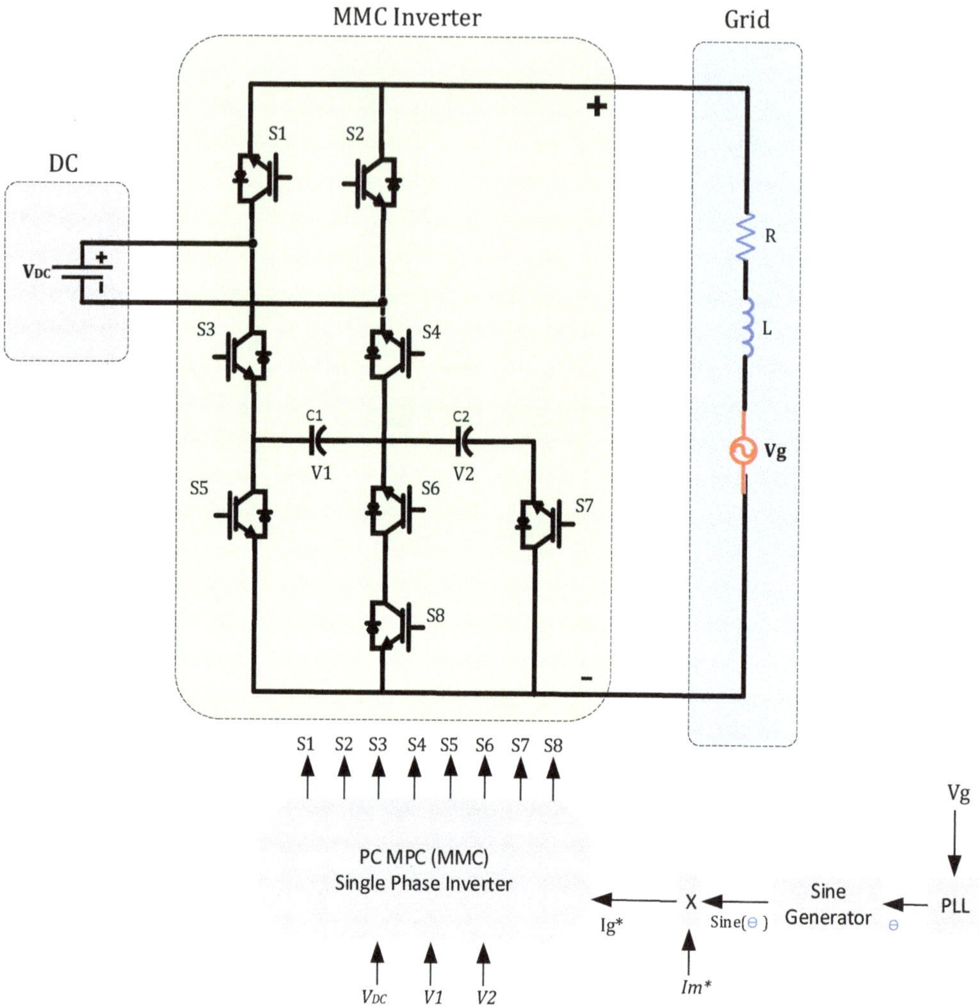

Figure 11. PC MPC for MMC grid-connected single-phase inverter.

A suggested PC MPC strategy in [49] aims to regulate a multi-level grid-connected inverter. The quasi-Z-source inverter topology is a single-stage inverter topology. One controlling loop is used for each model of the cascade system. Mathematical equations are obtained after examining the different operating modes of the investigated inverter topology. The multi-level operation is achieved by cascading three quasi-Z source inverters. The presented PC MPC is employed to operate each inverter individually. The current predictive controlling system maintains stable operation during different operation modes with low THD of the grid current. The discrete-time form of the grid current is calculated via the Euler approximation method to predict future current values. The introduced PC MPC method uses two-step horizons to predict the future values of objective variables. The deadband time is considered in this study to deal with natural delays when the inverter switches from one mood to another. The research suggests a new parametrization method to accurately calculate the grid side inductor and ensure a normal operation under faulty system scenarios. The simulation results and experimental tests are presented to examine the validity of the proposed controlling system [49]. Figure 12 shows cascaded quasi-Z-source MMC single-phase inverter with PC MPC.

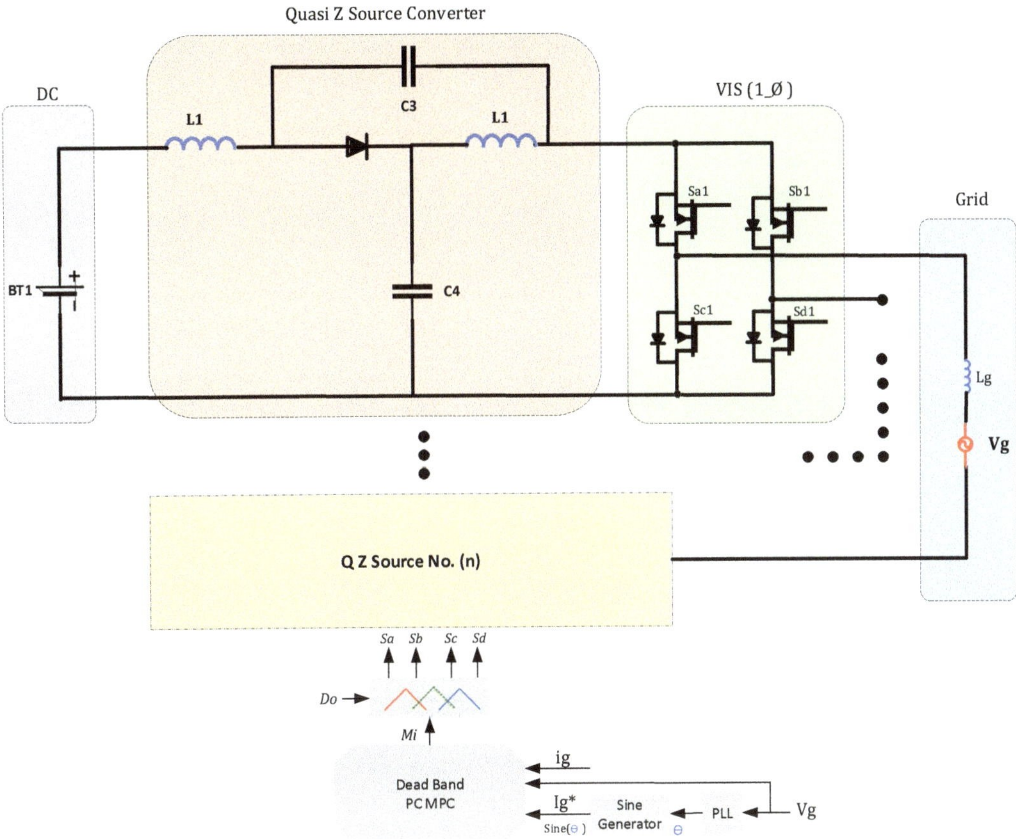

Figure 12. MPC for quasi-Z-source cascaded MMC single-phase inverter.

4. Limitations and Future of MPC for Grid-Connected PV Applications

The online computation effort needed to capture future behavior is one of the main issues in applying MPC approaches in the power electronic field, especially for applications that require a high switching frequency. Increasing the sampling frequency (fs) can overcome this challenge; however, this can lead to increasing steady-state efforts. This explains why several studies rely on hierarchical control methods to eliminate steady-state errors [105].

The lack of modulation process in the PC MPC approaches leads to the operation of the power converter with a variable switching frequency. While operating the power converter can be seen as an attractive feature in a simple implementation, the modulation process plays a crucial role in minimizing the switching losses and associated distortion of the grid current [106].

The steady-state error is one of the main drawbacks of employing the MPC for power electronics converters. The controlled variable in the system starts oscillating around the reference variable, causing a significant steady-state error. This issue occurs because the optimization process is performed corresponding to the sampling time. Thus, controlling performance is reduced between the two intervals of sampling time [107].

In recent years, model predictive control (MPC) has emerged as a viable control strategy for grid-connected renewable energy applications. Experimental and theoretical studies have demonstrated that MPC provides reliable performance in terms of system efficiency, robustness, and overall control quality for grid-tied projects. Compared to conventional control technologies, MPC offers significant advantages, including flexibility in regulating multiple objective variables without adding excessive complexity. However, as a relatively modern control approach in the field of power electronics, MPC presents several opportunities for further research. Below are some key directions for the future development of MPC in power electronic applications:

- Fixing the Switching Frequency in PC-MPC

Finite Control Set MPC (PC-MPC) is often preferred for power converter applications due to its effective control performance and straightforward implementation. However, one of its main challenges is the variable switching frequency caused by the absence of a modulation stage. Future research is expected to propose novel methods to stabilize the switching frequency in PC-MPC and address this critical issue.

- Enhanced Discretization Strategies

MPC is a nonlinear control strategy that relies on a precise model of the power converter. Typically, discrete-time models are used to predict the future values of objective variables. Improving the discretization strategies—particularly by adopting a longer prediction horizon—could enhance steady-state response and minimize steady-state errors, offering more reliable performance.

- Optimal Tuning of Weighting Factors

The cost function in MPC often includes multiple control variables, and a weighting factor is used to prioritize these variables based on design requirements. Selecting the correct weighting factor is crucial for achieving optimal controller performance. Future studies will likely focus on developing effective tuning strategies to determine the optimal weighting factor, leading to significant improvements in MPC performance.

5. Conclusions

This study highlights the recent advancements in MPC strategies for grid-connected PV applications. MPC has proven to be a valid and competitive choice for controlling grid-connected inverters for PV systems. Both CC MPC and PC MPC are applicable for controlling power electronic converters; however, PC MPC is preferred in power electronics applications due to its simple implementation and design flexibility. The evolution of modern control theory and digital signal processors facilitates the use of various MPC categories in grid-connected inverters. The rule to apply MPC approaches in multilevel converters requires a clear understanding of the grid-connected system needs and the dynamic behavior of the power converter circuits. This understanding is critical to design a practical MPC controller. A key feature of MPC techniques is their ability to control multiple objectives simultaneously without increasing complexity, making them a suitable control platform for grid-connected PV applications. Recent developments in power converter topologies using MPC control for power electronics in PV applications demonstrate the promising potential of MPC strategies for the future of power electronics.

Author Contributions: Conceptualization, Y.A. and A.D.; methodology, Y.A. and A.D.; software, Y.A. and A.D.; validation, Y.A. and A.D.; formal analysis, Y.A. and A.D.; investigation, Y.A. and A.D.; resources, Y.A. and X.M.; data curation, Y.A. and A.D.; writing—original draft preparation, Y.A.; writing—review and editing, A.D.; visualization, Y.A. and A.D.; supervision, A.D. and X.M.; project administration, A.D.; funding acquisition, A.D. All authors have read and agreed to the published version of the manuscript.

Funding: This research received no external funding.

Data Availability Statement: Not applicable.

Conflicts of Interest: The authors declare no conflict of interest.

References

1. Sarver, T.; Al-Qaraghuli, A.; Kazmerski, L.L. A comprehensive review of the impact of dust on the use of solar energy: History, investigations, results, literature, and mitigation approaches. *Renew. Sustain. Energy Rev.* **2013**, *22*, 698–733. [CrossRef]
2. Hernandez, R.; Easter, S.; Murphy-Mariscal, M.; Maestre, F.; Tavassoli, M.; Allen, E.; Barrows, C.; Belnap, J.; Ochoa-Hueso, R.; Ravi, S.; et al. Environmental impacts of utility-scale solar energy. *Renew. Sustain. Energy Rev.* **2014**, *29*, 766–779. [CrossRef]
3. Gao, C.-K.; Na, H.-M.; Song, K.-H.; Dyer, N.; Tian, F.; Xu, Q.-J.; Xing, Y.-H. Environmental impact analysis of power generation from biomass and wind farms in different locations. *Renew. Sustain. Energy Rev.* **2018**, *102*, 307–317. [CrossRef]
4. Uddin, S.; Kumar, S. Energy, emissions and environmental impact analysis of wind turbine using life cycle assessment technique. *J. Clean. Prod.* **2014**, *69*, 153–164. [CrossRef]
5. Darwish, A.; Abdelsalam, A.K.; Massoud, A.M.; Ahmed, S. Single phase grid connected curent source inverter: Mitigation of oscillating power effect on the grid current. In Proceedings of the IET Conference on Renewable Power Generation (RPG 2011), Edinburgh, UK, 6–8 September 2011; pp. 1–7.
6. Kibler, K.M.; Tullos, D.D. Cumulative biophysical impact of small and large hydropower development in Nu River, China. *Water Resour. Res.* **2013**, *49*, 3104–3118. [CrossRef]
7. Kömürcü, M.I.; Akpınar, A. Importance of geothermal energy and its environmental effects in Turkey. *Renew. Energy* **2009**, *34*, 1611–1615. [CrossRef]
8. Wang, G.; Konstantinou, G.; Townsend, C.D.; Pou, J.; Vazquez, S.; Demetriades, G.D.; Agelidis, V.G. A review of Power Electronics for grid connection of utility-Scale Battery Energy Storage Systems. *IEEE Trans. Sustain. Energy* **2016**, *7*, 1778–1790. [CrossRef]
9. Li, Y.; Ruan, X.; Yang, D.; Liu, F.; Tse, C.K. Synthesis of multiple-input DC/DC converters. *IEEE Trans. Power Electron.* **2010**, *25*, 2372–2385. [CrossRef]
10. Wai, R.-J.; Lin, C.-Y.; Liaw, J.-J.; Chang, Y.-R. Newly designed ZVS multi-input converter. *IEEE Trans. Ind. Electron.* **2010**, *58*, 555–566. [CrossRef]
11. Solero, L.; Lidozzi, A.; Pomilio, J. Design of multiple-input power converter for hybrid vehicles. *IEEE Trans. Power Electron.* **2005**, *20*, 1007–1016. [CrossRef]
12. Karanayil, B.; Ciobotaru, M.; Agelidis, V.G. Power flow management of isolated multiport converter for more electric aircraft. *IEEE Trans. Power Electron.* **2016**, *32*, 5850–5861. [CrossRef]
13. Al-Chlaihawi, S.J. Comparative study of the multiport converter used in renewable energy systems. In Proceedings of the2016 International Conference on Applied and Theoretical Electricity (ICATE), Craiova, Romania, 6–8 October 2016; pp. 1–6.
14. Mihai, M. Multiport converters—A brief review. In Proceedings of the 2015 7th International Conference on Electronics, Computers and Artificial Intelligence (ECAI), Bucharest, Romania, 25–27 June 2015. [CrossRef]
15. Rafiqi, I.S.; Bhat, A.H. Role of UPQC in addressing power quality issues in the power grid connected with renewable energy sources. In Proceedings of the 2022 1st International Conference on Sustainable Technology for Power and Energy Systems (STPES), Srinagar, India, 4–6 July 2022; pp. 1–4. [CrossRef]
16. Darwish, A.; Holliday, D.; Ahmed, S.; Massoud, A.M.; Williams, B.W. A single-stage three-phase inverter based on CUK converters for PV applications. *IEEE J. Emerg. Sel. Top. Power Electron.* **2014**, *2*, 797–807. [CrossRef]
17. Alharbi, Y.; Darwish, A. Control of CUK-based Microinverter topology with energy storage for residential PV applications. *Energies* **2023**, *16*, 2293. [CrossRef]
18. Darwish, A.; Finney, S.; Holliday, D. New Three-phase AC-DC rectifiers with reduced numbers of switches. In Proceedings of the 8th IET International Conference on Power Electronics, Machines and Drives (PEMD 2016), Glasgow, UK, 19–21 April 2016. [CrossRef]
19. Darwish, A.; Wang, Y.; Holliday, D.; Finney, S. Operation and control design of new three-phase inverters with reduced number of switches. In Proceedings of the 2016 International Symposium on Power Electronics, Electrical Drives, Automation and Motion (SPEEDAM), Capri, Italy, 22–24 June 2016; pp. 178–183. [CrossRef]

20. Barbosa, P.; Steimer, P.; Steinke, J.; Winkelnkemper, M.; Celanovic, N. Active-neutral-point-clamped (ANPC) multilevel converter technology. In Proceedings of the 2005 European Conference on Power Electronics and Applications, Dresden, Germany, 11–14 September 2005. [CrossRef]
21. Akagi, H. Large static converters for industry and utility applications. *Proc. IEEE* **2001**, *89*, 976–983. [CrossRef]
22. Sun, K.; Zhang, L.; Xing, Y.; Guerrero, J.M. A distributed control strategy based on DC bus signaling for Modular Photovoltaic Generation Systems with Battery Energy Storage. *IEEE Trans. Power Electron.* **2011**, *26*, 3032–3045. [CrossRef]
23. Meyer, C.; De Doncker, R.W.; Li, Y.W.; Blaabjerg, F. Optimized control strategy for a medium-voltage DVR—Theoretical investigations and experimental results. *IEEE Trans. Power Electron.* **2008**, *23*, 2746–2754. [CrossRef]
24. Stieneker, M.; Engel, S.P.; Stagge, H.; De Doncker, R.W. Optimization of the pulse-width-modulation strategy for redundant and non-redundant multi-level cascaded-cell converters. In Proceedings of the 2013 IEEE Energy Conversion Congress and Exposition (ECCE), Denver, CO, USA, 15–19 September 2013; pp. 101–108. [CrossRef]
25. Ota, J.I.Y.; Sato, T.; Akagi, H. Enhancement of performance, availability, and flexibility of a battery energy storage system based on a modular multilevel cascaded converter (MMCC-SSBC). *IEEE Trans. Power Electron.* **2015**, *31*, 2791–2799. [CrossRef]
26. Chanhom, P.; Sirisukprasert, S.; Hatti, N. DC-link voltage optimization for SOC balancing control of a battery energy storage system based on a 7-level cascaded PWM converter. In Proceedings of the 2012 9th International Conference on Electrical Engineering/Electronics, Computer, Telecommunications and Information Technology (ECTI-CON 2012), Phetchaburi, Thailand, 16–18 May 2012; pp. 1–4. [CrossRef]
27. Hillers, A.; Biela, J. Optimal design of the modular multilevel converter for an energy storage system based on split batteries. In Proceedings of the 2013 15th European Conference on Power Electronics and Applications (EPE), Lille, France, 2–6 September 2013; pp. 1–11.
28. Jimichi, T.; Fujita, H.; Akagi, H. Design and experimentation of a dynamic voltage restorer capable of significantly reducing an energy-storage element. *IEEE Trans. Ind. Appl.* **2008**, *44*, 817–825. [CrossRef]
29. Tao, H.; Duarte, J.L.; Hendrix, M.A.M. Three-port triple-half-bridge bidirectional converter with zero-voltage switching. *IEEE Trans. Power Electron.* **2008**, *23*, 782–792. [CrossRef]
30. Saito, W. A Future outlook of power devices from the viewpoint of Power Electronics trends. *IEEE Trans. Electron Devices* **2023**, *71*, 1356–1364. [CrossRef]
31. Buccella, C.; Cecati, C.; Latafat, H. Digital control of Power Converters—A survey. *IEEE Trans. Ind. Inform.* **2012**, *8*, 437–447. [CrossRef]
32. Kim, K.; Sung, M.; Jin, H.-W. Design and implementation of a delay-guaranteed motor drive for Precision Motion Control. *IEEE Trans. Ind. Inform.* **2011**, *8*, 351–365. [CrossRef]
33. Kazmierkowski, M.P.; Jasinski, M.; Wrona, G. DSP-based control of grid-connected power converters operating under grid distortions. *IEEE Trans. Ind. Inform.* **2011**, *7*, 204–211. [CrossRef]
34. Rafin, S.M.S.H.; Lipo, T.A.; Kwon, B.-I. Performance analysis of the three transistor voltage source inverter using different PWM techniques. In Proceedings of the 2015 9th International Conference on Power Electronics and ECCE Asia (ICPE 2015-ECCE Asia), Seoul, Republic of Korea, 1–5 June 2015; p. 1428-143.
35. Atalik, T.; Deniz, M.; Koc, E.; Gercek, C.Ö.; Gultekin, B.; Ermis, M.; Cadirci, I. Multi-dsp and -FPGA-based fully digital control system for cascaded multilevel converters used in facts applications. *IEEE Trans. Ind. Inform.* **2012**, *8*, 511–527. [CrossRef]
36. Monmasson, E.; Idkhajine, L.; Cirstea, M.N.; Bahri, I.; Tisan, A.; Naouar, M.W. FPGAs in industrial control applications. *IEEE Trans. Ind. Inform.* **2011**, *7*, 224–243. [CrossRef]
37. Leibl, M.; Kolar, J.W.; Deuringer, J. Sinusoidal input current discontinuous conduction mode control of the Vienna Rectifier. *IEEE Trans. Power Electron.* **2016**, *32*, 8800–8812. [CrossRef]
38. García, G.; Flores-Bahamonde, F.; Valderrama-Blavi, H.; Martínez-Salamero, L.; Maixé-Altés, J. Control of a three-phase AC/DC Vienna converter based on the sliding mode loss-free resistor approach. *IET Power Electron.* **2014**, *7*, 1073–1082. [CrossRef]
39. Lee, J.-S.; Lee, K.-B.; Blaabjerg, F. Predictive control with discrete space-vector modulation of Vienna Rectifier for driving PMSG of wind turbine systems. *IEEE Trans. Power Electron.* **2019**, *34*, 12368–12383. [CrossRef]
40. Molina-Martínez, E.J.; Roncero-Sánchez, P.; López-Alcolea, F.J.; Vázquez, J.; Torres, A.P. Control scheme of a bidirectional inductive power transfer system for electric vehicles integrated into the grid. *Electronics* **2020**, *9*, 1724. [CrossRef]
41. Fuentes, E.; Kennel, R. Finite-set model predictive control of the two-mass-system. In Proceedings of the 2011 Workshop on Predictive Control of Electrical Drives and Power Electronics (PRECEDE), Munich, Germany, 14–15 October 2011; pp. 82–87. [CrossRef]
42. Zhao, S.; Blaabjerg, F.; Wang, H. An overview of artificial intelligence applications for Power Electronics. *IEEE Trans. Power Electron.* **2020**, *36*, 4633–4658. [CrossRef]
43. Restrepo, J.; Viola, J.; Aller, J.M.; Bueno, A. A simple switch selection state for SVM Direct Power Control. In Proceedings of the 2006 IEEE International Symposium on Industrial Electronics, Montreal, QC, Canada, 9–13 July 2006; pp. 1112–1116. [CrossRef]

44. Eloy-García, J.; Arnaltes, S.; Rodríguez-Amenedo, J. Direct power control of voltage source inverters with unbalanced grid voltages. *IET Power Electron.* **2008**, *1*, 395–407. [CrossRef]
45. Monfared, M.; Sanatkar, M.; Golestan, S. Direct active and reactive power control of single-phase grid-tie converters. *IET Power Electron.* **2012**, *5*, 1544–1550. [CrossRef]
46. Baktash, A.; Vahedi, A.; Masoum, M. Improved switching table for direct power control of three-phase PWM rectifier. In Proceedings of the 2007 Australasian Universities Power Engineering Conference (AUPEC), Perth, WA, Australia, 9–12 December 2007; pp. 1–5. [CrossRef]
47. Malesani, L.; Mattavelli, P.; Buso, S. Robust dead-beat current control for PWM rectifiers and active filters. *IEEE Trans. Ind. Appl.* **1999**, *35*, 613–620. [CrossRef]
48. Han, J.-H.; Kim, I.-S. Double-Loop controller design of a single-phase 3-level Power Factor Correction converter. *Electronics* **2024**, *13*, 2863. [CrossRef]
49. Mayne. *Model Predictive Control: Theory and Design*; Nob Hill: Madison, WI, USA, 2009.
50. Froisy, J.B. Model predictive control: Past, present and future. *ISA Trans.* **1994**, *33*, 235–243. [CrossRef]
51. Baliga, B. Power semiconductor devices for variable-frequency drives. *Proc. IEEE* **1994**, *82*, 1112–1122. [CrossRef]
52. Kouro, S.; Cortes, P.; Vargas, R.; Ammann, U.; Rodriguez, J. Model predictive control—A simple and powerful method to control power converters. *IEEE Trans. Ind. Electron.* **2008**, *56*, 1826–1838. [CrossRef]
53. Mayne, D.Q.; Rawlings, J.B.; Rao, C.V.; Scokaert, P.O.M. Constrained Model Predictive Control: Stability and optimality. *Automatica* **2000**, *36*, 789–814. [CrossRef]
54. Borrelli, F. Discrete Time Constrained Optimal Control. Ph.D. Thesis, Automatic Control Laboratory ETH Zurich, Zurich, Switzerland, 2002.
55. Lei, S.; Jin, N.; Jiang, J. Current-prediction-controlled quasi-Z-source cascaded multilevel photovoltaic inverter. *Electronics* **2024**, *13*, 1824. [CrossRef]
56. Aurtenechea, S.; Rodriguez, M.A.; Oyarbide, E.; Torrealday, J.R. Predictive direct power control—A new control strategy for DC/AC converters. In Proceedings of the IECON 2006—32nd Annual Conference on IEEE Industrial Electronics, Paris, France, 6–10 November 2006; pp. 1661–1666.
57. Geyer, T.; Oikonomou, N.; Papafotiou, G.; Kieferndorf, F.D. Model predictive pulse pattern control. *IEEE Trans. Ind. Appl.* **2011**, *48*, 663–676. [CrossRef]
58. Quevedo, D.E.; Aguilera, R.P.; Geyer, T. Model Predictive Control for Power Electronics Applications. In *Handbook of Model Predictive Control*; Birkhäuser: Cham, Switzerland, 2018.
59. Besselmann, T.J.; Almer, S.; Ferreau, H.J. Model predictive control of load commutated inverter-fed synchronous machines. *IEEE Trans. Power Electron.* **2015**, *31*, 7384–7393. [CrossRef]
60. Silva, C.A.; Yuz, J.I. On sampled-data models for model predictive control. In Proceedings of the IECON 2010—36th Annual Conference of IEEE Industrial Electronics, Glendale, AZ, USA, 7–10 November 2010; pp. 2966–2971.
61. Vargas, R.; Ammann, U.; Rodríguez, J. Predictive approach to increase efficiency and reduce switching losses on matrix converters. *IEEE Trans. Power Electron.* **2009**, *24*, 894–902. [CrossRef]
62. Gamboa, P.; Silva, J.F.; Pinto, S.F.; Margato, E. Predictive optimal matrix converter control for a dynamic voltage restorer with flywheel energy storage. In Proceedings of the IECON 2009—35th Annual Conference of IEEE Industrial Electronics (IECON 2009), Porto, Portugal, 3–5 November 2009; pp. 759–764. [CrossRef]
63. Rodriguez, J.; Cortes, P. *Predictive Control of Power Converters and Electrical Drives*; Wiley: Chichester, UK, 2012.
64. Rodriguez, J.; Kazmierkowski, M.P.; Espinoza, J.R.; Zanchetta, P.; Abu-Rub, H.; Young, H.A.; Rojas, C.A. State of the art of finite control set model predictive control in power electronics. *IEEE Trans. Ind. Inform.* **2012**, *9*, 1003–1016. [CrossRef]
65. Alharbi, Y.; Darwish, A.; Ma, X. Cascaded multi-input single-output boost inverter for mismatch mitigation at PV submodule level. *Electricity* **2024**, *5*, 93–111. [CrossRef]
66. Lunardi, A.; Lourenço, L.F.N.; Munkhchuluun, E.; Meegahapola, L.; Filho, A.J.S. Grid-connected power converters: An overview of control strategies for renewable energy. *Energies* **2022**, *15*, 4151. [CrossRef]
67. Ahmed, K.H.; Massoud, A.M.; Finney, S.J.; Williams, B.W. A Modified Stationary Reference Frame-Based Predictive Current Control with Zero Steady-State Error for LCL Coupled Inverter-Based Distributed Generation Systems. *IEEE Trans. Ind. Electron.* **2010**, *58*, 1359–1370. [CrossRef]
68. Cortés, P.; Rodríguez, J.; Antoniewicz, P.; Kazmierkowski, M. Direct Power Control of an AFE Using Predictive Control. *IEEE Trans. Power Electron.* **2008**, *23*, 2516–2523. [CrossRef]
69. Fayyaz, M.M.; Syed, I.M.; Meng, Y.; Aman, M.N. Comprehensive Predictive Control Model for a Three-Phase Four-Legged Inverter. *Energies* **2023**, *16*, 2650. [CrossRef]
70. Abd-Elaziz, A.A.; Dabour, S.M.; Elmorshedy, M.F.; Rashad, E.M. Application of FCS-MPC for Split-Source Inverter-based Single-Phase Grid-Connected PV Systems. In Proceedings of the 2023 IEEE Conference on Power Electronics and Renewable Energy (CPERE), Luxor, Egypt, 19–21 February 2023; pp. 1–6.

71. Billinton, R.; Bagen. Reliability considerations in the utilization of wind energy, solar energy and energy storage in Electric Power Systems. In Proceedings of the 2006 International Conference on Probabilistic Methods Applied to Power Systems, Stockholm, Sweden, 11–15 June 2006; pp. 1–6. [CrossRef]
72. Sadamoto, T.; Chakrabortty, A.; Ishizaki, T.; Imura, J.-I. Dynamic modeling, stability, and control of power systems with distributed energy resources: Handling faults using two control methods in tandem. *IEEE Control Syst.* **2019**, *39*, 34–65. [CrossRef]
73. Bughneda, A.; Salem, M.; Richelli, A.; Ishak, D.; Alatai, S. Review of Multilevel Inverters for PV Energy System Applications. *Energies* **2021**, *14*, 1585. [CrossRef]
74. Kanouni, B.; Badoud, A.E.; Mekhilef, S. A multi-objective model predictive current control with two-step horizon for double-stage grid-connected inverter PEMFC system. *Int. J. Hydrogen Energy* **2022**, *47*, 2685–2707. [CrossRef]
75. Kaymanesh, A.; Chandra, A.; Al-Haddad, K. Continuous Control Set Model Predictive Control for Multilevel Packed E-Cell Inverter. In Proceedings of the 2021 IEEE Canadian Conference on Electrical and Computer Engineering (CCECE), Virtual, 12–17 September 2021; pp. 1–6.
76. Metri, J.; Vahedi, H.; Kanaan, H.Y.; Al-Haddad, K. Model predictive control for the packed U-Cells 7-level grid connected inverter. In Proceedings of the 2016 IEEE International Conference on Industrial Technology (ICIT), Taipei, Taiwan, 14–17 March 2016; pp. 1214–1219.
77. Kadhum, H.; Watson, A.J.; Rivera, M.; Zanchetta, P.; Wheeler, P. Model Predictive Control of a Modular Multilevel Converter with Reduced Computational Burden. *Energies* **2024**, *17*, 2519. [CrossRef]
78. Tarisciotti, L.; Zanchetta, P.; Watson, A.; Bifaretti, S.; Clare, J.C. Modulated Model Predictive Control for a Seven-Level Cascaded H-Bridge Back-to-Back Converter. *IEEE Trans. Ind. Electron.* **2014**, *61*, 5375–5383. [CrossRef]
79. Siami, M.; Khaburi, D.A.; Rivera, M.; Rodríguez, J. A computationally efficient lookup table based FCS-MPC for PMSM drivesfed by matrix converters. *IEEE Trans. Ind. Electron.* **2017**, *64*, 7645–7654. [CrossRef]
80. Yoo, H.-J.; Nguyen, T.-T.; Kim, H.-M. MPC with constant switching frequency for inverter-based distributed generations in microgrid using gradient descent. *Energies* **2019**, *12*, 1156. [CrossRef]
81. Geyer, T.; Papafotiou, G.; Morari, M. Model predictive direct torque control—Part I: Concept, algorithm and analysis. *IEEE Trans. Ind. Electron.* **2009**, *56*, 1894–1905. [CrossRef]
82. Rodriguez, J.; Pontt, J.; Cortes, P.; Vargas, R. Predictive control of a three-phase neutral point clamped inverter. *IEEE Trans. Power Electron.* **2007**, *24*, 2697–2705.
83. Tomlinson, M.; Mouton, H.d.T.; Kennel, R.; Stolze, P. A fixed switching frequency scheme for finite-control-set model predictive control—Concept and algorithm. *IEEE Trans. Ind. Electron.* **2016**, *63*, 7662–7670. [CrossRef]
84. Choi, D.-K.; Lee, K.-B. Dynamic performance improvement of ac/dc converter using model predictive direct power control with finite control set. *IEEE Trans. Ind. Electron.* **2014**, *62*, 757–767. [CrossRef]
85. Falkowski, P.; Sikorski, A. finite control set model predictive control for grid-connected ac–dc converters with LCL filter. *IEEE Trans. Ind. Electron.* **2017**, *65*, 2844–2852. [CrossRef]
86. Karamanakos, P.; Nahalparvari, M.; Geyer, T. Fixed switching frequency direct model predictive control with continuous and discontinuous modulation for grid-tied converters with LCL filters. *IEEE Trans. Control Syst. Technol.* **2020**, *29*, 1503–1518. [CrossRef]
87. Karamanakos, P.; Geyer, T. Guidelines for the design of finite control set model predictive controllers. *IEEE Trans. Power Electron.* **2019**, *35*, 7434–7450. [CrossRef]
88. Morari, M.; Lee, J.H. Model predictive control: Past, present and future. *Comput. Chem. Eng.* **1999**, *23*, 667–682. [CrossRef]
89. Cagienard, R.; Grieder, P.; Kerrigan, E.; Morari, M. Move blocking strategies in receding horizon control. *J. Process Control* **2007**, *17*, 563–570. [CrossRef]
90. Geyer, T. Generalized Model Predictive Direct Torque Control: Long prediction horizons and minimization of switching losses. In Proceedings of the 2009 Joint 48th IEEE Conference on Decision and Control (CDC) and 28th Chinese Control Conference (CCC), Shanghai, China, 15–18 December 2009; pp. 6799–6804.
91. Davari, S.A.; Khaburi, D.A.; Kennel, R. An improved FCS–MPC algorithm for an induction motor with an imposed optimized weighting factor. *IEEE Trans. Power Electron.* **2011**, *27*, 1540–1551. [CrossRef]
92. Cortes, P.; Kouro, S.; La Rocca, B.; Vargas, R.; Rodriguez, J.; Leon, J.I.; Vazquez, S.; Franquelo, L.G. Guidelines for weighting factors design in model predictive control of power converters and drives. In Proceedings of the IEEE International Conference on Industrial Technology, Gippsland, Australia, 10–13 February 2009.
93. Shadmand, M.B.; Jain, S.; Balog, R.S. Autotuning technique for the cost function weight factors in model predictive control for power electronic interfaces. *IEEE J. Emerg. Sel. Top. Power Electron.* **2018**, *7*, 1408–1420. [CrossRef]
94. Dragicevic, T.; Novak, M. Weighting factor design in model predictive control of power electronic converters: An artificial neural network approach. *IEEE Trans. Ind. Electron.* **2018**, *66*, 8870–8880. [CrossRef]
95. Stellato, B.; Geyer, T.; Goulart, P.J. High-speed finite control set model predictive control for power electronics. *IEEE Trans. Power Electron.* **2016**, *32*, 4007–4020. [CrossRef]

96. Azab, M. A finite control set model predictive control scheme for single-phase grid-connected inverters. *Renew. Sustain. Energy Rev.* **2020**, *135*, 110131. [CrossRef]
97. Errouissi, R.; Al-Durra, A.; Muyeen, S.M. A robust continuous-time MPC of a DC–DC boost converter interfaced with a grid-connected photovoltaic system. *IEEE J. Photovolt.* **2016**, *6*, 1619–1629. [CrossRef]
98. Lekouaghet, B.; Boukabou, A.; Lourci, N.; Bedrine, K. Control of PV grid connected systems using MPC technique and different inverter configuration models. *Electr. Power Syst. Res.* **2018**, *154*, 287–298. [CrossRef]
99. Matiushkin, O.; Husev, O.; Rodriguez, J.; Young, H.; Roasto, I. Feasibility Study of model predictive control for grid-connected Twisted Buck–Boost Inverter. *IEEE Trans. Ind. Electron.* **2021**, *69*, 2488–2499. [CrossRef]
100. Trabelsi, M.; Bayhan, S.; Ghazi, K.A.; Abu-Rub, H.; Ben-Brahim, L. Finite-control-set model predictive control for grid-connected packed-U-cells multilevel inverter. *IEEE Trans. Ind. Electron.* **2016**, *63*, 7286–7295. [CrossRef]
101. Bighash, E.Z.; Sadeghzadeh, S.M.; Ebrahimzadeh, E.; Blaabjerg, F. High quality model predictive control for single phase grid-connected photovoltaic inverters. *Electr. Power Syst. Res.* **2018**, *158*, 115–125. [CrossRef]
102. Liu, J.; Cheng, S.; Liu, Y.; Shen, A. FCS-MPC for a single-phase two-stage grid-connected PV inverter. *IET Power Electron.* **2019**, *12*, 915–922. [CrossRef]
103. Sajadian, S.; Ahmadi, R. Model Predictive-Based Maximum Power Point Tracking for Grid-Tied Photovoltaic Applications Using a Z-Source Inverter. *IEEE Trans. Power Electron.* **2016**, *31*, 7611–7620. [CrossRef]
104. Sebaaly, F.; Sharifzadeh, M.; Kanaan, H.Y.; Al-Haddad, K. Multilevel Switching-Mode Operation of Finite-Set Model Pre-Dictive Control for Grid-Connected Packed E-Cell Inverter. *IEEE Trans. Ind. Electron.* **2021**, *68*, 6992–7001. [CrossRef]
105. Wang, D.; Shen, Z.J.; Yin, X.; Tang, S.; Liu, X.; Zhang, C.; Wang, J.; Rodriguez, J.; Norambuena, M. Model predictive control using artificial neural network for power converters. *IEEE Trans. Ind. Electron.* **2021**, *69*, 3689–3699. [CrossRef]
106. Khalid, M.; Ahmad, F.; Panigrahi, B.K.; Al-Fagih, L. A comprehensive review on advanced charging topologies and methodologies for electric vehicle battery. *J. Energy Storage* **2022**, *53*, 105084. [CrossRef]
107. Harbi, I.; Rodriguez, J.; Liegmann, E. Model-predictive control of multilevel inverters: Challenges, recent advances, and trends. *IEEE Trans. Power Electron.* **2023**, *38*, 10845–10868. [CrossRef]

MDPI AG
Grosspeteranlage 5
4052 Basel
Switzerland
Tel.: +41 61 683 77 34

Electronics Editorial Office
E-mail: electronics@mdpi.com
www.mdpi.com/journal/electronics

www.ingramcontent.com/pod-product-compliance
Lightning Source LLC
Chambersburg PA
CBHW061501210926
51850CB00008B/1784
9783725852291